AF553631

PHYSIOLOGY OF FISHES

ENCYCLOPAEDIA OF FISH AND FISHERIES - II

PHYSIOLOGY OF FISHES

By

Dr. Arvind N. Shukla

School of Studies of Zoology & Biotechnology

Vikram University

Ujjain

DISCOVERY PUBLISHING HOUSE PVT. LTD.

NEW DELHI-110 002

Published by:

DISCOVERY PUBLISHING HOUSE PVT. LTD.
4383/4B, Ansari Road, Darya Ganj
New Delhi-110 002 (India)
Phone : +91-11-23279245; 23253475; 43596065
E-mail : discoverypublishinghouse@gmail.com
orderdphbooks@gmail.com
web : www.discoverypublishinggroup.com

***First Published:* 2009**
***Reprinted:* 2024**

ISBN: 978-81-8356-303-1 (Set)
ISBN: 978-81-8356-384-0

Physiology of Fishes

© Author

All rights reserved. No part of this publication should be reproduced, stored in a retrieval system, or transmitted in any form or by any means: electronic, mechanical, photocopying, recording or otherwise, without the prior written permission of the author and the publisher.

This book has been published in good faith that the material provided by authors/editors is original. Every effort is made to ensure accuracy of material, but the publisher and printer will not be held responsible for any inadvertent error(s). In case of any dispute, all legal matters are to be settled under Delhi jurisdiction only.

Printed at:
Infinity Imaging Systems
Delhi

Preface

The present title *"Encyclopaedia of Fish and Fisheries"* has been designed for undergraduate and postgraduate students of all Indian Universities. The text of the present title is organized on some major areas like fishing techniques, fish behaviour, and fish physiology, fisheries of important Indian forms. Fishes inhabit every kind of aquatic environment, and their wide distribution resulted in many different designs for their special mode of life. The present title, with a skilful mix of breadth combined with detail, reviews what is known about fishes. Biological inter-relationship are fully discussed, and fishes with features of special interest are highlighted, whether they are economically significant or not. The present text brings together many scattered observations as well as the results of recent investigations. It is hoped that the general biologist, the zoologist and the Icthyologist will find wealth of exciting information in it.

The present title gives a comprehensive overview of the fishery science and also offers diverse information on the subject of fisheries to make the readers up-to-date with the latest in this field where proliferation of scientific information has not only been fast but enormous too in the last decade. Written in scholarly yet easy to understand language. It combines the usefulness of a reference work with the readability of a browsing book, perfect for any one attuned to the splendor of our natural world. It is hoped that the book will prove indispensable to fisheries organizations, lecturers and the young zoologists engaged in teaching and research.

In the preparation of this book large number of books and research papers have been consulted. So no authenticity is claimed.

The author expresses his gratitude to Mr. Wasan and staff of M/s Discovery Publishing House for their whole hearted co-operation in the publication of this book.

The author tried hard to be accurate and upto date in statement and realises the impossibility of completely avoiding errors therefore, the author will greatly appreciate having his attention called to any questionable statement.

Author

Contents

1

INTRODUCTION

As in other vertebrates, the skin of a fish is the envelope for the body and is the first line of defense against disease. It also affords protection from, and adjustment to environmental factors that influence life, for it contains send receptors tuned to the surroundings of a fish. Furthermore, the skin has respiratory, excretory, and osmoregulatory functions. Housed in the skin too are the devices of colour and living light that either conceal the individual, advertise its presence, or afford sexual recognition. Also present, in the skin of the electric catfish *(Malapterurus),* are electric organs with prey-stunning discharge. The skin also contains venomous glands, as in the madtom catfishes *(Noturus),* and mucous glands, the secretions of which give the characteristic slimy touch and odor to a fish.

STRUCTURE

The skin of a fish consists of two layers. The outer is the epidermis, and the inner, the dermis or corium. Fish epidermis is similar in many ways to the lining of the human mouth. Typically it is composed superficially of several layers of flattened, moist epithelial cells. The deepest layers are a zone of active cell growth and multiplic-ation (stratum germinativum). Here cell multiplication goes on all the time to replace from within the outermost layer of cells as it is worn off, and to provide for growth. These epithelial cells from the epidermis are the first to close a surface wound.

The dermal layer of the skin contains blood vessels, nerves and cutaneoussense organs, and connective tissue. While a fish is being skinned, the fibers of the connective tissue (membranous skeleton) that bind the skin to the underlying muscle and bone are very evident.

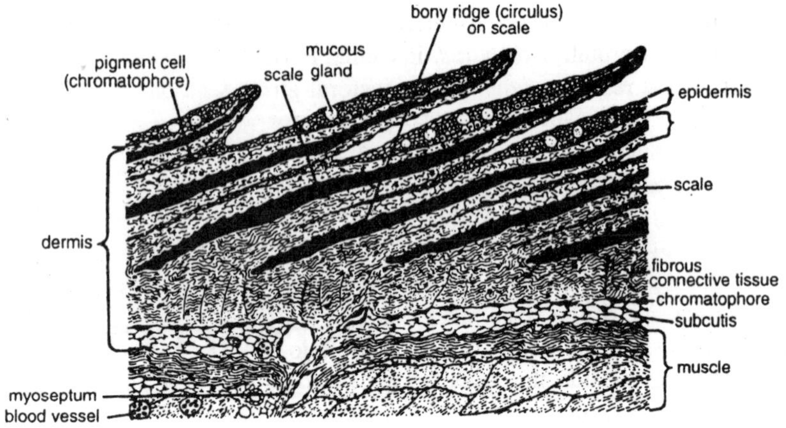

Figure 1.1 : Section of fish skin.

The dermis plays the main role in the formation of scales and related integumentary structures.

Scattered among the flattened cells of the epidermis are numerous openings of the tubular and flask-shaped mucous gland cells that extend into the dermis. These cells secrete the slippery mucus that covers most fishes. In some, as in a foot-long hagfish *(Myxine,* sometimes called slime-eel) these cells work so well that a single individual is said to be capable of secreting a pail-full of slime. Presumably mucus lessens the drag on a fish when it swims through the water. As mucus is sloughed off, it may be envisioned as carrying away microorganisms and irritants that might be harmful if they accumulated. In some species mucus coagulates and precipitates mud or other suspended solids in water. Fish odors are present in mucus. Mucous cells are also probably a source of chemical communication among fishes (e.g., alarm substance or Schreckstoffe).

SCALATION

Outstanding among the special features of the skin are its appendages, and outstanding among these, of course, are the very prominent scales that most fishes display. However, some kinds are "naked," in the technical sense of having no scales. Examples include the lampreys (Petromyzonidae) and North American freshwater catfishes (Ictaluridae). An intermediate structural category comprises fishes that are nearly naked and have scales only on a few places of the body, or have scales reduced to a few prickles. Some of these prickles are in very localised positions. Examples of partly scaled fishes include the paddlefish *(Polyodon),* a relative of the sturgeons, that inhabits streams in central North America and has also a near-relative, the freshwater swordbill, *Psephurus,* in China. *Polyodon* is essentially naked, except in the regions of the throat, pectoral girdle, and on the upturned base of its tail, where some scales are found. Some sticklebacks (Gasterosteidae), are naked including *Culaea* and *Pungitius;* others, notably the threespine stickleback, *Gasterosteus aculeatus,* have bony plates (rarely absent). Interestingly, the extent of bony plating is related to the degree of salinity of the water inhabited by this fish.

The sculpins (Cottidae) of the Northern Hemisphere, which inhabit streams, some of the large lakes, such as the Great Lakes, and Arctic seas, are essentially naked but may be variously adorned. Commonly among these cottids the axils of the pectoral fins and the head are scattered with little prickles that appear to, be vestiges of scales. Other derivatives of scales include teeth, the spiny "sting" of the st ni frays (Dasyatidae), and bony armoring plates such as in the seahorses and pipefishes (Syngnathidae) and the armored catfishes of South America (such as Loricariidae). One kind of the common carp *(Cyprinus carpio),* the mirror variety, has only a few large scales that are often separated by large areas of skin. Then there are a few fishes that have scales so small and/or so deeply im-bedded that the species appear to be naked although they are not; freshwater eels *(Anguilla)* provide a legendary example. Other examples include the brook trout *(Salvelinus fontinalis)* and the burbot *(Lota),* a freshwater member of the cod family (Gadidae).

Scale patterns are fundamentally associated with body segmenta tion as manifested initially in embryonic development by the vertebrae and myomeres. In arrangement, scales are most often imbricated and thus overlap like the shingles on a roof with the free margin directed toward the tail in a manner that minimizes friction with the water. Rarely, aberrant individuals exhibit total or partial reversal of this arrangement. In some fishes, such as the burbot *(Lota)* and the freshwater eels *(Anguilla)* the pattern is mosaic; rather than overlapping one another, the scales are minutely separated, or meet their neighbors only at their margins.

Scale Shapes

Although they comprise only a few basic structural types, scales exhibit many modifications that are often characteristic of groups or species. First, let us classify fish scales on the basis of prevalent shapes, and then repeat on the basis of structure. On the basis of

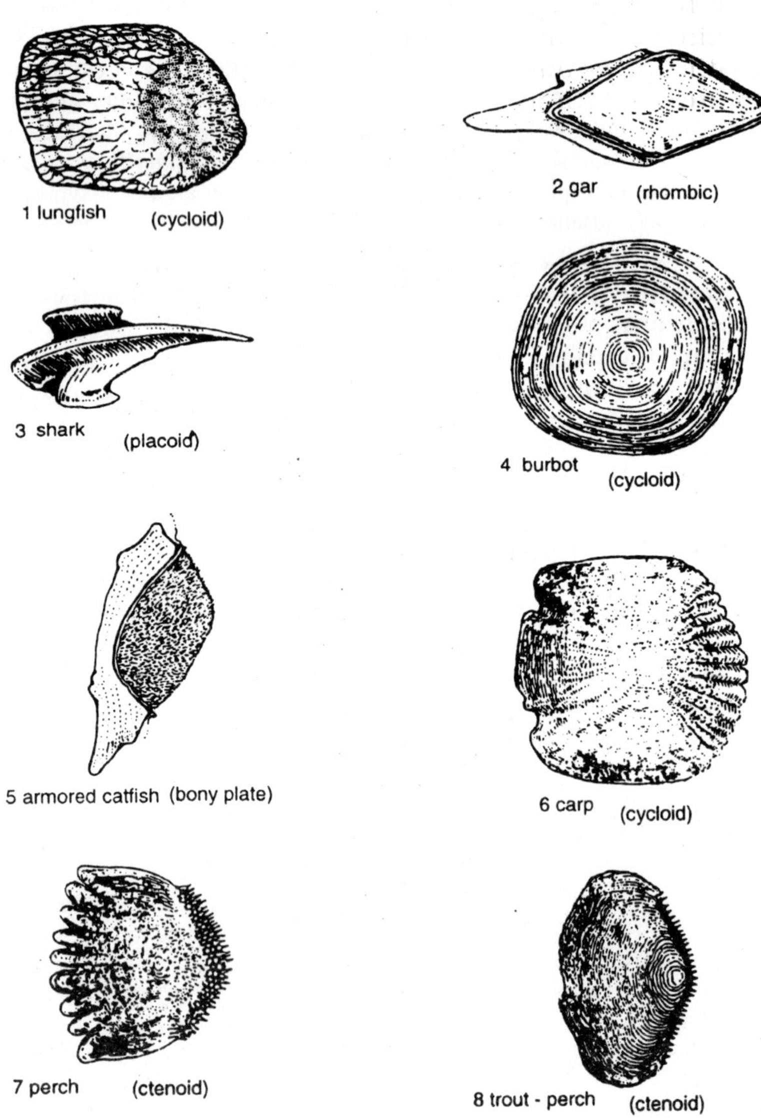

Figure 1.2 : Diversity in sacalation among fishes.

shape, one type is platelike (placoid), with each plate carrying a small cusp, as common among the sharks (Elasmobranchii). A second type is diamond-shaped (rhombic) and characterises the integument of the gars of North America (Lepisosteidae) and the bichirs *(Polypterus)* of the Nile. Such scales also occur on the tail of the sturgeons (Acipenseridae) of the Northern Hemisphere and of the American paddlefish *(Polyodon)*. *A* third type of fish scale on the basis of shape is cycloid. It gets its name because it is typically smoothly disc-like, and prevalently more or less circular in outline (the exposed, caudal surface or margin is entire, not toothed as it is in the next type of scale). In the fourth type, ctenoid, the posterior surface or margin is toothed or comblike.

Cycloid scales are found on most soft-rayed fishes (Malacopterygii); cteno d scales almost universally characterize the spiny-rayed bony fishes (Acanthopterygii).

However, some of the soft-rayed fishes have ctenoid-like contact organs on their scales, for example, a few of the characins (Characidae), killifishes (Cyprinodontidae), and livebearers (Poeciliidae). Some spiny-rayed fishes have cycloid scales exclusively, for example, the brook silverside, *Labidesthes*. Many spiny-rayed species exhibit both cycloid and ctenoid scales. As an example, the common basses *(Micropterus)* of North American fresh waters have mostly ctenoid scales, but on the cheeks, in the axils of the paired fins, about the vent, and elsewhere, there may be patches of scales that altogether lack teeth. These fishes may still be considered as predominantly ctenoid in their scalation, but the degree and extent of development of the teeth on the scales vary from place to place on the body.

Structural Types

Structurally, there are two types of fish scales, placoid and nonplacoid. Nonplacoid scales are basically of three kinds-cosmoid, ganoid, and bony-ridge. The last term is proposed to fill a gap in the story (in need of refinement) which has been developed concerning scales over the past century. In addition, bony fishes have many modifications of scales including reduction to tiny prickles or expansion and strengthening into strong armor, as previously described.

Placoid. Placoid scales, also called dermal denticles, have an ectodermal cap that is usually of an enamel-like substance (as on human teeth) termed vitrodentine. The cap is underlain by a body of dentine (as in vertebrate teeth generally) with a pulp cavity and dentinal tubules emanating from it. Each, scale has a disc-like, basal plate in the dermis with a cusp projecting outward from it through the epidermis. Placoid scales occur among sharks and their relatives (*Chondrichthyes*) *Cosmoid.* Cosmoid scales have a thinner, harder, outer layer than do placoid ones. Although the external material has a slightly different crystallographic

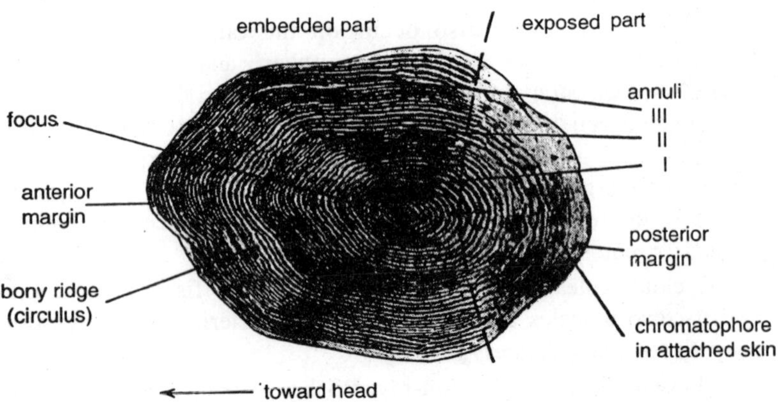

Figure 1.3 : Bony-ridge scale of the ctenoid type.

outer layer than do placoid ones. Although the external material has a slightly different crystallographic makeup from that of enamel of placoid scales it also has been termed vitrodentine. The particular layer beneath this enamel is hard and noncellular and has been called cosmine. There follows inwardly a vascularised mid-layer that is a zone of perforate bony substance (termed isopedine). A further distinctive feature of this type of scale is reportedly that growth is at the edge of the scale from beneath-not from without, because no living cells cover the surface. Cosmoid scales are found both in the living *(Latimeria)* and extinct lobefins. In *Latimeria,* the scales have a denticulate outer surface, almost ctenoid in gross aspect. The lungfishes (Dipnoi) have scales which appear to be cycloid scales because the basic cosmoid structure is highly modified.

Ganoid. In a ganoid scale, the outer layer is a hard inorganic substance (ganoine), differing from vitrodentine. Beneath the ganoin cap there is a cosmine-like layer. The innermost lamellar bony layer is isopedine. Besides differing in structure from a cosmoid scale, a ganoid scale is alleged to grow not only at the edges and from underneath, but also on the surface. Among living fishes ganoid scales are best represented in bichirs *(Polypterus)* and the gars (Lepisosteidae) where they invest the entire body. In rhombic shape they are also present on the upturned lobe of the tail of such chondrosteans as the sturgeons (Acipenseridae) and the paddlefishes (Polyodontidae), as shown previously.

Bony-Ridge. Bony-ridge scales are typically thin and translucent, lacking both dense enameloid and dentinal layers of the three other kinds. Bonyridge scales characterize the many living species of bony fishes (Osteichthyes) that have cycloid or ctenoid scales. The outer surface of such a scale is marked with bony-ridges that alternate with valley-like depressions. The inner part or plate of the scale is made up of layers of criss-crossing fibrous connective issue. Growth of these scales is both on the outer surface and f.om beneath.

For describing the structure and development of the bony-ridge scale, there is need to standardize the names of parts because of inconsistencies of past usage. For the ridges on the scales, two terms seem permissible: ridges and circuli. Changes in the growth pattern of the individual may be reflected in the character and distribution of the ridges. Natal-, metamorphic , breeding-, and year-marks (annuli) have been identified on this basis in many species. In most ordinary ctenoid or cycloid scales, a nuclear, central zone is recognisable. This zone may properly be called the' focus of the scale. It is the first part to develop and it is often central in position, although later differential growth in the forepart or after part of the scale may lead to the apparent (relative) displacement of the focus toward the posterior or the anterior margin. In many species, grooves (radii) radiate from at or near the focus toward one or more of the margins of the scale. On ctenoid scales, the most commonly used and acceptable term for the denticulations on the posterior scale margin is teeth.

In the use of scales for the classification and identification of fishes or for the study of life histories, it has sometimes been desirable

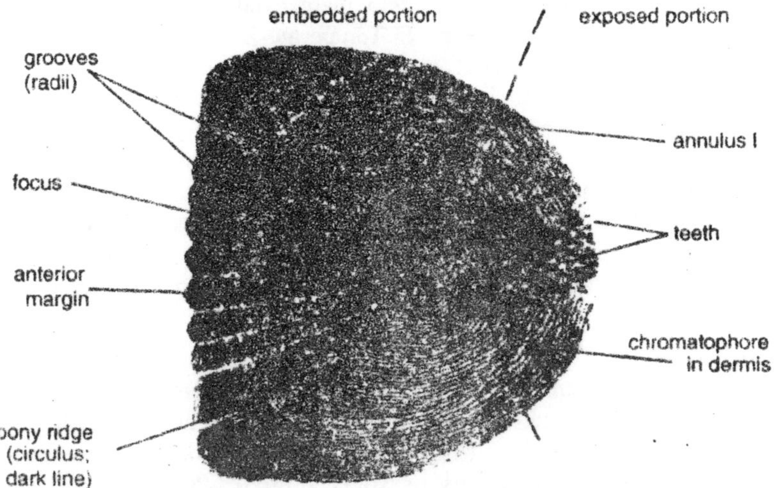

Figure 1.4 : Bony-ridge scale of the ctenoid type.

to describe various regions of a scale. For most cycloid and ctenoid scales there is a readily recognisable anterior field that in many scales roughly corresponds to the imbedded portion of the scale. There is also a posterior field, often approximating the part ofthe hscaleat can be seen without removing it from the body of the fish. By what characteristics may one ordinarily recognize, given a single fish scale, the imbedded or anterior field from the exposed or posterior one? In ctenoid scales, as indicated, the teeth are typically part of the posterior field. In cycloid scales, the posterior field may often be identified by the fact that the ridges are least distinct there, often appearing as if they had been eroded or abraded. Frequently also, pigment cells (chromatophores) adhere in this zone.

In development, bony-ridge scales first appear in the dermis as tiny aggregations of cells, most often forming first on the caudal peduncle and spreading from there. Such an aggregation soon forms a scale platelet, the focus of the definitive scale. These platelets make their first appearance at different sizes of individuals in different fishes; an ordinary size at which scales first appear might be something under an inch for common species. Soon ridges are deposited on the surface of the growing scale, a fact that contradicts an antiquated idea that the ridges are the margins of superimposed laminae or layers of which scales were presumed to have been composed. The only ridges, then, which appear on the surface of ordinary fish scales, are deposited on the outward surface during development, and they are laid down somewhat in relation to growth. In Temperate-Zone waters, deposition undergoes yearly seasonal changes which makes circuli partial indicators of the annulus on a fish scale. The deepest part of the scale, the plate (also "basal plate," "lamellar layer"), is made up of successive layers of parallel fibers; those in one layer run at sharp angles to those in adjacent layer(s). Some calcification of this fibrillary plate may occur to strengthen the scale. Certain fishes, however, lack the fibrillary plate, as in the eel *(Anguilla)* where the scale is a composite of more or less concentrically arranged loculi of calcareous deposition.

Derivatives

Scales are held to be the starting point for several structures in fishes. Jaw teeth in the sharks and relatives as well as those of higher vertebrates are modifications of placoid scales, so too are spines such as those of the dorsal fins of spiny dogfish *(Squalus)* and the chimaeroids and the "stinger" on the tail of the sting rays (Dasyatidae). The saw teeth of the sawshark *(Pristis)* are of like origin. The lancets of the surgeonfishes *(Acanthurus)* obviously originated from bony-ridge scales as did also the sawtooth-like belly scutes of herrings (Clupeidae). Without an ancestry in scales, soft-rays (lepidotrichia) may never have appeared to support fins in the Osteichthyes, nor superficial bones to invest the skull in these fishes and their vertebrate descendants. Both lepidotrichia and dermal-bones are derivatives of bony scales. Specially perforated and tubulated modifications of scales constitute a hardened channel for the lateral line sensory canal in many bony fishes. Armature too is various, ranging from spaced bony-plate bucklers in the sturgeons (Acipenseridae), through semirigid cases in pipe-fishes and seahorses (Syngnathidae), to encasements rivaling those of the most completely boxed turtles in the trunkfishes (Ostraciidae) for example, the cowfish *(Lactophrys quadicornis)*. Extensions of dermal armature beyond the regular body surface are seen in the porcupinefish *(Diodon hystrix)*.

Use in Classification and Natural History

Fish scales, which are useful to the fish for many purposes suggested above and for additional reasons, are also useful to ichthyologists for purposes of classification and natural history. How are scales useful in classification? In the major groups of living fishes the lampreys and hagfishes (Cyclostomata) are scaleless; the sharks and relatives (Chondrichthyes) are characterised by the placoid-scale type; the lobefins (Crossopterygii) and lungfishes (Dipnoi) have cosmoid scales; primitive bony fishes are associated with ganoid scales (Ganoidei of Agassiz); the higher bony fishes most often have bony-ridge scales. Besides being useful for broad aspects of fish classification, scales have characteristics that are useful for separating orders and families. If one is fortunate enough to be in an area where a good part of the fauna is made up by families represented by a single species (monotypic families), such as the bowfin (Amiidae) and the cod (Gadidae) families in the Great Lakes region, he or she can identify the species as easily by a single scale as by a whole individual. Scales have utility too, in classifying the remains of fishes in the waste heaps (kitchen middens) of prehistoric man, and in fossil deposits. Ability to classify on the basis of scales has proven worth also in the study of food habits of fish-eating animals.

The diagnostic value of scale morphology for the separation of species varies among families or genera. Where there are many closely related species, perhaps of recent origin, scale structure is likely to be as much alike as entire fishes. For example, in the whitefishes *(Coregonus)*, the many closely similar forms in the cisco subgenus

Leucichthys cannot be satisfactorily distinguished on the basis of scales. The same is true within the large genera of minnows (Cyprinidae), both in Asia and in North America. Thus, scale morphology is often relatively useless for effecting fine separations, although it may be indicative of affinities or of memberships in major groups. Highly useful, however, in taxonomic work have been scale counts such as numbers in the lateral line, and rows along or around the body.

In addition to values in identification and classification, scales have some utility in the interpretation of life history. Recall, for example, that from annuli on the scales one may determine age in years. From the spacing of annual rings and a knowledge of the length of the individual at capture, fish length at each previous year mark may be calculated, after proportionality of scale growth to body growth has been determined. Sometimes scales also disclose how many times a fish has spawned. In the Atlantic salmon *(Salmo salar),* for example, there are spawning marks on the scales. Thus, one can tell how old such a fish was when it first went to sea, its age at capture and at first spawning, and also how many times it has spawned. The time when the individual first went to sea can be discerned by an abrupt change in the distance between the circuli, for when growth is accelerated in the sea the circuli are more widely spaced.

BARBELS AND FLAPS

Besides smooth nakedness and some relatively minor divergences from basic scalation, fishes have evolved a wide variety in integumentation. Included are weird extensions of the skin into barbels and flaps. Barbels have appeared independently in many major groups as accessory feeding structures that carry sensory organs. Barbels of different structure and location are present on sturgeons *(Acipenser),* marine and freshwater catfishes (nematognath or siluroid Cypriniformes), goatfishes (Mullidae), and many others. The sargassum fish *(Pterophryne)* and the alga-resembling seadragon *(Phyllopteryx)* are the most frequently cited examples of the extension of the skin into flaps. The function ascribed in both these fishes is protective resemblance; at least, it cannot be denied that fishes with flaps are camouflaged by their adornment.

COLOURATION

From the large variety of colouration that fishes exhibit, certain generalisations may be drawn concerning basic pattern in relation to habit. Free-swimming, open-water fishes such as the many marine herrings and relatives (Clupeidae) as well as the early planktonic stages of many bottom fishes such as the flounders (Pleuronectidae) are mostly of simple colouration grading from a whitish belly, through silvery lower sides, to upper sides and back that are irridescent blue or green. Bottom dwellers and weed-bed occupants are often very strongly and intricately marked above and pale beneath. The most brilliantly, elegantly, and bodaciously coloured fishes are among the frequenters of the tropical coral reefs. Included are cardinalfishes (Apogonidae), butterflyfishes and angelfishes (Chaetodontidae), wrasses (Labridae), damselfishes (Pomacentridae), surgeonfishes (Acanthuridae), parrotfi-0shes (Scaridae), and triggerfishes (Balistidae). Indigenous to streams of eastern North America are the darters (Percidae; Etheostomatinae) the males of many of which are beautifully coloured, as are many of the small freshwater tropical fishes sold in the aquarium trade. Three types predominate in the colour of oceanic fishes: silver in the upper zones; red in the middle range; and violet or black in the greater depths.

Colouration in fishes is primarily due to skin pigments. Background colour or complexion is due, of course, importantly to underlying tissues, to body fluids, and even to gut content. Background colour is the essential hue of the blind cavefishes (Amblyopsidae). In other fishes, hues ranging from bright to dull that mask background colour are dermal in origin. The common ground of colouration in fishes is the prevalent lightness on the ventral body surface, darkness on the back, and gradual shading on the sides from light below to dark on the back as described above. This plan illustrates the primary principle of camouflage by obliterative countershading; students of animal colouration call this Thayer's principle. Beyond this general plan, there are many extraordinary features of colour dress in fishes. One of these is colour uniformity due either to lack or excess of pigment in albino or in melanistic fishes.

Lack of pigment and resultant transparency characterizes the pelagic, freeswimming young of many kinds of fishes. Another extraordinary feature is almost uniform colouration by some one hue or another. An example of this would be the revealing uniform gold colour that characterizes certain genetic strains of goldfish *(Carassius)* or the magnificent orange shown by the garibaldi *(Hypsypops)* of the southern California coast. Still other examples approaching uniform colouration would include the over-all blackness or melanism that characterizes many of the deep-sea fishes, although some are reversely countershaded. Other fish species, however, successfully rival the most colourful of the butterflies or birds in their combinations of multiple hues on single individuals.

Sources of Colour

Colouration in fishes is due to schemachromes (colours due to physical configuration) and biochromes or true pigments. White schemachromes are seen in the skeleton, gas bladder, scales, and testes; tyndall blues and violets are in the iris; and, iridescent colours are in the scales, eyes and intestinal membranes. The foregoing schemachromes are also to be seen in the integument. Biochromes include carotenoids (yellow, red, and other hues), chromolipoids (yellow to brown), indigoids (blue, red, and green), melanins (mostly black or brown), porphyrins and bile pigments (red, yellow, green, blue, and brown), flavines (yellow, often with greenish fluorescence), purines (white or silvery), and pterins (white, yellow, red, and orange). Carotenoids, melanins, flavines and purines appear in fish skin. The liver, eggs, and eyes have carotenoids. Melanins occur in the endoderm and skin. Muscle and blood have porphyrins whereas the skeleton and bile have bile pigments. Flavines are widespread in blood, muscle, spleen, gills, heart, kidneys, eggs, liver, and eyes. Purines are in the scales and eyes. Pterins are in the eyes, blood, liver, kidneys, and stomach.

The special cells that give colour to fishes are of two kinds, chromatophores and iridocytes. Chromatophores are assorted in hue and impart true colour. They are located in the dermis of the skin, either outside or beneath the scales. Also, they are often found in the peritoneum and even deeply around the brain and spinal cord. Cytoplasmic inclusions in chromatophores, called pigment granules, are the actual sources of the colour. The pigment migrations within the chromatophores—the granules can disperse through the cell or concentrate in the center—account for many of the colour changes that fishes exhibit. The pigment in the granules reflects some wave lengths of light and absorbs others-the ones that are reflected are the ones that are seen; those that are absorbed cannot be seen. The basic chromatophores according to the colours of their pigment granules are red and orange (erythrophores), yellow (xanthophores), black (melanophores), and white (leucophores). Red, yellow, and orange pigments are most often carotenoids that are acquired through food and are related to vitamin A. The black or brown pigments, melanins, are highly polymerised compounds derived from the amino acid tyrosine and other phenols. In addition to these colours, fishes exhibit others—such as green and brown. These and other hues come from various associations of the three basic kinds of chromatophores. Black and yellow chromatophores interspersed among one another, for example, give ostensibly green hues. Yellow and black make brown, but orange and blue may also associate to give brown.

Iridocytes could be called mirror cells, because they contain reflecting materials that mirror colours outside the fish. Both leucophores and iridophores contain purines, primarily guanine, but the former contain small crystals that can move back and forth in the cytoplasm, whereas in the latter there are large crystals incapable of movement and usually stacked in layers. Guanine itself is a breakdown product of nucleic acids (genetic material).

Significance

The functions of colouration and other visual signals (bioluminescence) are for communication with other members of the same species (intraspecific signals) and communication with other kinds of animals (interspecific signals). Intraspecific signals serve social (recognition, threat, warning) and sexual purposes, and may also offer cues by the host fish to other small fishes that browse on the body surface and clean it. Young cichlid fish travel in swarms near their parents, to which they flee when danger approaches. Recognition of their parents is based partly on colour. For example, in *Hemichromis,* which are brightt red during the period of caring for the young, the small fish have a pronounced preference for red and orange over all colours. Interspecific signals may be for warning or intimidating potential predators and other assailants or for decoying or masking purposes directed toward either prey or predators. Cott grouped principal functions of colouration under three headings*concealment, disguise,* and *advertisement*. The various kinds of concealing colour are suggested as being: (1) general colour resemblance; (2) variable colour resemblance; (3) obliterative shading; (4) disruptive colouration; and (5) coincident disruptive colouration.

Colour Resemblance

General colour resemblance between fish and the background is the basic characteristic of fishes to resemble the shades and hues of the habitats which they frequent. As previously stated, many coral-reef fishes are as very highly coloured as the coral heads that they frequent, but some at least may be revealingly coloured to advertise their presence to other fishes as removers of parasites from them or to afford recognition to congeners or species mates. Weed-bed inhabitants, living where there are zones of brightness and shadow, are commonly mottled. Littoral dwellers, living over lightshaded bottoms are light-coloured, but over dark bottoms (or at greater depths or at night) the same species are dark.

Variable colour resemblance is the ability of a fish (or other animal) to change colour, gradually or rapidly in order ostensibly to match its background more perfectly. Some such variations in colour resemblance occur with life history stage. The stream resident phase of a migratory rainbow trout *(Salmo gairdneri) is* multicoloured, including dark spots and parrmarks (young), and rosy sides (adult). In the sea, the same individual grades from steel-blue above, through silver, to white below. There is also variation in colour among fishes with the seasons, with day and night and even with momentary changes in habitat. Trout *(Salmo)* occupying bright, or partly shaded riffles in the summer have a mottled aspect. The same individuals in pools or in ice-covered waters in the winter would be rather evenly dark.

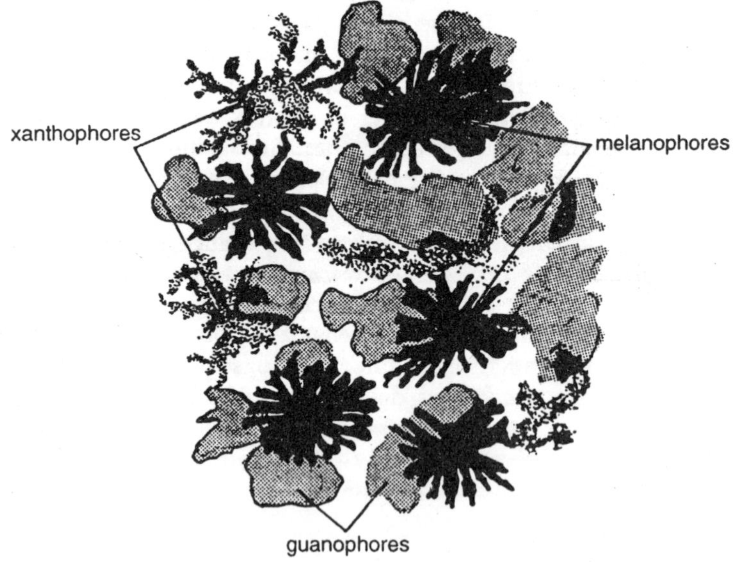

Figure 4.5 : Colour cells in the skin of a flounder, Paralichthys.

One form of variable colour resemblance alters with change in body form or structure. It is to be seen in the kind of a shift that takes place when the transparent leptocephalus larva of the eel *(Anguilla)* comes from the ocean into a stream. This eel hatches far at sea in the tropical Atlantic. The transparent, pelagic larvae migrate to coastal streams of North America and Europe. With arrival at the streams, glassy band-like individuals (leptocephali) change body shape to a pencil-like form and become opaque (elvers), pigmented following Thayei s principle.

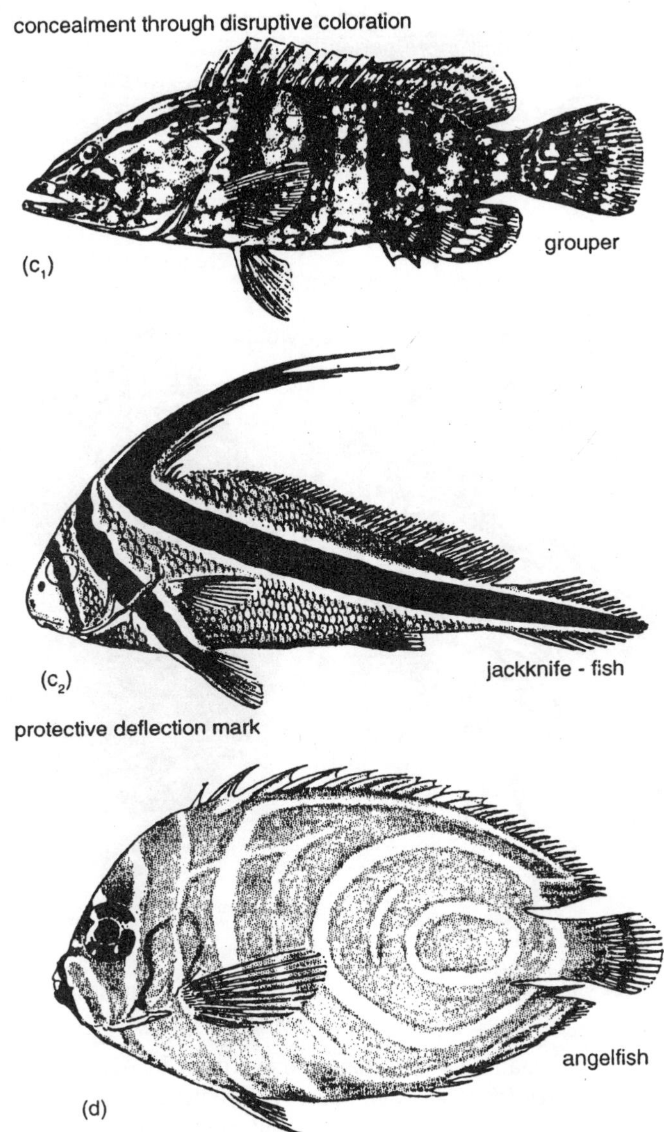

Figure 4.6. Camouflage in fishes (Figure Contd.)

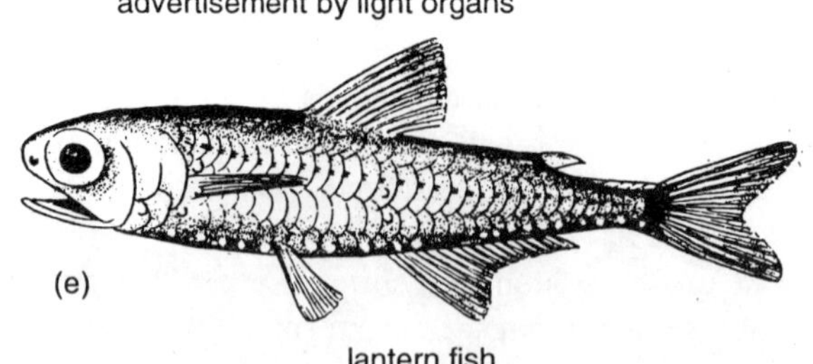

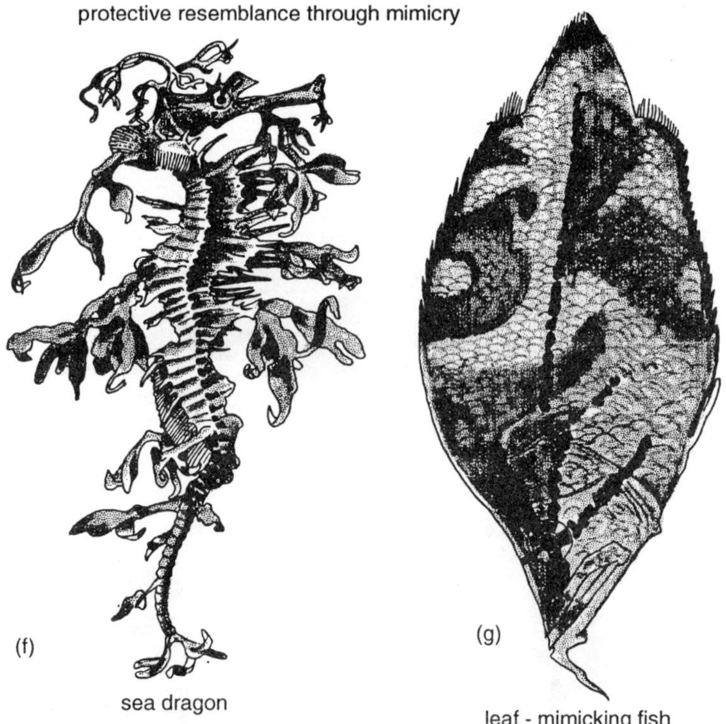

Figure 4.6 Continued.

Rapid, adjustable colour change may result in quickly matching a fish to changes in its surroundings. This type of colour adaptation is usually brought about by rearrangement of pigment granules within chromatophores with the chromatophores themselves becoming differentially concealed or prominent. Speedy alteration in appearance may be of special value for concealment when the fish moves over various backgrounds. This ability to change colour by fishes has been termed the most wonderful automatic cryptic device in existence. The internal regulation of these rapid changes is very complex. It involves reflex activities induced through the visual stimulation of the eyes and/or the pineal body, through hormones, and through direct action of light on skin and/or chromatophores.

Adjustment of colouration in fishes may be semi-permanent as well as rapid. A semi-permanent adjustment is brought about by an increase in the number of chromatophores—as an artist increases the number of stipples for shading. So, a fish, living in a darkened area, or over a dark substratum for a long time, may gradually add chromatophores. Such an adjustment takes time to mature, and though it is partly reversible by contraction of the chromatophores, the chromatophores themselves may not be destroyed.

In contrast, rapid adjustment is brought about through rearrangement of the pigment granules within the chromatophores by expansion and contraction. The changes that are achieved by fishes in this way exceed in rapidity and extent that in the famed chameleons. Some of the characins (Characidae) of aquarists flash light and then darken instantly when their nests are threatened.

The reaction is similarly instantaneous in certain reef fishes when they are moved from a brightly lighted and coloured area to subdued light with dull hues. A classical example of physiological colour change in fishes is among the flatfishes (Pleuronectiformes). These fishes are adapted to life on the bottom by the great morphological adaptation that results in perpetual living on one side of the body and in having the side turned toward the bottom unpigmented and that turned toward the light pigmented (rare anomalies are the reverse of this). Equally as striking as the physiological adaptation of flatfishes to rapid colour change is their matching ability. A flatfish placed on an illuminated checkerboard will, after some time, afford a background almost suited for a game! The mystery remains, however, as to how the retina in flatfishes (Pleuronectidae) can initiate this remarkable matching of pattern.

Thus the colouration of most fishes is not constant throughout life since there are both short-term and long-term changes. Such pigmentary changes or responses are both morphological and physiological. For melanophores, morphological change is slow and exhibits buildup of pigment and cells. Physiological change can be rapid (often within a few minutes) and involves a redistribution of pigment granules within the melanophores. Both of these changes are responses either to visual or nonvisual stimuli, with the latter being either coordinated by nerves and hormones or uncoordinated where the melanophore is its own receptor and effector (not known for fishes). In eyeless sharks *(Mustelus)* and blinded catfishes *(Ictalurus)*, darkening takes place in light and paling in darkness, whereas intact individuals darken by melanin dispersal when placed on a dark background. Similarly, by pituitary hormone responses, removal of the pituitary results in paling in another shark *(Scyl*lium) and pituitary implants result in darkening in a skate *(Raja)*. The role of nerve impulses in chromatic regulation has been demonstrated by cutting peripheral nerves and having the denervated area darken, and by electrically stimulating the peripheral nerves. In *Scyllium,* at least, darkening as a black background response involves MDH (melanin dispersing hormone) and paling on a white background involves MDH or secretion of MAH (melanin aggregating hormone), both from the pituitary. There is also good evidence for MDH in some teleosts *(Anguilla, Ictalurus,* and *Fundulus)*. In the relatively few species of teleosts studied, there is uneven information on the colour change mechanism but there is general agreement that coordination of colour change is by interaction of nervous and hormonal control.

Obliterative Shading

Yet another kind of colouration for concealment is obliterative shading. The optical principle upon which this type of concealing colouration depends is that of countershading. Light and shade give an observer the third dimension of objects seen. Shadow, then, is a serious matter when trying to reduce the visibility of any object, as one may try to do, say, in time of war, to obscure a vehicle, a ship, or an aircraft. Practical countershading is evidenced by most fishes-also, by most birds, mammals, reptiles and amphibians-since they possess dark dorsal surfaces and light bellies, tending to make them appear optically flat, like a shadow (Thayer's principle). The surfaces normally directed toward a source of light are countershaded by darkening, whereas those which would normally be in shadow are counterlighted, and properly graded tones between render the object flat and thus reduced in visibility. Countershading has disappeared in cave fishes *(Amblyopsis* and *Anoptichthys)* and reversed in the depths of the sea, as previously stated.

Disruptive Colouration

Yet another means of concealment that has evolved in fishes is disruptive colouration, a further means of camouflage. Since it is the continuity of surface, bounded by a specific contour or outline, that enables us chiefly to recognise an object with the shape of which we are familiar, disruption of this contour tends to conceal. Thus, for effective concealment, it is essential that the telltale appearance of form should be destroyed. The function of camouflage for a fish, then, may be thought to be to prevent, or to delay as long as possible, recognition on sight. When the surface of the fish is covered by irregular patches of contrasted colour and tones, these patches tend to catch the eye of the observer and to draw attention away from the shape that bears them. In contradicting the form, the patterns concentrate the attention upon themselves, and thus the patterns may cause the object that bears them to pass for part of the general environment.

Coincident disruptive colouration is a special kind of camouflage. It may appear to join together separate, unrelated parts of the body in order further to reduce chances of recognition. Appendages are concealed by this device and eyes are too. The eye, particularly its staring black pupil, is made to appear another shape by various forms of coincident disruptive eye masks and thus, joined to some other part of the head, ceases to resemble an eye. However, it has also been proposed that the eye-line-horizontal or tear-drop vertical-is a sight-line for aiming attack.

Related Forms of Concealment and Disguise

Associated with concealment by use of colour are other means of masquerade or disguise, often relating structure with habit as shown well by Breder. Habits such as cryptic attitudes are used to conceal position and may complete the role of colour disguise.

Gars *(Lepisosteus),* both young and adult, have the habit of basking motionless near the surface. In posture, form, and colouration they strongly resemble floating twigs or logs and are often mistaken for them. *Strongylura,* a genus of needlefishes (Belonidae), looks much like the stems of plants in its habitat. Young of some halfbeaks *(Hemiramphus)* and pipefishes (Syngnathidae) in profile look like leaves of the eelgrass *(Zostera).* Other fishes that resemble plant leaves as young or adults include the leatherjacket (Oligoplites *saurus),* the tripletail *(Lobotes surinamensis),* a filefish *(Monacanthus polycanthus),* certain spadefishes (Ephippidae; *Platax),* and the orange filefish *(Alutera schoep* fi). The young of some flyingfishes *(Cypselurus)* look very much like the blossoms of a plant *(Barringtonia)* in their habitat.

Resembling algal fronds (i.e., thalli) or fragments of them, are several fishes including seadragons (Syngnathidae; *Phyllopteryx),* the dwarf wrasse *(Doratonotus megalepis),* and the giant kelpfish *(Heterostichus rostratus).* In addition, the Atlantic spadefish *(Chaetodipterus faber)* resembles a seed pod *(Rhizophora),* and the lumpfish *(Cyclopterus lumpus)* looks like an algal flotation capsule.

Disguise is also accomplished by various conspicuous localised characters, which simply tend to reduce the resemblance of the fish to itself. Deflective and directive marks are important here. Deflective marks are those which have been thought of as deflecting the attack of an enemy from a more or less vital part of the body to some other part. A simple illustration is the dark spot on the tail of the bowfin *(Amia)*; another is in the young of certain angelfishes. During ritualised territorial fights of one angelfish *(Pterophyllum scalare),* there are changes in the darkness of the vertical bar pattern and of the eye spot on each of the gill covers. A directive mark, contrariwise, has been thought of as diverting the attention of prey from the most dangerous part of the predator's body. Stargasers (Uranoscopidae) have fringed mouths that may well obscure the organ when the fish is mostly buried in the sand. Prey fishes move into the region of the obscured mouth, perhaps directed or attracted there by fleshy flaps that resemble food, and are engulfed (even though the flaps may serve primarily to keep sand out of the mouth). The luring of prey to the mouth region is another function of directive marks, usually a combination of structure and colour. Goosefishes and anglers (Lophiiformes) have one or two slender, stalked appendages, each a virtual "bait-rod" (illicium), extending forward over the snout, sometimes with a little luminous bulb or variously tasseled lure at the tip of the structure. One stargazer *(Uranoscopus)* of the Mediterranean is said to have a little red "worm" in its mouth that lures prey to their demise.

Advertisement

Some forms of colouration of fishes appear to advertise or to reveal rather than conceal their presence. Outstanding examples of this are among the darters (Etheostomatinae) of American streams. Among these small members of the perch family (Percidae) are the most brightly hued freshwater fishes of this continent. Colouration of this type may be of significance for sexual recognition; hence, it is a form of advertisement. Experiments with sticklebacks *(Gasteros-teus)* suggest value of colour in sexual recognition for some fishes, but trials with other species show that colour may not be discriminated or may be valueless for this purpose.

Use in Classification

Colour pattern in fishes is used often as a character to distinguish smaller taxonomic units such as species and subspecies. In spite of variation among individuals of one size and kind, and variability with sex, age, habitat, and so on, fundamental colour pattern, often the exact placement of individual chromatophores, is under genetic control. Not uncommonly, therefore, colour characters appear in the descriptions of the smaller taxonomic categories.

LIGHT ORGANS

Although absent in freshwater fishes, living light (bioluminescence), like that in the common fireflies or glowworms (Lampyridae), is not uncommon in the oceans and appears in many groups of animals, including many marine fishes. It is most extensively developed in many of the midwater and bottom dwelling deepsea species. The suspicion is that light organs may serve in species and mate recognition by the fishes themselves. Certain light organs on the head have been implied to act as lures to attract prey. Most of the luminescent lights are blue or green. The organs are useful as taxonomic characters, for example, in the lanternfishes and relatives (Myctophiformes).

Luminescence in fishes is of two types: *(a)* that which results from the presence of luminous bacteria living on the fish in a symbiotic relationship; *(b)* that which arises from self-luminous cells on the fish, the photophores. Self-luminescence may be either light generated by the animal within its own tissues (intracellular luminescence) or by discharge of luminous secretion (extracellular luminescence). An example of the latter is a searsiid, *Searsia schnakenbecki*.

Fish families in which luminescence is of bacterial origin are represented in both shoal and deep waters and include the grenadiers (Melanonidae), cods (Gadidae), pine-cone fishes (Monocentridae), anomalopids (Anomalopidae), luminous cardinalfishes (Acropomatidae), slipmouths (Leiognathidae), sea basses (Serranidae), cardinalfishes (Apogonidae), swallowers (Saccopharyngidae), and deepsea anglerfishes and some of their relatives in the order Lophiiformes. Elasmobranchs with self-luminescence include some sharks *(Spinax, Centroscyllium, Etmopterus)* and an electric ray *(Benthobatis moresbyi)*. The greatest number of self-luminous species among the bony fishes belong to the deepsea scaly dragonfishes (Stomiatidae) and to the lanternfishes (Myctophiformes). Photophores are also present in the toadfishes and midshipmen (Batrachoididae).

Many hypotheses have been advanced to explain the value of

bioluminescence among fishes. Evidence supports the importance of light organs in illuminating the dark waters of fish such as the lantern fish *Gonichthys* and stomiatoids. Bioluminescence is a vital part of the courting behaviour of the midshipman. Other hypotheses awaiting confirmation are the use of light as lures for prey, to deter or confuse predators, for advertising, or to obliterate the fish's silhouette. Further possibilities also include the spacing out of the territories of midwater fishes. The ventral prevalence of luminous organs and tissues in such fishes suggests protection from the larger number of predators above.

POISON GLANDS

Poisonous glands are present in many fishes as skin derivatives, ostensibly as adaptations of mucous glands. They secrete a venom that when injected by puncture into man may be painful or, rarely, lethal. The study of these glands and their secretions is a part of the field of ichthyotoxism so well advanced by the work of Halstead. The field of ichthyotoxism includes the various forms of intoxication resulting from eating poisonous fishes (ichthyosarcotoxism) or being stung by venomous fishes (ichthyoacanthotoxism).

As an integumentary derivative, venom glands have evolved independently several times in widely separated families of fishes. Ichthyoacanthotoxism is known to result from the stings of many fishes including the following (selected to show spread through several orders), for which the stinging apparatus is briefly described and venomous examples given.

a. Sharks, fin spine with venomous glandular epithelium in groove-horn shark *(Heterodontus francisci)* and spiny dogfish *(Squalus acanthias)*.

b. Rays (Dasyatidae and Myliobatidae), caudal stinger with glandular skin of its sheath-stingrays *(Dasyatis)*, spotted eagle ray *(Aetobatis narinari)*, bat stingray *(Myliobatis californicus)*.

c. Chimaeras (Chimaeridae), dorsal fin spine with venom from glandular epithelium composing its sheath and lining its *groove-Chimaera* and *Hydrolagus* (the ratfish).

d. Scorpionfishes, lionfishes, and rockfishes (Scorpaenidae), dorsal, anal, and *pelvic fin spines* with venom glands in their grooves-including species of scorpionfishes proper *(Scorpaena)*, bullrouts *(Notesthes)*, lionfishes *(Pterois)*, and stonefishes *(Synanceja)*.

e. Weeverfishes (Trachinidae), opercular stinger and spines of dorsal fin, all with venom glands in their grooves-Trachinus.

f. Toadfishes (Batrachoididae), opercular stinger and dorsal fin spines with glands at their *bases-Batrachoides, Thalassophyryne,* and *Opsanus.*

g. Catfishes, spines of dorsal and pectoral fins with glands beneath skin, opening through pores at bases of spines-stonecats and madtoms *(Noturus),* bullheads *(Ictalurus),* seacatfishes *(Galeichthys felis* and *Bagre marinus),* and Indo-Pacific catfishes *(Heteropneustes, Clarias,* and *Plotosus).*

h. Surgeonfishes (Acanthuridae), stinger on each side of caudal peduncle presumably with venom glands in sheath of *spine-Acanthurus, Naso.*

i. Dragonets (Callionymidae), fin spines with venom *glands-Callionymus.*

j. Rabbitfishes (Siganidae), dorsal, anal, and pelvic fin spines with venom glands on *them-Siganus.*

k. Stargasers (Uranoscopidae), shoulder stingers with venom glands at bases.

Venom from stings of the following fishes may be fatal as well as having painful symptoms to humans: rays, chimaeras, scorpionfishes, weeverfishes, and stargasers. Nonfatal but nevertheless painful to humans (sometimes very much so) are stings of the venomous sharks, toadfishes, catfishes, surgeonfishes, rabbitfishes, and dragonets. All sting puncture wounds convey to a recipient human the chance for secondary infection including gangrene and tetanus. Nothing is known of the biological significance of the venom apparatus in fishes although it is easy to surmise roles in food getting, offense, or defense.

2

FEEDING AND DIGESTION

Fishes, as all animals, require adequate nutrition in order to grow and survive. Through observation in the field and examination of the contents of digestive tracts, and through physiological studies in the laboratory, researchers have learned much concerning feeding behavior and the kinds of organisms that are eaten as well as the mechanisms that have developed for digestion. Experiments have also disclosed dietary values of various classes of foods and have analysed factors that affect growth. Particularly significant has been the information on nutrition gained as a corollary of man's attempts to propagate fishes in the most efficient manner possible.

The gross anatomy of the digestive tract is described in elsewhere in this chapter.

NATURAL FOOD OF FISHES

Varieties of Foods

Nature offers a great diversity of foods to fishes including nutrients in solution and hosts of different plants and animals. Little is known of the direct uptake of soluble nutrients by fishes, although there is some evidence of direct uptake of glucose from ambient water. Many essential and inessential compounds and ions in the water are absorbed directly (as through the gills) or are swallowed with food and then absorbed in the digestive tract. For example, extra-enteral absorption exists for calcium, vital to scale and bone formation.

It is to be expected that a group as diversified as the fishes has become adapted to a wide variety of foods. Some fishes feed exclusively on plants, others feed only on animals, whereas a third and larger group derives its proteins, carbohydrates, and fats, as well as vitamins and most minerals necessary for growth and upkeep, from both plant and animal sources. Specialised parasitic fishes such as the sea lamprey (*Petromyzon marinus*) live on the blood and the tissue fluids of other fishes. Many young fishes, with tiny mouths and often still with yolk in their yolk sacs, begin to feed on plankton, the microscopic plant and animal life of water, including bacteria. Some fishes, adapted with numerous, elongate, and close-set gill rakers such as the gizzard shads (*Dorosoma*) feed for their entire lives principally on phytoplankton. Other fishes that have cutting teeth, such as parrotfishes (Scaridae) and surgeonfishes (*Acanthurus*), as well as the freshwater tilapia (*Tilapia melanopleura*) and the grass carp (*Ctenopharyngodon idella*) take parts of plants as food, nibbling off leaves, fronds and small stems. The parrotfishes are particularly well fitted for this habit with incisor-like teeth that are more or less fused into a cutting-edged beak. The beak serves to bite off pieces of algae whereas a set of pharyngeal grinding teeth mills the pieces to break open the cells and expose their contents to the action of digestive juices. The grinding teeth also mill pieces of coral.

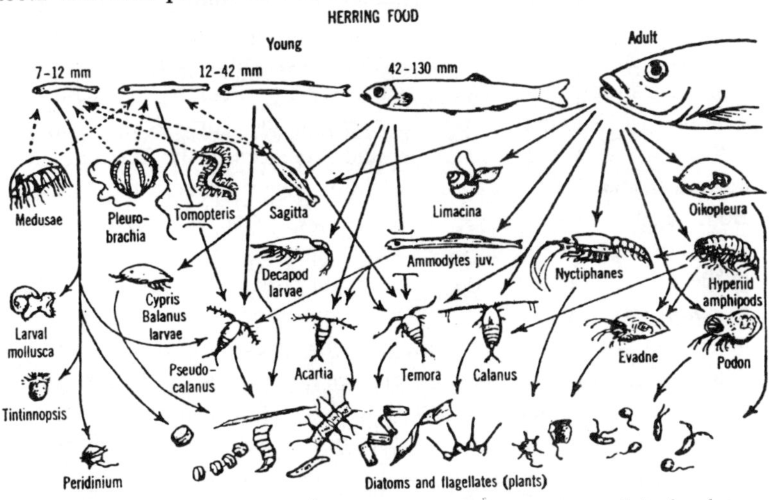

Figure 2.1 : Diagrammatic summary of the food relationships of the herring (**Clupea harengus**) *at different sizes.*

A wide range of kinds and sizes of animals is important in the food chains of fishes. Among the earliest animal foods to be consumed are animal plankton organisms, or zooplankton. Zooplankton includes many different kinds of protozoans, microcrustaceans and other microscopic invertebrates, and the eggs and larvae of many animals including those of fishes themselves.

Of outstanding importance as fish food among the larger invertebrate animals are: annelid worms (Annelida); snails, mussels, and clams (Mollusca); and crustaceans and insects (Arthropoda). All of the classes of vertebrates are prey for fishes-birds, mammals, reptiles, amphibians, and, of course, fishes themselves. It is not uncommon to find a small rodent, snake, or turtle in the stomach of a bass (*Micropterus*), a pike (Esox), a gar (*Lepisosteus*), or a bowfin (*Amia*). Sometimes a duckling or a hapless songbird will be found in one of the foregoing fishes as well, or in a goosefish (*Lophius*). As every angler knows, frogs are a natural food of predatory freshwater fishes; hence, they make good bait for many game species. Not even humans escape, for they are attacked by several fishes, including barracudas (*Sphyraena*), certain sharks, and piranhas (*Serrasalmus*) of South America.

Abundance of Food

Because of natural fluctuations in abundance, any one food organism is not of constant numerical availability to fishes. Such fluctuations of forage organisms are often cyclic and due to factors of their life histories or to climatic or other environmental conditions. Fish migrations often reflect a search for a particularly abundant source of food. The herring (*Clupea harengus*) in the North Sea, for example, follows plankton concentrations, and predators of the herring, such as certain mackerels (Scombridae), follow the fish about the North Sea. Both growth and survival in fishes have been shown repeatedly to reflect changes in abundance in food organisms. Availability of plankton at the time fish hatch is important to survival of the year class. Trouts (*Salmo; Salvelinus*) in American streams accomplish most of a year's growth in a few weeks of the late spring or early summer, when there is a peak of insect emergence. This emergence takes place at the time when the aquatic stages of many different kinds of insects metamorphose and leave the water preparatory to reproduction. During this interval the prey is seasonally concentrated and particularly vulnerable to predation.

Abundance of a potential food species often determines whether or not it will be eaten by fishes, for indeed availability is a key factor in determining what a fish will eat. Most fishes are highly adaptable in their feeding habits and utilise the most readily available foods. Relatively few kinds approach being strict herbivores or carnivores, and perhaps none at all feed solely on any one organism. Some, such as the Bermuda angelfish (*Holacanthus bermudensis*), may even change their diet with the season and may be quite herbivorous in winter and spring and become predominantly carnivorous in summer and early fall.

Food Chains, Food Webs, and Food Pyramids

Fishes are tied to other forms of life in their environment by food webs. If one considers the relative positions of the eaters and the eaten, herbivores and carnivores are at different vertical positions (trophic levels) of a food pyramid. Usually the largest carnivore or top predator can be placed at the apex of the pyramid. Such a pyramid represents the decreasing numbers of these animals in a community as their individual size increases; it may also represent decreasing biomass in the different levels of the pyramid with the least at the top, (the largest predator).

The bottom level is, of course, occupied by green plants that bind the sun's energy for further transfer through the living world. Then comes an intermediate level composed of the herbivores-mollusks, certain insect larvae, and many crustaceans and fishes. Finally there is the highest trophic level, occupied by carnivores, in which there may be several tiers of fishes that prey on fishes that prey on fishes, and soon.

To complete the round of the wheel of life or to close the food cycle, excretions of living fishes and other organisms and their remains when dead are acted upon by bacteria. The materials composing them are returned into solution to give again to plants the raw materials to harness radiant energy into new protoplasm during the process of photosynthesis.

Food relationships may appear relatively simple when we look only at the adult stage of a fish, say a black bass (*Micropterus*) or a bluegill (*Lepomis macrochirus*), but when the entire life cycle of a fish is considered, from fry to juvenile to adult, food webs are likely to be very complex. They appear as a web of interrelations of eating and being eaten involving many phyla of aquatic organisms as the food changes with the life history stage.

A food cycle is fairly easy to visualise in the upper sunlit waters of lakes, rivers, and the oceans, as having a producing or synthetic phase, possessing consumer links and having return of raw materials to the synthetic phase through excreted wastes, death, and decay to be used again. But what of the dark, deepwater abyssal zones of the oceans, the deeps of great inland lakes such as Superior and Baikal? Here the organisms, including the fishes, depend on items that move into the deeps from above, either by migration or in the form of a rain of carcasses. Furthermore, in the deepwater zone of standing waters, the transfer may involve a great series of layers through which nutrients are relayed by organisms from the manufacturing zone of the upper waters to the purely consuming zones of the deepwaters. In this connection, one may visualise an upper zone in which synthesis is going on, followed downward by a deeper zone in which there are some crustaceans and other invertebrates and fishes that browse up into the lower part of the synthetic level.

Beneath the level of the browsing, pelagic crustaceans may be another one where there is so little light that no synthesis is possible but in which perhaps there are some small fishes that feed upon the crustaceans, to bridge the light and dark areas, and so on through successive fishes and other animals to the deeps. Here all organic matter that sinks accumulates into concentrations of ooze to become food for deep sea bacteria and food webs built upon them. At each successive downward level (the sea bottom excluded) the total biomass becomes smaller and smaller.

FEEDING HABITS

In any discussion of the feeding of fishes, both the manner and the stimuli for feeding need to be included. The feeding habits or the feeding behavior of fishes is the search for and ingesting of food. These should be distinguished from food habits and diet, which are the materials habitually or fortuitously eaten.

Major Feeding Types

As for the manner of feeding, only one broad common characteristic prevails the food is taken into the mouth. Other than this the feeding habits and adaptations of fishes are very diverse. Nevertheless, certain very broad types of food getting are recognizable either by species or by life'history stages. On this basis, fishes can be classified, although somewhat arbitrarily, according to their feeding habits as predators, grasers, food strainers, food suckers, and parasites.

Predators. Fishes that feed on macroscopic animals have certain adaptations in common. They usually have well-developed grasping and holding teeth, as in many sharks (Elasmobranchii), the barracudas (*Sphyraena*), the pikes, pickerels, and muskellunge (*Esox*), or the gars (*Lepisosteus*). In predatory fishes there is well-defined stomach with strong acid secretions, and the intestine is shorter than that of herbivores of comparable size. Many predators such as the voracious bluefish (*Pomatomus saltatrix*) and many deepsea fishes actively hunt their prey, whereas others, like the groupers (*Epinephelus*) often lie in wait till an animal passes and then dart out to grasp it. The anglerfishes (Lophiidae and Antennariidae) have developed an anterior ray of the first dorsal fin into a lure to attract their prey. The archerfish (*Toxotes jaculator*) of Southeast Asia even shoots down insects from plants at the water's side by spitting at them. The accurate aim of this fish testifies to its well-developed aerial vision. Some predatory fishes hunt by sight whereas others, including many sharks (Squaliformes), nocturnal fishes (such as bullheads, *Ictalurus*) and morays (Muraenidae) rely largely on smell, taste and touch and probably also on their lateral-line sense organs to locate and catch their prey.

Grasers

In grazing, the actual taking of the food is by bites, often by individual small ones. Sometimes organisms are taken singly or at other times in small groups in a rather continual type of browsing much like cows or sheep in a pasture. Grazing characterises many fishes that feed on plankton or on bottom organisms. For example, a bluegill (*Lepomis macrochirus*) feeds along the bottom of a lake, taking this and that, often individual dipteran midge larvae. Many young fishes that grow to be predators of other fishes feed on plankters which they hunt individually.

Parrotfishes (Scaridae) or butterflyfishes (Chaetodontidae) browsing on a coral reef, may scrape off coral particles to ingest polyps or bite off pieces of algal fronds as previously indicated. A very specialised kind of grazing is that in which fish browse on parts of one another. Scales that are plucked from others and ingested (lepidophagy) constitute the bulk of food of an Indian catfish (Schilbeidae) and an African cichlid. Furthermore, it is not uncommon in the crowded waters of fish hatcheries to see trouts (*Salmo; Salvelinus*) nip parts of fins from one another.

Strainers

The straining of organisms from the water is a generalised type of feeding, inasmuch as food objects are selected by size and not by kind. Plankton filtration and swallowing of the pea-soup-like concentrate illustrates this feeding act as in many herring-like (clupeoid) fishes including the gizzard shads (*Dorosoma*), as previously indicated. The menhaden (*Brevoortia tyrannus*), which occurs in large schools along the American Atlantic coast, swims mouth agape through the

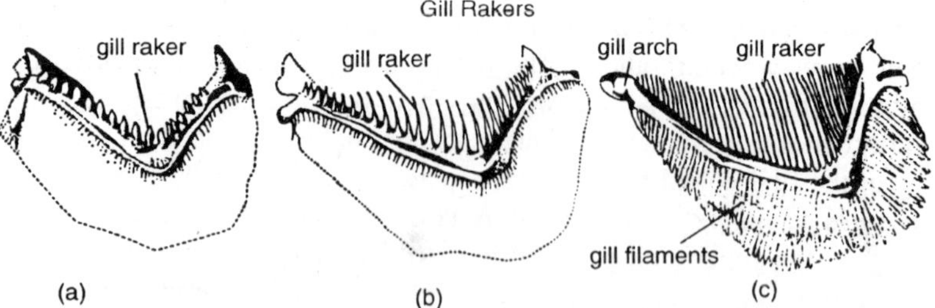

Figure 2.2 : Relationship of gill rakers to feeding habits, showing increasingly efficient sieve potential from left to right: (a) *round whitefish,* Prosopium cylindraceum; (b) *lake whitefish,* Coregonus clupeaformis; (*c*) *blackfin cisco,* Coregonus nigripinnis.

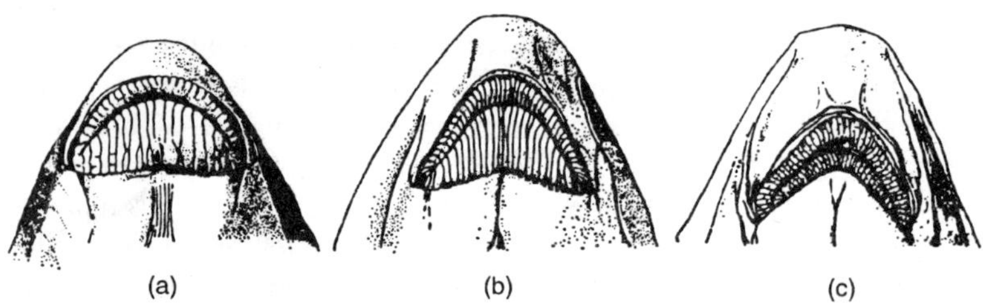

Figure 2.3 : Variations in the development of the suctorial lips and mouth in three redhorse suckers (Moxostoma); (a) shorthead redhorse, M. macrolepidotum; (b) golden redhorse M. erythrurum; (c) silver redhorse, M. anisurum.

rich plankton beds. An adult menhaden is capable of straining as much as 1-2 gallons of water a minute through its gill rakers. The fish can thus swallow during the same short time, several cubic centimeters of plankton concentrate, mainly diatoms and crustaceans. Other fishes, some attaining large sizes, such as the paddlefish (*Polyodon*), the basking shark (*Cetorhinus*) and the whale shark (*Rhincodon*) also have efficient food straining or filtering adaptations. A principal adaptation of the strainers is the development of numerous, close-set, and elongated gill rakers.

Suckers

The sucking into the mouth of food or food-containing material is often practiced by bottom feeding fishes such as the sturgeons (Acipenseridae) and suckers (Catostomidae). Old World minnows (such as fringelips, *Labeo* and *Osteochilus*), with inferior mouths and sucking lips, have similar habits. In a few carp (Cyprinidae) species in Southeastern Asia the sucking response depends so strongly on the stimulus of touch on the frilled, fleshy lips that fishermen bite the lips off these fishes before crowding them into holding pens with other species. The concern is that the other fishes will be damaged by being mistaken for the algae- or moss-covered stones on which the minnows usually feed by scraping and sucking.

Fishes that suck in mud to extract the organisms in it may or may not get a good mouthful of food with each ingestion. In some, the food items are separated from the sediments before being swallowed but in others, such as some oriental catfishes (Schilbeidae, Siluridae), remains of flocculent bottom deposits can be found in the digestive tract together with high concentrations of bottom organisms. This feeding habit indirectly plays havoc with machinery for oil extraction from fish wastes because metal parts become covered with a sticky emulsion of fish fats and clay.

Parasites

Parasitism is perhaps the most unusual and highly evolved feeding habit among animals. In the fish world, an outstanding example of this practice is represented by parasitic lampreys (some Petromyzonidae) and hagfishes (Myxinidae) that suck body fluids from the host fish after rasping a hole in the side of the body. A species example is the sea lamprey (*Petromyzon marinus*) of the western North Atlantic and adjacent continental waters. Another is the Pacific lamprey (*Lampetra tridentata*) that includes whales among its hosts. A deepsea eel (*Simenchelys parasiticus*) is also parasitic. The males of some of the deepsea anglerfishes (*Ceratias*) are obligatory parasites on the females of the same species. Shortly after hatching, the male finds a female and attaches by his mouth to her body. The female obligingly responds by developing a fleshy papilla from which the male fish can absorb nutrients since subsequent to attachment he takes no free-living food at all. The male remains relatively small, little more than an animated gonad. The intra-uterine absorption of nourishment (and respiratory exchange) by the embryos of certain livebearing fishes represents an even greater nutritional specialisation than the foregoing extreme of parasitism.

Feeding Adaptations

The diversity in feeding habits that fishes exhibit is the result of evolution leading to structural adaptations for getting food from the equally great diversity of situations that have evolved in the environment.

Lips

A significant advance in vertebrate evolution was the appearance

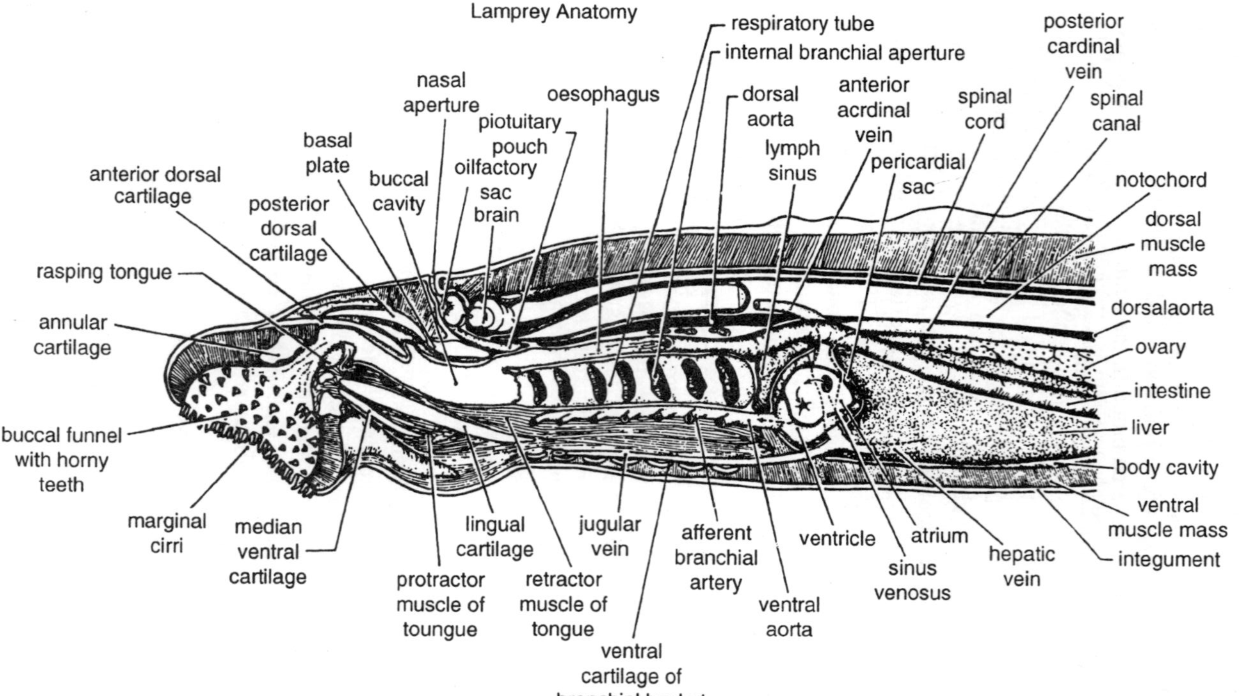

Figure 2.4 : Jawless mouth, foregut and related structures of an adult sea lamprey (Petromyzon marinus).

of true jaws to border the mouth opening. Most generally the jaw-equipped mouth has a biting function and fishes that swallow large morsels of food usually have unmodified relatively thin lips. Suctorial feeders have an inferior mouth and fleshy modification of lips. Notable among these are the sturgeons (Acipenseridae) of the North Temperate Zone and the suckers (Catostomidae) of North America and Asia. The lips of the sturgeons and suckers are mobile and described as plicate (having folds) or papillose (having small tufts of skin or papillae). Many suctorial feeders also have well developed barbels more or less bordering the mouth, as in the sturgeons (*Acipenser*) and Asiatic hillstream loaches (Homalopteridae). The barbels have many sensory end organs and help to locate food grubbed from soft bottom materials.

Suctorial lips of free-living fishes may also serve as holdfast organs in fast flowing mountain streams, as in one of the Southeastern Asiatic sisorid catfishes, *Glyptosternum*, some of the armored catfishes of South America (Loricariidae), and in the loach-like gyrinocheilid (*Gyrinocheilus*) of Southeastern Asia. The last fish has an extremely specialised suctorial adaptation in having developed in addition to suctorial lips, an opercular structure with a separate water inhalent and exhalent device; for these purposes the fish has two branchial openings on each side. The mouth is relieved thereby of taking in respiratory water when engaged in suction for holdfast or feeding. The mouth is small and the inner surface of the lips has rasp-like folds to facilitate the scraping of algae from the stones to which the fish adheres.

In the parasitic members of the lamprey family (Petromyzonidae) and in the hagfishes (Myxinidae) the jawless suctorial mouth serves both as a holdfast for attachment to the host and as a food-remover from the host. The sucking disc of lampreys is also used to dislodge and transport stones from the nest pit in streams. Later the disc serves to anchor spawning individuals into a side-by-side position with head directed upstream and genital pores located over the nest excavation.

Modifications in the Shape of the Mouth

Among the grasers and suctorial feeders there exist not only specially developed lips but also adaptations of other mouth parts. The trumpetfishes (Aulostomidae), the cornetfishes (Fistulariidae), and the pipefishes (Syngnathidae), as well as many butterflyfishes (Chaetodontidae) of coral reefs, have mouths that resemble elongated beaks. This adaptation is achieved by a protraction of the hyomand-

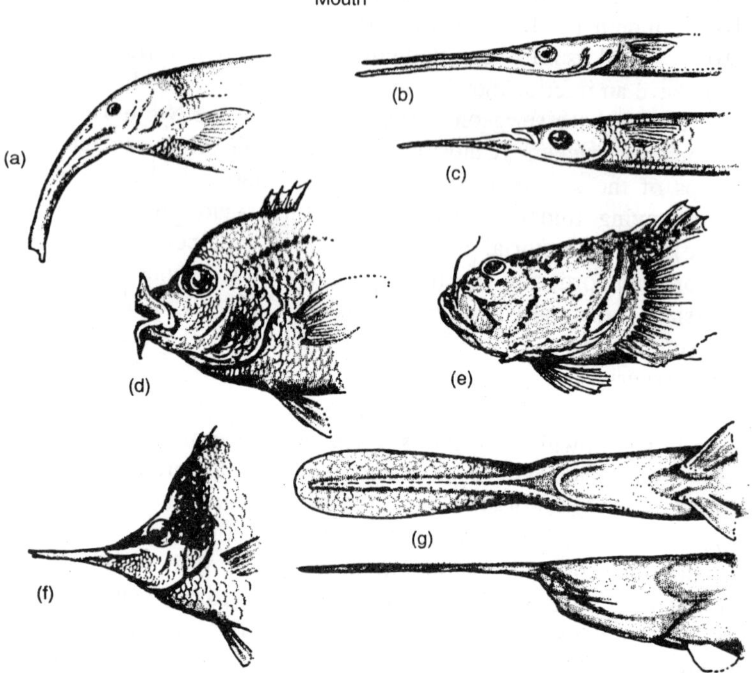

Figure 2.5 Some structural adaptations of the mouth among fishes: (a) elephantfish, Gnathonemas elephas; (b) garfish, Strongylura longirostris; (c) halfbeak, Hyporhamphus unifasciatus; (d) thick-lipped mojarra, Cichlasoma lobochilus; (e) stargaser, Zalescopus tosae;· (f) longnose butterflyfish, Forcipiger longirostris; (g) paddlefish, Polyodon spathula.

ibular bone rather than by a lengthening of the lower jaw bones (dentaries) themselves.

Among the fishes, the method of feeding may be by suction such as a syringe would exert in the case of the trumpetfishes, cornetfishes, and pipefishes, or it may be a selective grazing action with sharp teeth where the long snout enables the butterflyfish to reach into small crevices of the coral.

Some predators, such as the dories (Zeidae), certain wrasses (Labridae), and the European bream (*Abramis brama*, Cyprinidae), can form temporary tubes in which to engulf their prey from close range by forward extension of the jaws enabled by special articulation of the premaxillaries and other skull bones. Other adaptations, also of skull bones and their articulations, to increase the gape of the mouth exist among predatory deepsea fishes such as the deepsea viperfish (*Chauliodus*).

A peculiar structure among mouth modifications has arisen in the halfbeaks (Hemiramphidae) where the lower jaw projects into a beak, often a third of the length of the fish itself, with the mouth opening above it. Halfbeaks are usually surface-feeding fishes, and it has been surmised that the "beak," as well as being an aid in food getting, serves for steering and maintenance of equilibrium in conjunction with a posteriorly inserted dorsal fin.

Teeth

Outstanding among the obvious adaptations for feeding in fishes are the teeth. They are thought to have arisen from scales covering the lips, as represented in living sharks (Squaliformes), where the placoid scales of the skin visibly grade into teeth on the jaws. In the bony fishes (Osteichthyes), teeth are of three kinds, based on where they are found: jaw, mouth, and pharyngeal. Jaw teeth are variously those on the maxillary and premaxillary bones above, and on the dentaries below. In the roof of the oral cavity, teeth are variously borne by the median vomer and by the palatine and ectopterygoid bones on each side. In the floor of the mouth, the tongue often has teeth on it. Pharyngeal teeth occur as pads on various gill arch elements in many species. In the carps (Cyprinidae) and suckers (Catostomidae), the only teeth are those deep in the pharynx that develop from modifications of lower elements of the last gill arch. Tooth-like modifications of gill rakers and dermal bones (perhaps scales) are not uncommon supplemental Zor'nments found on the inner surfaces of the pharyngeal arches in many predacious fishes such as the northern pike (*Esox lucius*).

Based on their form, some major kinds of jaw-teeth are the following: cardiform villiform canine, incisor, and molariform. Cardiform teeth are numerous, short, fine, an 'poiinted. A pad of them on a bone resembles the multiple-toothed wool card of spinning-wheel days, hence the term cardiform. Such dentition with variations is found in very many fishes that have multiplerowed teeth; included would be North American catfishes (Ictaluridae), perches (Percidae), and many seabasses (Serranidae). Villiform teeth are more or less elongated cardiform ones, in which the length-to-diameter relationship resembles that of intestinal villi (as in the needlefishes, *Belone*, and lionfishes, *Pterois*). Canines are dogtooth-like, often even quite fang-like. They are elongated and subconical, straight or curved and are adapted for piercing and holding; they are possessed by the walleyes (*Stizostedion*), among many others. In certain fishes, such as the morays

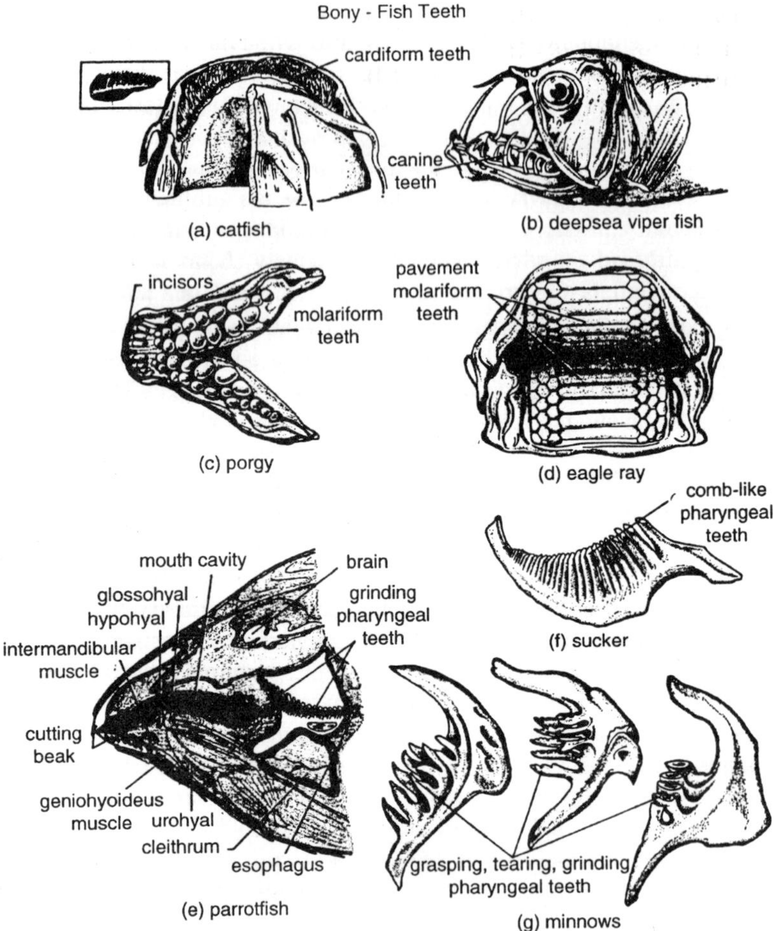

Figure 2.6 : Some variations of teeth in fishes by structure and location: (a) *upper jaw of flathead catfish*, Pylodictis olivaris; (b) *deepsea viperfish*, Chauliodus sloani; (c) *lower jaw of a porgy, Sparidae;* (d) *pavement teeth on roof and floor of mouth of an eagle ray*, Myliobatis; (e) *sagittal section of a parrotfish (Scaridae) to show "beak" from fusion of jaw teeth and to show pharyngeal teeth;* (f) *pharyngeal teeth of a golden redhorse sucker*, Moxostoma erythrurum; (g) *types of minnow pharyngeal teeth, from left to right, showing pointed, grasping teeth of a barb* (Barbus), *crenulated, tearing teeth of a rudd* (Scardinius), *and molariform grinding teeth of the tench* (Tinca).

(Muraenidae) and the pelagic marine predator, the so-called handsawfish (*Alepisaurus*), the canines are hinged, yield to backward pressure, but snap into a locked position when pushed toward the mouth openings-obviously an adaptation to retain living, moving prey.

Sharply edged cutting teeth are called incisors. In some fishes such as the sea bream (*Archosargus*) they look almost human; in others they are crenulated (sometimes saw-edged), and in still others they have become variously fused into cutting "beaks" as in the parrotfishes (Scaridae). Molariform teeth are for crushing and grinding, and hence have flattened, often broadly occlusal surfaces; they are generally characteristic of the bottomdwelling skates (including *Raja*), the chimaeras (Holocephali), and several drums (Sciaenidae).

There is strong correlation among kind of dentition, feeding habits, and food eaten. Predacious fishes, such as the pike (*Esox*), gars (*Lepisoteus*), and deepsea gulpers and swallowers (Saccopharyngoidei) have sharply pointed teeth of apparently great use in grasping, puncturing, and holding prey. Feeders on plankton and scrapers of encrusting periphyton characteristically have toothless jaws, although by special adaptation the jaw itself may have a cutting edge as in the stoneroller minnow (*Campostoma anomalum*) where the dentaries have a hardened, gristly leading edge that is used in scraping surfaces for food. In skates (Rajidae) and in drums (Sciaenidae) there are grinding (molariform) teeth in oral or pharyngeal cavities in association with a diet that includes snails, clams, and hard-bodied crustaceans. Razor-like cutting teeth have developed in predacious fishes such as the formidable piranha (*Serrasalmus*) of the Amazon and the barracuda (*Sphyraena*) of warm seas. Some vegetable feeders, such as many parrotfishes (Scaridae), chop their food into a veritable salad with cutters resulting from the fusion of individual teeth into sharp-edged ridges on the jaws. Initial cutting is then followed-up with fine trituration by means of pharyngeal grinding plates.

Within a single group the diversity of dentition on identical bones may be almost as great in principle as the foregoing gamut. In the carps and minnows (Cyprinidae), the pharyngeal teeth range from sharp in the carnivores such as the creek chub (*Semotilus*) to molariform in the common carp (*Cyprinus carpio*) and have almost disappeared in some species. In these and many other fishes, teeth are of value in classification. Many species of fossil sharks have been described solely on the basis of teeth.

In general, teeth are absent in plankton feeders and in some of the more generalised omnivores. They are present on increasing numbers of bones in the more and more predacious fishes. The premaxillary bones are toothed in most fishes that have any jaw teeth at all. This is true of many such softrayed species as the bowfin

(*Amia*), the gars (*Lepisosteus*), the salmons and trouts (Salmoninae) and of tooth-bearing spiny-rayed fishes in general (for example, the large order Perciformes).

The maxillae are typically toothed in those soft-rayed fishes that carry premaxillary teeth. However, the maxillae are characteristically toothless in otherwise tooth-bearing, spiny-rayed fishes. In the process of evolution from soft-rayed fishes to spiny-rayed ones, the maxillae are tooth-bearing bones only in some soft-rayed kinds. In those with spiny rays, the maxilla is edentulous, and is excluded from the gape, no longer forming part of the margin of the mouth.

Gill Rakers

Besides protecting the tender gill filaments from abrasion by ingested materials that are coarse in texture, gill rakers are also specialised in relation to food and feeding habits, as previously described. They are very stubby and unadorned in omnivores such as the green sunfish (*Lepomis cyanellus*) and the pumpkinseed (*L. gibbosus*). In many plankton feeders, the gill rakers are elongated, numerous, and variously lamellated or ornamented, presumably to augment efficiency in straining. Simple, but very numerous rakers are possessed by gizzard shads (*Dorosoma*) and the paddlefish (*Polyodon*). Ornamented structures, however, are found in flatfishes (Pleuronectidae) and are of taxonomic use. Here each raker resembles a feather in having' a main axis with lateral processes (and even branches on these lateral elements). When these processes on adjacent rakers overlap, a sieve is formed that can strain very finely.

The Digestive Tube

Another adaptation that fishes have for feeding is the great distensibility of the esophagus. Seldom does an individual choke to death because it cannot swallow something that it got into its mouth. Only occasionally is a predacious fish found in mortal distress because of a prey fish lodged in its throat. Catfishes (Ictaluridae) and sticklebacks (Gasterosteidae) are frequently offenders to their predators in this regard. Their spinous fin rays, when erected, cause them to become stuck. Yet most often if these spines had been depressed, the engulfed and lodged fish would have slipped down the esophagus very easily. In general the esophagus is so distensible that it can accommodate anything that the fish can get into its mouth and can sometimes even accommodate the item if it happens to double on itself two or three times on its way to the stomach.

The stomach, too, shows various adaptations, one of which is shape. In fish-eating fishes, the stomach is typically quite elongate as in the gars (*Lepisosteus*), bowfin (*Amia*), pikes (Esox), barracudas (*Sphyraena*), and the striped bass (*Morone saxatilis*). In omnivorous species, the stomach is most often sac-shaped, similar to that in humans. A very special adaptation is the modification of the stomach into a grinding organ, as in the sturgeons (*Acipenser*), gizzard shads (*Dorosoma*), and mullets (Mugil). Here the stomach is reduced in overall size but its wall, greatly thickened and muscularised. The lining too is heavily strengthened with connective tissue, and the lumen (the space within), made very small. The organ is not a bin for mixing and primary digestion but rather a food grinder, not unlike the gizzard of chickens and other fowl. Great distensibility is the adaptation of the stomach in the predatory deepsea swallowers (Saccopharyngidae) and the gulpers (Eurypharyngidae) enabling these fishes to take relatively huge prey.

A remarkable modification of the stomach exists in the puffers (Tetraodontidae) and porcupinefishes (Diodontidae) which can inflate themselves with water or air to assume often an almost globular shape. Either the stomach itself or an evagination from its anterior portion are filled by action of cardiac and pyloric sphincters and by another sphincter in the evagination itself. When inflated, this evagination or gastric sac is so distended that its walls are paper thin. Emptying of the sac proceeds, again by sphincter action, assisted by the abdominal body musculature. The presence of a distensible stomach in these families goes hand in hand with the absence of pelvic fins; the pelvic bones may be absent also. The adaptive value of this modification of the digestive tract is probably mainly one of defense, for many puffers and porcupinefishes have spines all over the body which can thus be erected. It ought to be mentioned that the gastric sac of puffers, porcupinefishes and some of their relatives is neither related to, nor derived from the gas bladder.

Not all fishes have a stomach, that is a portion of the digestive tube with a typically acid secretion and a distinctive epithelial lining different from that of the intestine. In the plant-feeding roach (*Rutilus rutilus*), a cyprinid of the Old World, for instance, epithelial tissue of the esophagus grades directly into that of the intestine. In other grasers such as the parrotfishes (Scaridae) analogous conditions are found. Some carnivores also have lost their stomachs, as the sauries (*Scomberesox*), as have certain plankton-feeding species in the pipefish-seahorse family, Syngnathidae. The primary criterion for being able to do without the stomach does not seem to be whether a fish is an

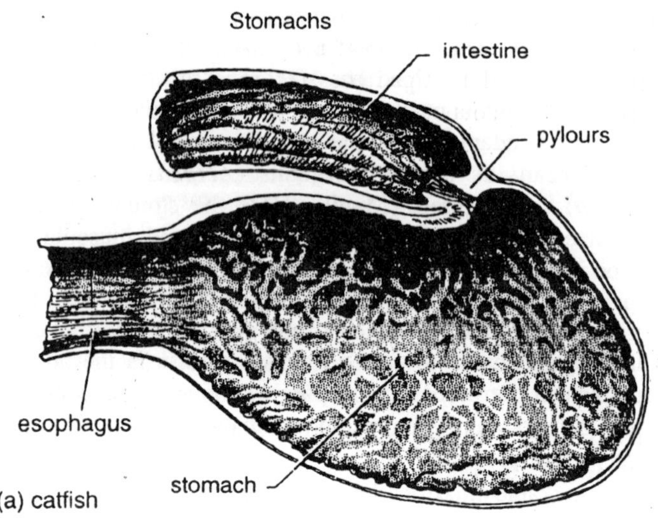

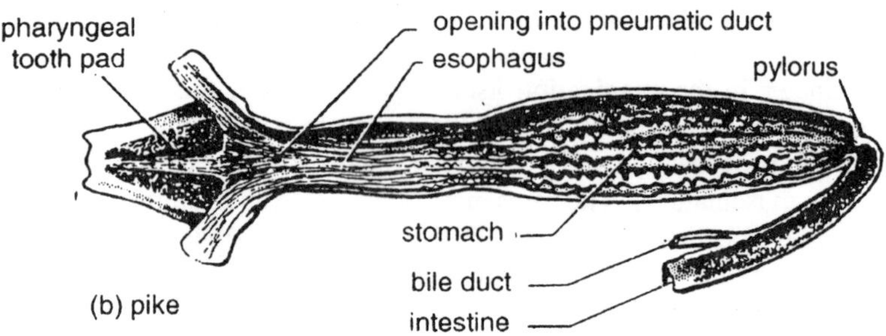

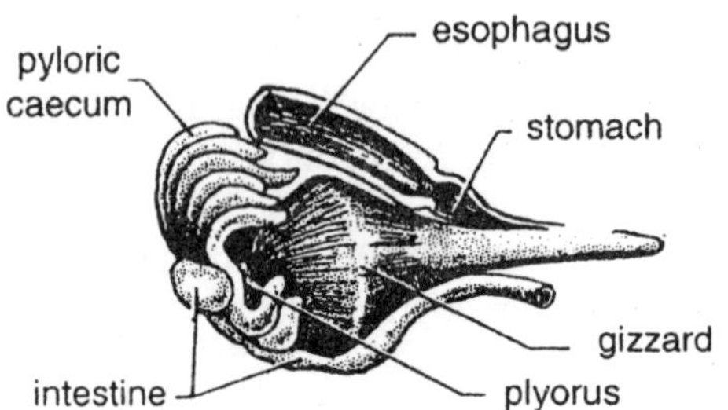

(c) mullet

Figure 2.7 : Variations in shape, appendages, and lining of the anterior portion of the digestive tract in three fishes.

herbivore or a carnivore but whether accessory adaptations for trituration and very fine grinding of food exist either in the form of teeth or a grinding apparatus such as a gizzard. Where stomachs exist, most pronouncedly in carnivores, they are characterised by a low pH and the prominent presence of pepsin among other digestive juices.

The intestine too, has many variations. It is shortened in essential carnivores such as the pike (*Esox lucius*) perhaps because meaty foods can be digested more readily than vegetable ones. In opposite fashion, it is often elongated and arranged in many folds in predominantly herbivorous species, as in certain loricariid catfishes. The sharks and relatives (Chondrichthyes), and a few other fishes have substituted a spiral valve or a large fold of absorptive tissue loosely wound in a roll for folds of the intestine and have ostensibly improved efficiency of digestion and/or absorption and visceral compactness thereby. The intestine itself seems to undergo digestion (autolysis) in fishes that cease feeding as sexual maturity and breeding arrive. The once-functional digestive tract of an adult sea lamprey (*Petromyzon marinus*) becomes a mere thread with practically no lumen by the time spawning is over and death approaches. Similar reduction has taken place in migrant Pacific salmons (*Oncorhynchus*) when they have reached their freshwater spawning grounds.

Stimuli for Feeding

Of interest both to ichthyologists and anglers are considerations of the stimuli to feeding in fishes. However, the mechanism of feeding behavior is a very complicated one, as we shall see. The stimuli to feed are of two kinds: (a) factors affecting the internal motivation or drive for feeding, including season, time of day, light intensity, time and nature of last feeding, temperature, and any internal rhythm that may exist; (b) food stimuli perceived by the senses like smell, taste, sight, and the lateral-line system, that release and control the momentary feeding act. The interaction of these two groups of factors determines when and how a fish will feed and what it will feed upon.

One important factor in feeding is time of day. Some fishes, such as the bullheads (*Ictalurus*) that find food by smell and taste, are predominantly night feeders. Still others, such as the pikes (*Esox*) and other predators that feed largely by sight, are most active during daylight hours. Season, strongly influencing water temperature in the nontropical areas and water levels in the tropical region, seems to have something to do with feeding also. Some fishes cease feeding

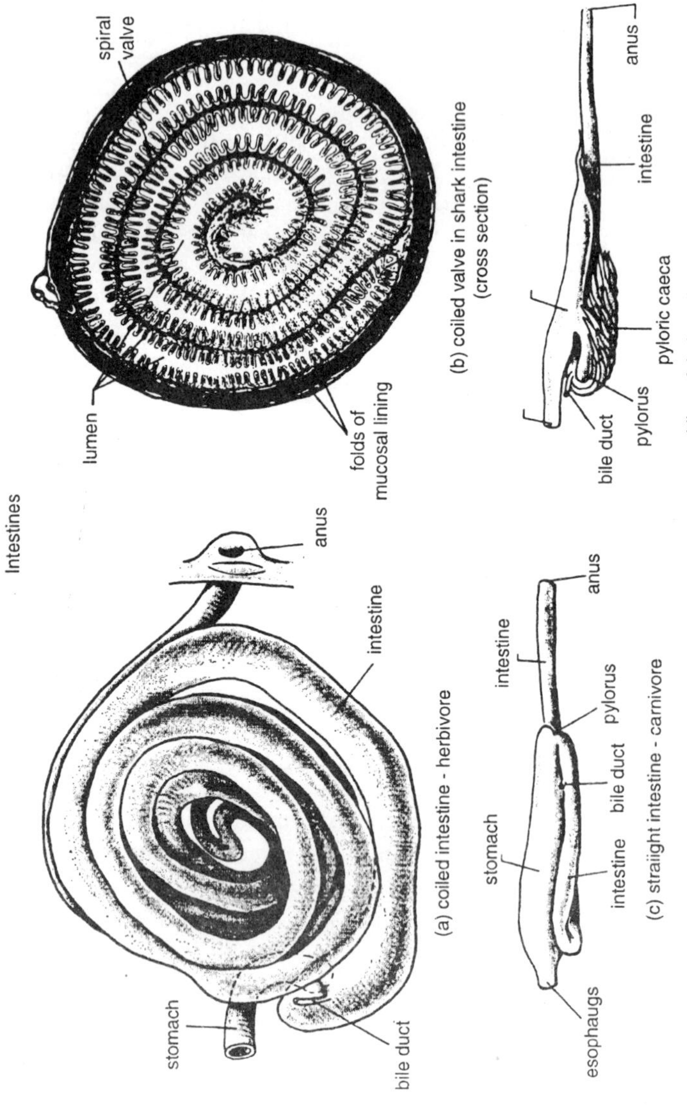

Figure 2.8 : Variation in intestinal length and other features among carnivourous and herbivorous fishes.

altogether during their spawning seasons, including salmons (*Salmo*, *Oncorhynchus*) and lampreys (Petromyzonidae). Others such as the swamp eels of southeastern Asia (Synbranchidae), while estivating in moist burrows in the mud, live for weeks on accumulated fat. Most fishes of the Temperate Zones feed very actively as environmental conditions change in the spring, near the beginning of a new growing season. Fishing lore is replete with contradictory statements regarding climatic conditions or stimuli that cause fish to bite. However, all that can be said truthfully is that very much remains to be discovered regarding the regulatory role of climatic factors in feeding activity.

Since fishes have senses of smell and taste, many workers have tested chemical factors as determinants of feeding. For many species, the chemical senses do indeed first cue the feeding act. The dogfish sharks (*Squalus*) as well as some of the man-eating sharks (*Carcharodon*) are led to feed importantly by chemical attraction, and moray eels (*Gymnothorax*) have been shown to find and select food primarily by smell. Catfishes (Ictaluridae) and goatfishes (Mullidae) feed mainly by taste and also by touch.

Visual stimuli trigger the feeding act in fishes with movement, color, and the shape of objects perceived each having its role as releasers of the act. Among schooling fishes of one species, as well as in aggregations of several species in a common environment, nonfeeding individuals may be triggered to feed by the discovery of food by one individual. In fact, it has been suggested that one adaptive value of the schooling habit is to facilitate feeding. Aquarium fishes can be trained secondarily to associate feeding with some auditory, olfactory, or visual stimuli. Similarly, fishes learn to avoid certain foods because of their texture, taste, or other quality. A pike (*Esox*), for example, soon learns to avoid a sharp-spined stickleback (*Gasterosteus*) as prey although the innate response of the pike may be to feed on an object with the dimensions, movement, and color of a stickleback.

Rapid changes in light intensity are also important in feeding. The yellow perch (*Perca flavescens*) and many other fishes exhibit circadian rhythms in feeding activity and movement with peaks at dawn and dusk. There are many subtle changes in the environment tied to day and night or tidal conditions; included are not only light, but also temperature, salinity, pH, and current. The variations of any one of these factors by itself or in combination with others are discernible by fishes and may influence activity and feeding pattern.

Innate timing mechanisms, termed physiological clocks, independent of, or unrelated to environmental stimuli may exist in fishes. In nature, however, timing landmarks such as diurnal light, pH, temperature, and other changes are rarely absent and usually trigger levels of activity.

Promptings for feeding may include those of actual contact and textural feel. In aquaria, when both the bluegill (*Lepomis macrochirus*) and the pumpkinseed (*L. gibbosus*) are offered small snails, only the pumpkinseed will crush the shells, eject them through their gill openings, and swallow the soft parts. However, if soft parts of small snails alone are offered the bluegill, they will be swallowed. This may be evidence of the significance of texture in feeding and a sense of touch in the mouth. It should be pointed out in this connection that although the jaw teeth are much the same in these two sun- fishes, the pharyngeal teeth of the pumpkinseed alone are truncated and thus adapted for a crushing function.

In addition to factors such as the foregoing, there are others that doubtless lead fishes to feed. These include hunger, gluttony, curiosity, and other less defined factors. After a long interval of enforced deprivation between feeding bouts in sticklebacks, the feeding that follows is in some respects more rapid, suggesting increased motivation. Hunger and satiation may be controlled by the hypothalamus. Regarding gluttony, the inclination to eat long after the normal capacity of the animal has been reached has been described for the bowfin (*Amia*) in aquaria. During some food utilisation experiments, small serranids (*Epinephelus guttatus*) gorged themselves far beyond a sustained need for optimal growth. Gluttonous feeding results in the passage of apparently less well-digested material in the feces than is usual. It may, of course, be prompted by the shortage of some nutrient in the food.

Even under constant environmental conditions experimental brown trout (*Salmo trutta*) show cyclic fluctuations in annual growth rates, reflecting variations in their feeding and metabolism. Growth has a spring maximum, is rapid in summer, and is slow or negligible in autumn and winter. Another aspect of changes in metabolism, apparently determined by internal factors, concerns the occurrence of periods of rapid growth in length alternating with periods of rapid growth in weight of this trout. The periods in question are of 4 to 6 weeks in duration, but the reason for the fluctuations remains unexplained.

To temper the action of all stimuli to feed, there is the relation between internal motivation or drive for feeding and the degree of selectivity controlling the proper feeding act, such as the capture of prey. For example, the less active the fish is, the more the stimuli emanated by the prospective prey regarding such things as dimensions, form, color, and movement must correspond to the ideal prey experience, perhaps also innately, or both.

DIGESTION, ABSORPTION, AND UTILISATION OF FOODS

The basic function of the digestive system is to dissolve foods by rendering them soluble so that they can be absorbed and utilised in the metabolic processes of the fish. The system may also function to remove dangerous toxic properties of certain food substances.

Movement of Food in the Tract

The movement of foodstuffs and their undigested remains in the digestive tract of a fish is accompanied, as it is in man and other vertebrates, by peristaltic waves of muscular contraction. In the anterior part of the tract, at least, the movement is voluntary due to the presence of skeletal muscles in the wall of the gut. Elsewhere it is involuntary and involves smooth muscles. The tongue, in contrast to that in higher vertebrates, is not mobile by itself; it has no skeletal muscle components and moves only when the fish moves its underlying visceral skeleton. The typical route of a bolus of food in a fish is essentially that of higher vertebrates, as already described.

Many predatory fishes appear to regurgitate large food items from the stomach with great facility. It has been suggested that this is made possible by the pronounced development of striated muscles in the walls of the esophagus extending to the stomach.

Intestinal Surfaces

The mouth cavity, esophagus and stomach are lined with a soft mucous membrane as is the rest of the tract. There are no salivary glands (except in some specialised fishes such as the parasitic lampreys of the family Petromyzonidae). However, the tract wall is liberally supplied with glands that secrete mucus which lubricates passage of food materials, and protects the gut lining. The gut is highly elastic and permits the passage of outsised masses of food. The linings of the intestines, small and large, are highly absorptive and pick up digestive products in solution. The absorptive capacity of these areas is increased by throwing the walls into lengthwise folds which as the typhlosole in lampreys, Petromyzonidae), transverse folds (rugae), or multitudinous fingerlike projections (villi) in such fishes as mullets (*Mugil*). Along the course of the tract, there are many gland cells that contribute digestive enzymes.

Glands and Digestive Enzymes

The general, widespread distribution of the mucous glands both outside and inside the fish has already been described. Additionally, gastric glands occur in the stomach at least in predacious fishes. These secrete hydrochloric acid and pepsinogen, effective in combination to split large protein molecules. In typically carnivorous fishes such as the northern pike (Esox *lucius*) gastric acidities of pH 2.4 to 3.6 have been measured. Consequently there exists in the stomach a suitable substrate for protein-splitting enzymes or peptidases that have been demonstrated in many fishes. Evidences for stomach enzymes other than peptidase are not clear-cut. Some minnows (Cyprinidae) lack gastric glands and, on this basis, may not possess a true stomach. Similarly the fish gizzard does not have digestive glands.

Pyloric caeca, when present, may have a digestive function, an absorptive one, or both. The caeca are finger-like pouches that extend outward from the intestine near the pylorus; the enzyme lactase has been found in them in trout (Salmoninae). A higher level of saccharase (invertase) activity has been found in the pyloric caeca and intestine of carp and bluegill compared to pickerel. The former ingest a considerable amount of vegetable matter whereas the pickerel ingest practically none. The pyloric caeca and the intestinal mucosa are sources of lipase which breaks down fats into fatty acids and glycerine. Absorption of fats may occur in the anterior part of the intestine. The copepods upon which anchovies, sardines, and herring prey contain up to 50 percent wax. The pyloric caeca of these fish produce a bile-activated wax lipase, an enzyme that catalyzes the metabolism of wax. The morphology of the pyloric caeca as single or branched (bifid) as well as their numbers are characters of value in the classification of some fishes.

Several discrete or diffuse glands are associated with the digestive tract. The thyroid, the thymus, and the ultimobranchial are of endocrine function but arise embryologically from the developing digestive tract and have no direct bearing on digestion.

The liver, as in other vertebrates, arises in the embryo as a ventral evagination of the developing intestine. The anterior portion develops into the liver proper and the posterior portion into the gall bladder and its ducts. Some holostean and some teleostean fishes have more than two liver lobes but there are always only two hepatic ducts from the liver to the cystic duct which terminates in the gall bladder. The gall bladder is a temporary storage organ for secretions of the liver. From the junction of the hepatic and cystic ducts there arises the common bile duct which enters into the beginning of the intestine near the pyloric region.

Some deepsea fishes have a vestigial gall bladder only and in others, such as the burbot (*Lota Iota*), the gall bladder is completely absent. The liver adjusts its form to that of the body cavity and consists of lobules of tubular glands where the blood bathes tiers of cells on one side while bile is secreted into the bile canals on the other side. The bile contains the fat-emulsifying bile salts along with the bile pigments, biliverdin and bilirubin, that originate from the breakdown of red blood cells and hemoglobin. The bile salts may not only help to hydrolise fats but also to adjust the digestive juices of the intestine to the proper alkalinity for the action of digestive enzymes.

Besides its role in digestion, the liver also acts as a storage organ for fats and carbohydrates (glycogen). It further has important functions in blood cell destruction and blood chemistry, as well as other metabolic functions such as the production of urea and compounds concerned with nitrogen excretion. Fishes vary greatly in the amount of fat that is stored in the liver but two general types can be distinguished. In flatfishes (Pleuronecti-formes) and cods (Gadidae), for example, fat is stored predominantly in the liver. Fat is to a great extent stored in the muscle of fishes such as tunas (Scombridae) and herrings (Clupeidae). Sharks and bony fishes differ in the relative amounts of unsaturated fatty acids in their liver oils. In some sharks the liver may take up as much as 20 percent of the body weight. Besides storing fat, the livers of fishes also store vitamins A and D. The content of these vitamins in the tunas for instance, is so high that persistent eating of their liver may lead to severe disturbances in human metabolism, which presumably result from hyper-vitaminosis.

The pancreas is really two organs rather than one; it has both exocrine and endocrine functions. It varies greatly in type of development and position in the various groups of fishes. In sharks and rays (Elasmobranchii) the pancreas is relatively compact, often two-lobed, but in bony fishes (Osteichthyes) the pancreas is often diffuse. There are smaller or larger nodules of pancreatic tissue in the omentum, the suspensory tissues of the intestine, in addition to dispersion of the pancreas into the liver to form a hepatopancreas among many spiny-rayed fishes (Acanthopterygii). The arrangement of endocrine cells that secrete insulin, the pancreatic islets or the islands of Langerhans, also varies in different groups of fishes which possess a discrete pancreas.

In most fishes the surface folds inside the intestine show a rectangular pattern, apparently instrumental in slowing down the passage of the food. The enzymes secreted by the small intestine, as well as bile and pancreatic secretions that pour into this part of the gut, work best at a pH ranging from neutral to alkaline. Different proteases, affecting terminal and inner bonds uniting the amino acids of proteins, are secreted either from the intestinal mucosa, pancreas, or pyloric caeca. Alkaline protease has been found in the intestine of rainbow trout and acid protease in the stomach of this species. Intestinal enzymes are secreted in an inactive form as zymogens, a general term for inactive enzymes, before chemical changes in the lumen of the intestine make them active in digestion. This change is brought about by other enzymes such as enterokinase. This adaptation prevents self-digestion (autolysis) of the intestinal mucosa.

Various enzymes that digest specific carbohydrates have been found in the intestine as well as in the pancreatic juice of fishes. In the predominantly herbivorous *Tilapia*, amylase activity was distributed throughout the gastrointestinal tract, but in carnivorous perch, the pancreas was the only source of amylase activity. There is evidence from several fishes, for example menhaden (*Brevoortia*), silverside (*Menidia*), and silverperch (*Bairdiella*) that fishes have endocommensal bacteria possessing cellulase which can break down the cellulose plant materials both for their own nutritive value, small as that may be, and to make the contents of the plant cells available for utilisation by the fish. Still herbivorous fishes largely rely on mechanical breakdown of plant cell walls, hence the extensive development of cutting, tearing, and grinding teeth among them, and the habit of feeding almost incessantly during daylight. Examination of the feces of an herbivore such as the Asiatic grass carp (*Ctenopharyngodon idella*) reveals that cells not broken down by the teeth appear intact and even retain their green pigments. Whole algal cells are also passed intact through the gut of some plankton feeders such as the gizzard shad (*Dorosoma cepedianum*).

The innervation of digestive organs is both sympathetic from paired ganglia lateral to the spinal cord and parasympathetic through branches of the vagus nerve. Experiments with sharks and rays have shown that sight and smell of food do not cause gastric secretion. This is in contrast to observations on mammals that have so-called psychic or appetite flow of digestive juices mediated through the vagus nerve. The pancreatic, and probably also intestinal secretion and bile flow are under both hormonal and nervous control. Several hormones have been isolated from fish intestinal tissue and shown to act on the flow of zymogens as well as other fluids such as bile and pH-buffering pancreatic secretion.

Absorption of Digested Materials

Absorptive Process

In order for digested foods to be absorbed, they must be in aqueous solution; hence, they themselves must be soluble. The component molecules must further be of a size that will enable them to cross the membranes of the cells lining the tract, pass into the circulatory system, and ultimately be carried to and enter cells that need them or store them. It is interesting to note in this context that fat absorption is intensified in the pyloric stomach of some fishes (as in the basking shark, *Cetorhinus*) and the pyloric caeca of others (as in the genus *Salmo*). Fats have been shown to enter into the lymph ducts in these regions without being split into their component fatty acid and glycerol molecules upon which intestinal absorption depends.

Although most digested proteins are absorbed in the intestines, some absorption of protein derivatives has been shown to occur in the shark stomach. It has also been shown that the smallest units into which proteins are broken down in intestines are dipeptides (compounds consisting of two amino acids). This apparently leaves to intracellular digestion the final breakdown of proteins into single building stones (amino acids) from which the fish proteins are resynthesised. Thus evidence accumulates to render less distinct one of the earlier suggested differences between lower invertebrates and vertebrates, that based on the presence of intracellular digestion in one and not the other group of animals.

Toxification of Flesh

Apparently because of the absorption of materials from foods, the flesh of a great many fishes is toxic to man (ichthyosarcotoxism). Some fish species are usually so, whereas others may be consumed safely hundreds of times and suddenly cause grave illness or even death when eaten again. The strongest evidence as to the origin of ichthyosarcotoxism now points to the feeding habits of fishes. The poison is thought to originate in marine plants and to be conveyed to man when the fish is eaten, perhaps after accumulation, concentration, or even alteration in the fish. Herbivorous fishes may transmit the poisons to the flesh of carnivorous fishes with no effect on the predator. Yet, when the predator is consumed by man, toxicity may result. Although widely distributed throughout the world, the greatest numbers of poisonous fishes are known from tropical waters. Temperate waters, however, have many puffers (Tetraodontidae), among others, which may be very poisonous with substances that they themselves produce. The Greenland shark (*Somniosus microcephalus*), sometimes poisonous, inhabits Arctic seas.

Poisonous fish flesh is known to exist in several sharks, including some requiem sharks (Carcharhinidae), cow sharks (Hexanchidae), dogfish sharks (Squalidae), and mackerel sharks (Lamnidae). In the bony fishes it occurs frequently among reef-dwellers. Families with known toxic representatives include the morays (Muraenidae), the mackerels and tunas (Scombridae), trunkfishes (Ostraciidae), puffers (Tetraodontidae), and porcupinefishes (Diodontidae). More than three hundred kinds of fishes have been incriminated in the ciguatera-type of fish poisoning, which is fatal to about 7 percent of the humans who eat the flesh of these fishes. There is no specific treatment, but a ciguatera antibody is being developed. The reactions are mostly nervous, with a reversal of cold and heat sensations being common in severe cases. Fish families periodically involved include the bonefishes (Albulidae), herrings (Clupeidae), anchovies (Engraulidae), goatfishes (Mullidae), basses (Serranidae), snappers (Lutjanidae), jacks (Carangidae), porgies (Sparidae), wrasses (Labridae), parrotfishes (Scaridae), barracudas (Sphyraenidae), surgeonfishes (Acanthuridae) and trunkfishes (Ostraciidae). Unfortunately, the occurrence of ciguatera is unpredictable and may arise very suddenly.

NUTRITION OF FISHES

Once food has been ingested and digested it may enter into various functions in the body of a fish. It may provide energy for life processes or provide material for the restoration or replacement of worn out cell components, or for growth and reproduction. A few fishes (tunas) can regulate their body temperature; in all others it fluctuates with that of the environment (poikilo-thermous condition). Thus their rate of metabolism and hence their nutritional requirements are, above all, dependent on the temperature of the surrounding water. Since waters in the Temperate Zones may have seasonal fluctuations of more than 25°C (77°F), a wide range of metabolic rates and food consumption can occur in one and the same individual throughout the year.

Most of our knowledge of food requirements of fishes stems from man's endeavors to raise fishes for food and for stocking in lakes and rivers. Detailed information is therefore restricted to relatively few species, mainly trouts and salmons of the family Salmonidae. Only here and there are the basic nutritional data on these salmonids supplemented by studies from natural freshwater and marine environments.

Food Conversion and Efficiencies

Fishes, like other animals, can be compared to complex machines with regard to the efficiency with which they utilise their food. One may simply measure the weight of food that brings about the increment of a weight unit of fish (be it pounds, grams, or kilograms) and arrive at a "conversion factor." For instance, hatchery-raised rainbow trout (*Salmo gairdneri*) may, under certain conditions and with certain foods, require 3.5 pounds of food for each pound of weight increase; this particular kind of food would, with these fishes, have a conversion factor of 3.5.

Values for conversion factors differ with the nature of the diet, the species and size of the fish, temperature, and other variables. Conversion factors range from 1.5 when certain dry artificial diets are fed to trout (Salmoninae) in hatcheries to 8 or more when diets with a high vegetable component such as corn and lupines are fed to carp (*Cyprinus carpio*) or other omnivorous fish species. The conversion factor is 2.5 to 3.0 for most hatchery diets. Weight-conversion factors in nature often tend to be higher than those measured in hatcheries because of the presence of much indigestible roughage, such as insect or crustacean shells, combined, perhaps, with greater energyexpending activity of fish in the wild than in confinement. The relation between food and weight gain can also be expressed as a weight conversion coefficient. For the example given in the above paragraph, this would be

$$\frac{1 \text{ weight unit gained}}{3.5 \text{ weight units fed}} = 0.2857$$

or a weight-conversion efficiency of 28.57 percent, the above ratio times 100.

Knowledge of efficiency of food conversion becomes more useful to the nutritionist, the physiologist, or the ecologist when it is expressed in calorie equivalents of the weights of food and fish flesh respectively. Carbohydrates, fats, and proteins can be burned to furnish heat energy as they combine with oxygen in the burning process. Metabolic processes can be thought of as similar to slow combustion or other chemical transformation where energy is derived from carbohydrates, fats, protein, and other compounds. Heat energy is expressed and measured as calories. One calorie is the amount of heat required to warm one gram of water one degree on the Centigrade scale. After suitable transformations other forms of energy such as chemical, mechanical, or electrical can also be expressed in terms of calories.

A fish transforms one compound into another when it produces flesh, scales, bones, and conducts its movements and other vital functions. The energy used for these purposes is derived from the breakdown of various compounds that all have different energy equivalents depending on their chemical composition and structure. For example, 1 pound of average fish flesh has about 600 large Calories (a large Calorie, written with a capital C, is a thousand times larger than the small calorie referred to above), sugar has 1695 and butter, a fat with a high energy equivalent, has 3260 Calories.

It is well known that some of the energy of gasoline exploding in the cylinder of the engine of an automobile is dissipated in overcoming the friction of the different parts of the engine and becomes lost as an effective forward propellant. Thus, the efficiency of an engine is never 100 percent. By very loose analogy we may compare such losses to those we find in animals when a portion of the energy they liberate in their metabolism goes into maintenance and such processes as respiration, digestion, excretion, and other vital functions and is not to be found in weight gain.

We can measure or weigh the amount of new material a fish adds to its own body during growth as well as we can measure or weigh the amount of food it eats. We can also express both of these quantities in calories instead of in pounds or grams, and compare them, namely

$$\frac{\text{calories contained in flesh added per unit of time}}{\text{calories contained in various foods eaten per unit of time}} \times 100$$

calories contained in various foods eaten per unit of time
This fraction, expressed as a percent, is called the over-all or gross efficiency coefficient or the first-order efficiency coefficient. For pike, between 14 percent to 33 percent of the calories consumed were found to be contained in the flesh. We can often go further and break down the denominator by determining the calories used in cell maintenance, respiration, excretion, and other metabolic processes that go on while

new flesh is laid down and therefore represent a part of the intake of energy not returned in growth (growth being the transformation of ingested material into new tissue beyond that required for repair or replacement). Then,

$$\frac{\text{calories in flesh added per unit of time}}{\text{(calories contained in food eaten per unit of time)} - \text{(calories used in metabolism, maintenance, etc.)}} \times 100$$

becomes the net or second-order efficiency coefficient that is always larger than the first-order one and provides a more exact measurement of the nutritive properties of various foods than the first-order efficiency coefficient.

Thus the first-order, or over-all coefficients, should be distinguished from the second-order coefficients or efficiencies. Second-order efficiencies are those in which the gain in weight, potential energy, or protein accretion is compared to the quantities of material or energy from which this gain is realised. In the denominator, the term *assimilated* may be substituted for *eaten* to reflect only food that passes through the gut wall and excludes egestion of materials not digested. Comparisons on the basis of calorie equivalents permit appraisal of the effects on food utilisation of the age of fishes, temperature, the amounts and kinds of food, and other variables.

Overall energy efficiencies decline with age and size of fish. They range from 58 percent for fish embryos (as in the mummichog, *Fundulus heteroclitus*) to 10 percent and less in large, old fishes (such as a flounder, *Pleuronectes*). Most energy coefficients lie between 15 percent for old fishes and 45 percent for young ones. Negative values are possible for short periods such as during the winter when many fishes of the Temperate Zones lose weight.

The building of body proteins is influenced by the kinds of food a fish eats. This was shown in trouts (Salmoninae) by a comparison between lots fed artificial food or natural food. Despite the fact that the artificial diet was superior to natural food in proteins and vitamins, the trout fed natural foods (insects, etc.) were twice as efficient as those on the artificial diet in the hatchery in converting food proteins into fish flesh, and, in addition, these natural-food fed trout had meat of a lower-water and higher-protein content.

Values for assimilation, expressed as percentage utilisation of foods retained in the body, have a mode between 80 and 90 percent; low water temperature and environmental stress may reduce these efficiencies. Provided diets are of the same composition, feeding experiments with the brown trout (*Salmo trutta*) and the northern pike (*Esox lucius*) indicated that fish utilise food more efficiently after fasting than when they are satiated and also that food is absorbed less completely when fish are overfed.

Quantity of Food Consumed

Predominantly herbivorous reef browers, such as parrotfishes (Scaridae) and surgeonfishes (Acanthuridae), feed more or less uninterruptedly throughout the day. In this way they, and many plankton feeders as well, ingest quantities of food much greater than do many carnivores. However, most attached algae have only half the caloric value and less than a quarter of the protein content of animal flesh. Furthermore in spite of the fact that unicellular planktonic algae and diatoms, and even parts of rooted aquatic plants may have adequate nutritive values, they contain a higher percentage of water. For these reasons, herbivores usually consume greater weights of food than carnivores.

Carnivores of different feeding habits also vary in the amounts they eat because their prey are often of very different compositions and sizes. Macrocrustacean feeders ingest relatively more indigestible roughage than piscivores. Some fishes may, at times, get a square meal of such a size that it lasts them for a week or longer. A northern pike (*Esox lucius*) after having swallowed a fish of nearly its own size may have the tail of the prey hanging out of its mouth for many hours while the anterior portion is being digested in the stomach. The aptly named gulpers (Saccopharyngidae) and swallowers (Eurypharyngidae) of the deep seas often take prey fish that are larger than themselves, having provided ample storage in their highly distensible foregut.

Fishes have a remarkable capacity for surviving long periods of starvation since they first mobilise stored glycogen and fats before body protein. Lungfishes (Dipnoi) and swampeels (Synbranchidae) survive a prolonged season of estivation by such resorption of bodily substances. Freshwater eels (An*guilla*) can survive for more than a year without taking food. The intestines of such starved fishes become nonfunctional and partly degenerate.

Quantities of food eaten by fishes over a certain period are often expressed as average rations per day or per a longer interval. Such an average can be a good illustration of the primary influence of temperature on feeding activity. The bluegill (*Lepomis macrochirus*) may consume average weekly rations up to 35 percent of the body weight (about 5 percent per day) in the summer at a mean water temperature of 20°C (75°F) and less than 1 percent (about 0.14 percent per day) during the winter when the water averages between 2 and 3°C (36°F). Not only the amount of food eaten but also the assimilation efficiencies are influenced by temperature. The metabolism of fishes increases with temperature up to an optimum temperature, above which the metabolism decreases. Marine fishes also compensate in their feeding rates for the higher energy and repair requirements of higher temperatures; the red hind (*Epinephelus guttatus*) more than doubles its feeding rates per day between 19° and 28°C.

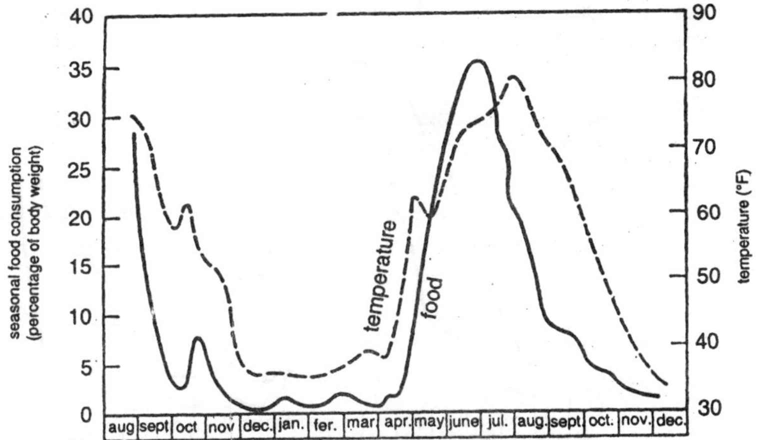

Figure 2.9 : Dependence of food intake on temperature as shown in the bluegill (Lepomis macrochirus).

Daily rations also vary with the size of fishes. Small fishes, like other small animals, have a higher metabolic rate than large ones of the same species. They, therefore, require relatively more food to maintain a unit of weight of their bodies than do large individuals. The daily maintenance ration of the red hind (*Epinephelus guttatus*) varies from 1.7 to 5.8 percent of the body weight for specimens that average 250 grams and between 1.3 and 3 percent for fish that average 600 grams as the water temperature rises from 19° to 28°*C*. Individuals of a species that average smaller in size than the red hind may have still higher daily ration requirements than the foregoing.

Availability of food to a fish also influences the amounts consumed. The red hind gorges itself with daily intakes of over 10 percent of the body weight at the onset of satiation feeding experiments but soon tapers off to more conservative rates of feeding. Yearlings of the carp

(*Cyprinus carpio*) in ponds may have a daily ration as high as 16 percent of their body weight when starting to feed at the onset of a growing season. Unlike most terrestrial animals, the majority of fishes experiences severe depletion for a part of every year of their lives.

Food Quality

True growth is the accretion of flesh (protein) and bone; it does not include the less permanent fat and water deposits in fish tissues. Nevertheless, the diet of fishes must be balanced and contain the primary or basic food components-proteins, carbohydrates, and lipids (fats)-in requisite, though differing amounts for different species of fishes. Vitamins and minerals are also required for growth, sustenance and replacement of tissues as well as for normal metabolism.

Basic Foods

Evolutionary adaptations among fishes have led to differing capacities in utilizing foods with varying amounts of proteins, carbohydrates, and fats. Trout (Salmoninae) for instance, develop fat-infiltrated livers when the fat content of their diet exceeds *8* percent (dry weight) whereas the Atlantic menhaden (*Brevoortia tyrannus*) normally feeds on algae that contain at least 25 percent fat. The common carp (*Cyprinus carpio*) and tilapias of the family Cichlidae can be raised on a diet of even higher lipid content. Not only do relative amounts of fat in the food affect food utilisation, but the composition of the fat also has an effect; unsaturated fatty acids with a low melting point (high iodine number) are digestible more readily than saturated fatty acids.

Permissible carbohydrate content in artificial diets, which, after all, are manufactured to be similar if not superior to a natural food of fishes, vary similarly for different species. When trouts are fed a diet with more than 12 percent digestible carbohydrate (wet basis), the results are abnormal deposition of liver glycogen and excessive mortalities. By contrast, the various minnows (Cyprinidae) raised for food throughout the world often grow well on food mixtures with 50 percent or more of starchy matter. Different carbohydrates vary in their digestibility by trout, for example, 99 percent of glucose was digested, whereas 73 percent of sucrose was digested. The most rigorous requirements of the three basic foodstuffs probably pertain to proteins. A certain protein content is mandatory in the diet of any animal. The protein requirement changes with changes in the fish's life cycle, being greatest for small, fast growing fish and for prespawning fish that are forming eggs and sperm. Invertebrate fish food averages approximately 11.5 percent protein. It appears that food with less than 6 percent protein (wet weight) cannot sustain any fish, let alone produce growth.

Not only are amounts of protein ingested important but so also are their relative amino acid content as intimated by the highest conversion efficiencies obtained when a fish eats others of its own kind. Like all animals, fishes can synthesise certain amino acids readily, but those of others only at rates too slow for satisfactory replacement. Therefore, these have to be supplemented in the diet. Yet a third group of amino acids is absolutely essential and has to be supplied in the food because fishes cannot synthesise them. Detailed studies of this aspect of fish nutrition have been made with the chinook salmon (*Oncorhynchus tshawytscha*). Arginine, histidine, isoleucine, leucine, lysine, methionine, phenylalanine, threonine, tryptophane and valine were the amino acids that this salmon either did not synthesise at all or manufactured in such small amounts that no growth was accomplished when they were absent from the diet. In this respect the chinook salmon resembles the rat (*Rattus*) rather than man who can dispense with two of the above compounds in his diet (histidine and arginine) because they can be synthesised in the human body.

Water, though not a basic food proper, is necessary in fishes, as in other animals, for metabolic processes including extracellular digestion. An animal may lose practically all of its fat and half of its protein and live, but a loss of only 10 percent of its water causes death. The amount of water taken in by fishes incidental to feeding is not known precisely. The presence of a set of sphincter muscles at the entrance to the stomach in some fishes and the straining and concentrating adaptations in the mouth and pharynx of others such as the plankton feeders suggest that accidental water intake may be insignificant. Freshwater fishes absorb water through exposed semipermeable membranes of the gills and other exposed surfaces. Marine fishes counteract osmotic water losses in part by swallowing sea water and in part by deriving water from food. It is probable that much of the water required for digestive processes by freshwater fishes is also extracted from food. The range of water content of foods taken in by fishes lies between 70 and 90 percent of wet weight of the foods, hard parts such as shells and bones not included.

Vitamins

Knowledge of the role of vitamins in fish nutrition is derived primarily from feeding experiments with several species of trouts and salmons (Salmoninae) and also channel catfish and carp. In the natural diet of animals, vitamins occur in sufficient amounts to supply the very limited quantities required for normal metabolic functions. However, when fishes are raised on artificial diets, the selection of food constituents, the property of rapid deterioration of some vitamins, and other factors may lead to specific deficiencies with their attendant impairments of body functions. Trout and salmon require ten members of the vitamin B complex. For carp the requirement has been established for pyridoxine, riboflavin, and pantothenic acid and for channel catfish some of the fat- and water-soluble vitamins. Trout need water-soluble vitamin C. Requirement of fat-soluble vitamins A, K, and E, but not D has been established for some fishes.

Minerals

Minerals, like vitamins, occur in most natural diets in sufficient quantities to satisfy the metabolic requirements of fishes; even in fish hatcheries mineral deficiencies are rare and easily alleviated when discovered.

Mineral and other elements can be classified according to their three major functions in the body: (*a*) *structural-calcium*, phosphorus, fluorine, and magnesium (tooth and bone formation); (*b*) *respiratory-iron*, copper, and cobalt (function and formation of hemoglobin); (*c*) *general metabolism-many* elements (body and cell functions). In the last category, sodium, potassium, calcium, and chlorine regulate osmotic balance and cell turgor. Boron, aluminum, zinc, arsenic and other trace minerals may also be required here; chlorine ions are needed for the manufacture of gastric juice; and magnesium, phosphorus, and chlorine ions are needed to activate digestive enzymes. Copper and zinc are required cofactors for some enzymes and sulfur and phosphorus occur in phospholipids and other nuclear constituents. Traces of phosphorus are important in adenosine triphosphate (ATP) and other high energy compounds. Bromine figures in the maturation of the gonads whereas iodine is a constituent of thyroid hormones. A deficiency of iodine can lead to goiter-like conditions in trouts at hatcheries and is alleviated by its addition to the artificial foods. Radioactive isotopes of calcium are absorbed directly from the water in increasing amounts as the respective mineral is lacking in experimental diets. Lithium, sodium, chlorine, and bromine also enter a fish through the gills, but most elements are adequately represented in the natural food of fishes and appear to be taken in through the food.

Table 2.1 : Effects of Vitamin Deficiencies on Fishes

Substance	Deficiency Symptoms
Vitamin B_1 (Thiamine)	Poor appetite; muscle atrophy; convulsions; nervous malfunctions; death
Vitamin B_2 (Riboflavin)	Eyes become opaque; growth ceases
Vitamin B_6 (Pyridoxine)	Nervous disorders; ataxia; anemia; growth reduction; death (in 4 to 6 weeks in trouts)
Vitamin B_{12} (Cobalamine) (Unnumbered vitamins in B complex)	Erratic hemoglobin and erythrocyte counts; growth reduction
Biotin	Mucous production impaired; convulsions; "blue slime disease"
Choline	Growth reduction
Folic Acid	Hemorrhagic kidney and intestine; growth reduction; anemia
Inositol	Growth reduction and probably anemia
Niacin	Increased sensitivity to ultraviolet light; "sunburn"
Pantothenic Acid	Gill disease; mucous production impaired; "blue slime disease"; sluggishness

The Chemical Composition of Fishes

Most data on the elementary composition of fishes refer to the composition and nutritive values of the flesh among commercially important species, some 350 to 400 in number, mostly marine bony fishes. The data are in terms of water, fat, proteins, carbohydrates, and minerals per unit of weight.

Water content of fish flesh averages 80 percent, slightly more than for birds and mammals. For fishes, extreme values of 53 and 89 percent have been recorded for certain species, seasons, and localities.

Water content of fish flesh varies with life history stage and species. Sex differences in water content are small as reported for a few species, as for example the herring, Clupea harengus. Males of this herring often contain 1 to 2 percent more water than the females. This difference may be correlated with the higher water content of testes than of ovaries. Sac fry (larvae with yolk sacs) embryos have a lower percentage of water than adults at maturity. The leptocephalus larvae of freshwater eels (*Anguilla*), however, contain more water than after transformation into elvers of adult body form and structure. Among adults, however, progressive dehydration follows with age. Anadromous species such as the salmons (*Oncorhynchus* and *Salmo salar*) and marine-run races of the sea lamprey (*Petromyzon marinus*) increase their water content perceptibly at spawning as a result of having replaced with water large amounts of fat and protein used for energy to reach the spawning sites in rivers. Cods (Gadidae), flounders (Pleuronectidae) and goosefishes (Lophiidae) may contain over 80 percent water, whereas freshwater eels (*Anguilla*) in the sea, the marine herrings and sardines (Clupeidae), and the tunas and mackerels (Scombridae) generally have a water content below 75 percent. The first group, represented by the cods, flounders, and goosefishes, is demersal, and its members have specific gravities slightly above that of sea water, to be raised, if necessary, with the help of their gas bladders. The second group represented by the herring and mackerel families, is largely pelagic, has flesh of a higher fat content than the first, and has some representatives with only small or vestigial gas bladders. The specific gravity of the latter is approximated to that of sea water by enlarged fat deposits in the muscle.

Fat content of fish flesh varies not only according to families but also with the season. Plankton feeders of Temperate seas such as the herrings and sardines (Clupeidae) have more fat and less water in their tissues after the vernal plankton pulses than after spawning, the time of their leanest condition. Water and fat content of the tissues of individual fish vary inversely to one another and fat content even varies from place to place on one fish. Muscle of the herring (*Clupea harengus*) has unusually high fat content in the abdominal region counterbalanced by another place of concentration near the dorsal fin.

Freshwater fishes also show group differences in relative fat content and generally have fats of different chemical characteristics than marine species. Tropical freshwater catfishes such as the labyrinthine catfishes (Clariidae) and the schilbeids (Schilbeidae) contain more fat than other groups in the same region, reflecting specific feeding habits and adaptations to utilise certain foods. Most freshwater fishes of the Temperate Zones store fat for the winter and/or in preparation for spawning. Important storage sites are the connective tissue and mesenteries of the intestines.

Only even-numbered fatty acids occur in the fats and oils of fishes. These fatty acids are less saturated than those occurring in land animals and in vegetable oils. Their oleic acid content, a precursor of cholesterol, is relatively low. For this reason they have been held to be beneficial in the diet of humans in relation to pathological hardening of the arteries.

Proteins make up between 14 and 23 percent of the wet weight of fishes. Here again, variations occur in species, seasons, and life history stage. The total nitrogen in the fish body is usually between 2 and 3 percent of the living weight. Shark-like fishes (Elasmobranchii) contain more nitrogen than bony fishes (Osteichthyes) because elasmobranch tissues have much osmoregulatory urea in them which bony fish tissues lack. On a dry weight basis, nitrogen is most highly concentrated in the liver, followed by the skin and then by the flesh.

Carbohydrates present in muscle for immediate energy release make up less than 1 percent of the wet weight of fishes; they are more concentrated than this in the liver where they are stored as glycogen.

Some of the many chemical elements found in fishes are more concentrated in marine than in freshwater species. The elevated content of these elements is thought to reflect their richness in sea water and their subsequent passive incorporation by marine fishes through concentration in all links of their food chains. Ash content in soft tissues of fishes is about 1 percent of their living matter. When bones and scales are included in the analysis, ash amounts to 3 percent or more because calcium phosphate, $Ca_3(PO_4)_2$, and calcium carbonate, $CaCO_3$, make up the bulk of these hard parts of fishes. Scales are about 30 to 35 percent ash on a dry weight basis.

GROWTH OF FISHES

The growth pattern of fishes is remarkable inasmuch as most of them have the capacity of sustained though diminishing growth (indeterminate growth) throughout their entire lives if sufficient food is available. Thus members of one species may be of variable sizes at the same ages in contrast to more definite sizes at any one age (determinate growth) as among a few fishes, such as males of the guppy (*Lebistes*), and generally among the individuals of terrestrial vertebrate groups. In other words, most fishes, in contrast to birds and mammals, do not cease growth after they have reached sexual maturity. Whereas the growth of fishes is thus much more variable and flexible than that of birds or mammals, there still exists an ultimate genetic limit to the growth of a species that is most evident in the smallest kinds. It has been proposed in explanation of the foregoing contrast that fishes living in a fluid medium that supports them mechanically can continue growth throughout their lives because there are more biotic than mechanical limits imposed on their maximum sizes.

Growth in animals has been defined above as the addition of structural or fleshy, hence protein, elements; fish growth in particular is of great importance to man as the exploiter and potential manager of fish populations for sport or sustenance. In fact it has been shown in controlled calorimetry experiments with trout and domestic livestock that fish are far superior in protein-building than the latter. This is because birds and mammals have to expend part of their caloric intake for the maintenance of body temperature and in supporting themselves; fish in contrast are poikilotherms and get such support from the surrounding water. Certain special aspects of growth and factors that influence it are discussed below.

Age and Growth Determinations

When the accretion of fish length (or weight) is plotted against time throughout the life span, the growth rate is represented by a curve. Seasonal fluctuations in growth are often disregarded and usually empirical data on growth are available only for the postembryonic or even postjuvenile periods of the life history.

Since many commercial fishes are available for measurements only once or a few times in their life cycle, man has sought to

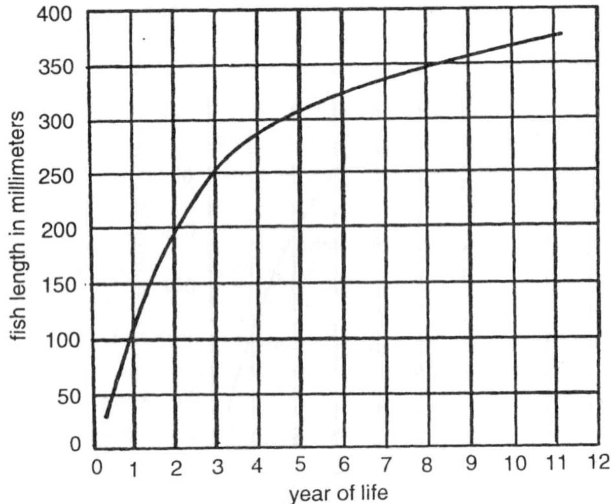

Figure 2.10 : Curve of growth in length as shown by the shallowwater cisco (Coregonus artedii).

reconstruct their rates and patterns of growth. Most useful for this endeavor-at least for fishes of the Temperate Zones-are the year marks on bony parts, such as scales, otoliths, spines, and opercular bones. Yearly zones of growth on these parts are due to a slowing down of temperature-dependent processes during the winter and a resumption of a more rapid metabolism at a time in the spring. Certain freshwater fishes of the monsoon tropics also show seasonal growth marks that correspond to the onset of the dry seasons. The growth zones can be measured and the correspondence between body growth and growth of bony parts can be established. For example; in certain positions on the body, scales grow at the same rate as the body (isometric growth) at least in the postjuvenile period. The growth history of an indjyidual can be reconstructed by calculations of proportionality. Although sometimes large numbers of individuals of different sizes are available for construction of life-table-like computations (length-frequency diagrams) and although sometimes the growth of fishes of known ages can be observed, the construction of growth patterns based on reading of year marks on hard body parts has come to be the prevalent method of growth determination among fishes of the Temperate Zones. In the uniformly warm waters of equatorial oceans, however, fish growth shows little or no seasonal fluctuation and the age of fishes there is extremely difficult to assess.

The length of a fish is often easier obtained than its weight or age. The following relation exists between the length and the weight of a fish: $W = a \cdot L^n$, where a is constant and the exponent n

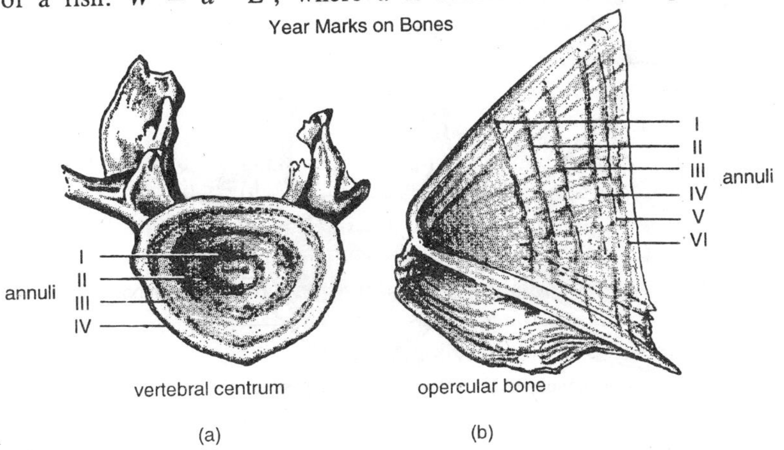

Figure 2.11: Annual growth marks (annuli) on (a) *the vertebra of a catfish* (Noturus) *and (b) the operculum of a perch* (Perca).

fluctuates between 2.5 and 4, most frequently near 3, since growth represents an increase in three dimensions whereas length measurements are taken in one dimension. This relationship also lends itself to comparison of individuals within and between different fish populations. The value sought then is the constant a (above), expressed as "ponderal index" or condition factor K in the formula $K = W/L^n$ which varies to provide the index when the exponent n is fixed at 3.

The Longevity of Fishes

Unconfirmed accounts mention that carp in ponds have reached an age of 200 or even 400 years, yet authenticated records of ages of captive fishes suggest that the most venerated old carp do not exceed 50 years.

By and large few fishes in the wild will live longer than 12 to 20 years and even in the shelter of an aquarium 30 years are rarely exceeded. Generally species with large individuals may be expected to exhibit older ages than species composed of little fish. Most fishes that grow longer than a foot (30 centimeters) have a life span of at least 4 or 5 years. Many minnows (Cyprinidae) of the Temperate Zones and other small fishes of the Temperate and Tropical Zones frequently have a life span of less than *2* years including some that are annual.

Genetic Factors

In nature the presence of genetic variations in growth potential in different populations of the same species is usually masked by environmental factors such as temperature or food supply. Nevertheless, some selected strains of the brook trout (*Salvelinus fontinalis*), rainbow trout (*Salmo gairdneri*), and the carp (*Cyprinus carpio*) have growth that is decidedly superior to that of other strains.

Fish growth usually slows down after the onset of sexual maturity when large amounts of nutrient material periodically go into egg or sperm forma-' tion. The gonads of the yellow perch (*Perca flavescens*) before spawning may make up more than 20 percent of the body weight of the female and more than 8 percent of the body weight of the male; after spawning the ovaries or testes may shrink to about I percent or less of the body weight.

The onset of sexual maturity and with it the change in growth rate in the life cycle of a fish may also be genetically determined as it is in the brown trout (*Salmo trutta*). Egg size also varies with genetic strains and with it varies the initial size of the fish at hatching. Large eggs produce large larvae that may also have the greatest growth potential of the lot. Advantages in growth pattern due to larger egg size may be noticeable for a short time until other factors such as food supply intervene; the advantages, however, may be lifelong.

Frequently there are size differences between mature male and female fish within a species. Size differences between the sexes that affect growth rates and growth patterns may be due to genetic factors, often correlated with inherited behavior patterns. Among the minnows (Cyprinidae) of North America, for example, the majority of species have larger females than males but in some species such as the hornyhead chub (Nocomis *biguttatus*) and the fathead minnows (Pimephalinae), the males are the largest. It is very likely that reproductive habits such as nest building effected these size differences between the sexes. The greatest size appears most often to be associated with the most "important" parent.

Seasonal and Temperature Factors

Fishes with a wide latitudinal distribution such as the largemouth bass (*Micropterus salmoides*) commonly have differences in rate of growth and in age at which they reach sexual maturity in northern versus southern waters. In Louisiana and Arkansas, for example, the largemouth bass may spawn after one year, but in southern Michigan or Wisconsin only after an average of three years. In Ontario and other comparable latitudes and climate, sexual maturity of this bass may be delayed until the fourth or fifth year. The ultimate size and age of the fish may not differ between northern and southern representatives of the species, but the shape of the curve that represents the plotting of weight against age will reflect a temperature-dependent difference in the growth pattern. Temperature also induces seasonal differences in growth since temperature affects metabolism and food consumption.

The seasonal rise in the temperature of fresh waters coincides with an increase of natural food for fishes but there are also seasonal and temperature dependent variations in activity that influence the amount of food diverted to growth and/or maintenance requirements. In one experiment, the brown trout (*Salmo trutta*) grew optimally between 7° and 9°C, and also between 16° and 19°C. Appetite was high once the water warmed to 10°C. Maintenance requirements also increased most rapidly between 9° and 11°C, then tapered off and became almost constant when 20°C was reached. Thus, optimum temperatures for rapid growth are those at which the appetite is high and maintenance requirements low whereas minimum growth occurs at intermediate temperatures when maintenance requirements are high because fish are most active. Below 7°C, maintenance requirements are low, but so is appetite; above 19°, appetite again falls off while , maintenance requirements remain at the same level. Thus, the brown trout achieves optimum growth outside of the temperature range where the fish show maximum activity.

Light, mostly through day length, may affect the rate and pattern of fish growth. Since changes in day length coincide with seasonal rise or fall in temperature, it is difficult to appraise the respective effects of these environmental factors. Illumination is known to act indirectly on thyroid metabolism and peaks in thyroid activity coincide with maximum day length but these peaks also cause increased swimming activity, resulting in a reduction in growth rate in experiments with the brown trout (*Salmo trutta*) in midsummer. If food is plentiful, shorter hours of day length may be most advantageous for rapid growth because of dark-induced rest patterns, but where strong competition occurs and feeding proceeds by sight, long days can be of advantage.

Biotic Factors

These factors have been called density-dependent variables but it would also be valid to term them "social." Among them is included competition or the influence of numbers on available food, nest sites, feeding behavior, and so forth. Since the expression of the biotic factors is most often studied in nature, it is very frequently confounded with the effects of environmental (physical and chemical) density-independent factors. Relatively few controlled experiments exist where density-dependent factors have been successfully segregated for scrutiny, either one by one or in combination.

The Indirect Effect of Fish on Fish

Experiments with the goldfish (*Carassius auratus*) grown in water previously inhabited by other goldfish have shown that such experimental fish grew better than their control running mates in "unconditioned" water. The improved growth may have been due to the uptake of regurgitated food from the conditioned water. More likely, however, it was due to the presence of a growth-promoting substance that could be isolated from the slime in the tanks with conditioned water. The substance was found to be effective in stimulating growth at dilutiorls as low as 1.2 parts per million (ppm).

Direct Effects of Fish on Fish (Group Effects)

Order of dominance or peck order has been observed as a factor in growth among captive fishes including sunfishes (Centrarchidae), trout-like fishes (Salmonidae), and cichlids (Cichlidae). One individual is at the top of the order (often the largest individual) and will regularly secure the best and/or most morsels when food is supplied to the group. If there are but a few fish in the holding tank, these hierarchal effects are often broken down, but there are indications that they persist in nature. The lake trout (*Salvelinus namaycush*) of northern lakes and rivers, among other fishes, has shown definite dominance of one individual over others. Whether growth differences due to peck orders result directly from increased food intake of the dominant individual or whether it comes about indirectly through the psychological stress placed on the lower members of the order is not known.

It has also been suggested that there may be a group effect in growth, over and above that of the hierarchy among individual fish. The suggestion is based on the following experiment. When different numbers of trout fry were placed in tanks of the same size and reared in groups of 25, 50, 80, 100, and 150 individuals with each group given an oversupply of food, the most successful group was the one which contained 80 fish. It appeared that the growth rate of individuals was influenced by group size as well as by the numbers of larger and smaller fish present.

The most pronounced group effect on growth is to be found in crowded fish populations. Fish grown in crowded conditions may release a chemical "crowding substance," which depresses growth. Many instances are known where fish in a body of fresh water become so numerous that their specific growth rate is reduced as compared to that of the average for the area or species. The condition has been grouped under the term "stunting." When some of the stunted fish or fish of a competing species are removed, the remainder responds to the new conditions and the specific growth rate increases. Among the fishes for which such stunting has been described are the yellow perch (*Perca flavescens*), European perch (*Perca fluviatilis*), bluegill (*Lepomis macrochirus*) and other members of the sunfish family (Centrarchidae), and a tilapia (Cichlidae-Tilapia *mossambica*).

3

BLOOD CIRCULATION

Mott reviewed the literature on the cardiovascular system of fishes and noted the fragmentary nature of available information. Since Mott's review many of the techniques of mammalian circulatory physiology have been applied to fishes, and much more information is now available. There are, however, still some areas in which there is a marked paucity of data; for example, very little is known about the hemodynamics of blood flow in arteries and veins in fish.

Recently, Robb has reviewed the anatomy of the cardiovascular system and Johansen and Martin have discussed some comparative aspects of cardiovascular function in vertebrates. Randall and Johansen and Hanson have discussed the functional morphology of fish and dipnoan hearts, respectively. Hunn has compiled a bibliography on the blood chemistry of fishes.

Fishes, including the Agnatha and Dipnoi, as well as the Chondrichthyes, Osteichthyes, Choanichthyes, and teleosts, represent a diverse group including over 20,000 species, extending in space over a wide variety of environments, and in time over widely divergent evolutionary pathways. The requirements of an aquatic environment enforces limitations on various systems and maintains morphological and physiological similarities between the groups. These similarities, however, may be superficial, and there is a danger of assuming conformity in function simply because of a general structural resemblance.

There is an apparent conformity of structure in the circulatory systems of fishes, with the exception of cyclostomes and lungfishes, blood pumped by the heart passes through the gill and systemic circulation before returning to the heart. Variations in the general scheme are, first, the large blood volume, low blood pressure, and the development of accessory hearts in the hagfish, and, second, the presence of a pulmonary circulation and a partial separation of oxygenated and deoxygenated blood in the heart of lungfish, associated with the transition from an aquatic to an aerial environment. A number of air-breathing fish have a well-developed circulation to respiratory surfaces other than the gills. The blood supply to these accessory respiratory organs is in parallel with the systemic circulation but in series with the gill circulation.

Of the small number of species of fish used in studies of the circulation, only a few have been investigated extensively. These include hagfish, dogfish, skate, ratfish, Port Jackson shark, trout, salmon, carp, tench, cod, lingcod, eel, and lungfish. The data have been collected using a variety of techniques under a variety of conditions. In some instances the fish were not intact and were either out of water, restrained, or anesthetised; in other experiments the recordings were obtained from intact, unanesthetizcd, and relatively unrestrained fish. The circulatory system is of necessity very responsive to changes in the environment and often adjusts rapidly to an experimental situation. The use of telemetry and other techniques has permitted experimentation on the cardiovascular system with relatively little interaction between the animal and the experimental procedure. However, the type of experimental procedure used can affect the results obtained. Differences in results reported in the literature may, therefore, result from the techniques used or represent interspecific or intraspecific variations in function. Thus, in discussing the cardiovascular system of fishes, one must be aware, first that generalisations rest on studies of only a small number of fish, and, second, that variations in results may reflect either species differences or differences in the techniques used.

BLOOD PRESSURES IN ARTERIES AND VEINS

The branchial or systemic heart in fish is usually the main propulsive organ for circulating the blood. Contractions of the heart serve to convert chemical energy into mechanical energy in the form of pressure and flow. The dissipation of energy as blood flows through the vascular network results in a fall in blood pressure. The hydraulic resistance to flow is inversely proportional to the fourth power of the radius of a blood vessel; therefore, the major portion of the drop in pressure occurs across the smallest blood vessels.

In fish, blood from the heart passes first through the ventral aorta into the gill circulation and then via the dorsal aorta into the general body circulation. In the majority of fish, the capillary bed of the gills represents a considerable proportion of the total vascular resistance to *flow*, and it is in series with all other capillary networks. In the lungfish, blood passes from the heart through the branchial vessels into ' the systemic and pulmonary circulations. The pulmonary vascular bed is in series with the gill circulation but in parallel with the systemic circulation. In the lungfish *Neoceratodus forsteri* there are capillaries in all gill arches, but in other species of lungfish, *Protopterus aethiopicus* and *Lepidosiren paradoxa*, there are no capillaries in the first and second gill arches, and branchial vessels connect the ventral and dorsal aortae directly.

In all fish, except the hagfish, the highest blood pressures have been recorded from the ventricle during systole: The resistance to *flow* between ventricle and ventral aorta is low; therefore, when the intervening valves are open, pressures in the ventricle and ventral aorta do not differ from each other by more than 1 mm Hg. Arterial pressures reported by earlier workers are sometimes unreliable because the fish were either restrained, out of water, ventral side up, anesthetised, or not intact. Johansen recorded ventral aortic pressures of 30 min Hg in the cod, *Gadus niorhua;* his fish were restrained and ventral side up, but not anesthetised. Robertson *et al.* measured pressures of 82/50 mm Hg in the ventral aorta and 44/37 mm Hg in the dorsal aorta of the spring salmon. There was no water flow over the gills when records were taken, initiating a reflex bradycardia as indicated by the low heart rates in these fish. Blood pressures have been recorded in unrestrained trout, salmon, and carp. Hanson has recorded blood pressure in a number of unrestrained Chondrichthyes (skate, dogfish, and ratfish). Arterial and intraventicular pressures have been recorded in immobilised elasmobranchs by Sudak, Satchell, and Satchell and Jones. All values were in the range of 30-70 mm Hg but show variability within a species as well as between species. In general, arterial blood pressures appear to be lower in elasmobranchs than in telcosts.

During diastole, ventral aortic pressure falls, valves in the corms or bulbus close, and pressure declines as the blood leaves the aorta. The pulse pressure in the ventral aorta of fishes is between 10 and 30 mm Hg, increasing to values as high as 40 mm Hg during hypoxia.

The first appreciable drop in blood pressure occurs across the gills. Hanson recorded dorsal aortic diastolic pressures which were 15% in skate *Raja binoculata*, 20% in dogfish *Squalus suckleyi*, and 26% in ratfish *Hydrolagus colliei*, of the ventral aortic blood pressure. The respective reductions in systolic pressure were 25, 25, and 40% of the ventral aortic pressure. The recorded pressure drop in trout, *Salmo gairdneri*, carp, *Cyprinus carpio*, and lungfish was between 40 and 50% of the respective ventral aortic pressure. Hypoxia probably increases the resistance to flow through the gills in the trout and dogfish, *Squalus acanthias*, whereas exercise and catecholamines cause a decrease in resistance to blood flow through the gills.

The dorsal aorta of fishes is a long tube, running the length of the body, which distributes blood to the systemic circulation. It is much less elastic than the ventral aorta. Blood pressure and flow in the dorsal aorta of the cod, *Gadus morhua*, is pulsatile even though blood flows through the capillaries of the gills before entering this vessel. Pulse pressures in the dorsal aorta are usually less than 50% of those in the ventral aorta.

The total peripheral resistance to blood flow probably decreases during swimming in the trout but increases during hypoxia. The pulmonary circulation in the lungfish has a lower resistance to blood flow than the systemic circulation. When the lungfish, *Protoptei!:s aethiopicus*, breathes, blood flow to the lungs increases, whereas during motor activity, systemic blood flow increases.

Venous pressures in fish range from slightly below atmospheric to about 10 mm Hg. Contractions of the heart within a noncompliant pericardium in elasmobranchs and lungfish create subatmospheric pressures in the venous system near the heart, producing aspiratory forces which increase venous return to the heart. Thus blood flow in veins near the heart in these animals is determined by the pumping action of the heart and skeletal muscles and aspiratory forces resulting from contractions of the heart within a noncompliant pericardium. The relative importance of these factors will be determined by the size of the vascular resistance to flow between the heart and the vein in question and the magnitude of the subatmospheric pressure created in the venous system by contractions of the heart. Flow in the vena cava of the lungfish, *Protopterus aethiopicus*, is largely determined by aspiratory forces when the animal is resting and by the pumping activity of skeletal muscle during exercise. Flow in the pulmonary vein of lungfish is largely dependent in the pumping action of the heart, but aspiratory forces do produce small increases in blood flow.

Pressures in the cardiovascular system of hagfish, are very low and are maintained by the action of the branchial and several accessory hearts. The hearts work independently of each other and circulate blood to a specific area of the body. The branchial heart pumps blood from the body and liver into the ventral aorta. The portal heart pumps blood from the gut and anterior cardinal vein to the liver. The action of skeletal muscles on a cartilaginous plate in the caudal region propels blood forward from large subcutaneous sinuses. This cartilaginous plate, the associated skeletal muscles, and a pair of lateral sacs with valves to regulate the direction of blood flow constitute the caudal heart. The activity of the skeletal muscles of this heart is initiated by impulses from the central nervous system. Both the branchial and portal hearts are aneural. Systolic arterial pressures are of the order of only a few mm Hg, and are determined by the interaction of the activity of the various hearts. Chapman *et al.* working on the Pacific hagfish, *Eptatretus stoutii*, noted that the branchial heart often stops and occasionally the portal hearts contract while the branchial heart is inactive. Johansen recorded large oscillations in the dorsal aortic blood pressure in *Myxine glutinosa* that were associated with contractions of the gill musculature and suggested that the gills take an active part in the propulsion of blood. Chapman *et. al.* did not observe any effect of gill contractions on blood flow in the Pacific hagfish.

The velocity of the pressure pulse through different parts of the circulatory systcm has not been measured in fish. Nothing is known about the reflection of pressure waves from arborisations in the circulation or the contribution of reflected waves to the pulse pressure at any point in the circulation.

THE HEART

Anatomy

The systemic or branchial heart of fishes is the main propulsive organ for circulating the blood. It consists of four chambers in series. Venous blood enters the sinus venosus from the liver and the ducts of Cuvier, passes into the atrium, and is then pumped into the ventricle and finally into the ventral aorta, via either a conus or bulbus (telcosts) arteriosus. All chambers with the exception of the bulbus are contractile, and a unidirectional flow through the heart is maintaincd by valves at the sinoatrial and atrioventricular junctions and at the junction of the ventricle and bulbus or conus. The contractile conus wall consists of a thin layer of cardiac muscle

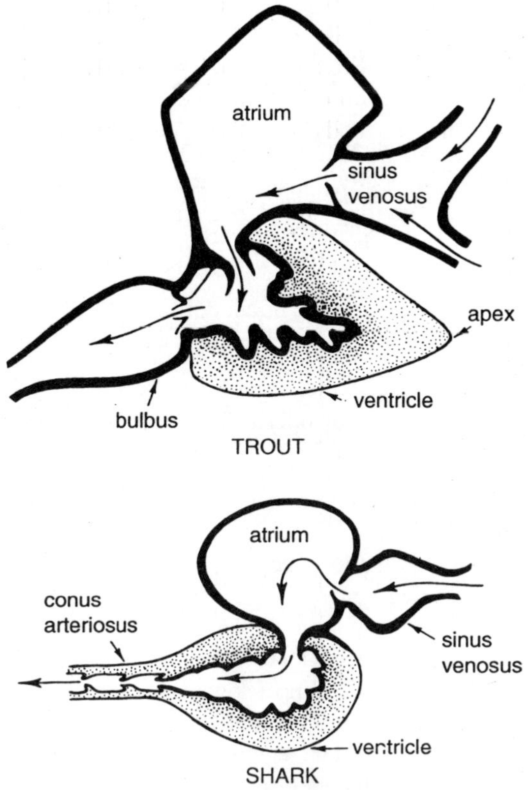

Figure 3.1 : Diagrams of the heart in a teleost (trout) and an elasmobranch (shark).

covering a fibrous elastic sheath, and it may have up to seven transverse rows of valves along its length. The bulbus, elastic and noncontractile, has only a single pair of valves guarding the exit of the ventricle.

The heart of the cyclostome is similar to that of other fish except that there is only a single duct of Cuvier in the adult (paired in other fish), and the conus is poorly developed. The exit of the ventricle is guarded by a pair of pocket valves, but there are no other valves in the conus. In *Petromyzon*, however, there are two longitudinal ridges similar to those found in the embryonic conus of other forms.

The heart of the lungfish is somewhat different from the typical fish pattern described above. The heart lies behind the pectoral girdle; the sinus and atrium dorsal to the Ventricle. The atrioventricular valves are absent and are replaced by an atrioventricular plug. There is a partial subdivision of the heart by the incomplete atrial and ventricular septa and the spiral folds in the conus arteriosus (bulbus chordis). The ventral aorta is either very short or nonexistent, and, as in amphibia, the branchial arteries arise from the anterior end of the conus arteriosus. The degree of subdivision of the heart varies between different genera of lungfish and is correlated with the prevalence of air breathing. *Neoceratodus*, the Australian lungfish, breathes air only when the oxygen content of the water is low, the African lungfish, *Protopterus aethiopicus*, and the South American lungfish, *Lepidosiren paradoxa*, are predominantly air breathers and spend several months each year on land. Division of the heart is least complete in *Neoceratodus* and best developed in *Lepidosiren*. The pulmonary vein and sinus venosus represent separate pathways for venous blood returning to the heart. Blood from the lungs is directed toward the left side of the atrium, whereas blood from the systemic circulation is directed into the right side. The partial separation of right and left atrium and ventricle aids the selective passage of blood from the lung into the left side of the ventricle. There are no conal valves but there are two folds or ridges, one large the other small, arising from opposite sides of the internal surface of the conus arteriosus (bulbus cordis). The free edges of the folds almost touch and act as a barrier which partially divides the conus into two outflow channels. The conus twists and turns, and in the anterior region the two ridges fuse to form a horizontal septum that completely divides the lumen into a dorsal and ventral channel. Blood from the left side of the ventricle is preferentially directed into the ventral channel and subsequently into the systemic circulation, whereas blood from the right side of the ventricle preferentially enters the dorsal channel which directs blood into the two posterior pairs of branchial arches, one pair of which directs blood to the pulmonary arteries.

The heart in all fishes, with the exception of that of the hagfish, has a coronary blood supply. The origin of the coronary supply is variable in lungfish; *Neoceratodus*, as in teleosts and elasmobranchs, has a coronary vessel derived from the anterior hypobranchial system. In *Lepidosiren* the coronaries arise from the second afferent artery. This transition from an efferent to an afferent origin for the coronary supply is associated with the absence of a capillary network in the second gill arch. In addition, this gill arch presumably contains oxygenated blood returning via the left side of the heart from the pulmonary circulation.

The hagfish branchial heart has no coronary supply, and blood perfusing the heart is almost totally deoxygenated. The hagfish heart is only sporadically active and the pressures developed are small; hence, the work done by the branchial heart is also small compared with that done by other vertebrate hearts. It is possible that heart work in the hagfish is limited by the absence of a coronary supply. The extensive development of accessory hearts in hagfish may represent an attempt to spread the load of circulating the blood in the absence of a heart capable of prolonged activity at high work rates. However, the hagfish branchial heart is capable of producing systolic pressures of the order of 20-30 mm Hg. The heart obeys Starling's law and when venous pressures are increased peak intraventricular pressures rise to 30 mm Hg, and the work done by the heart increases. Such increases in heart work, if large, may be anaerobic. The hagfish "usually lies virtually immobile in lightless, frigid marine canyons for very long periods of time". When food appears it swings into violent and effective action. Nothing is known of the changes in blood flow that may occur during these periods of activity.

The heart of fish is composed of typical vertebrate cardiac muscle fibers. The reported fiber diameters are smaller than those of mammals and may account for the relatively small number of intracellular potentials recorded from fish hearts. Jensen reported atrial and ventricular fiber diameters in the hagfish heart as 6.1 and 7.1 μ, respectively. These are about half the size of atrial and ventricular fibers in the dog.

Electrical Properties

Action potentials and electrocardiograms recorded from fish hearts are similar to those recorded from other vertebrate hearts. ensen recorded cardiac resting potentials of between 40 and 60 mV in the hagfish portal heart and in atrial and ventricular fibers in the branchial heart of three species of hagfish, an elasmobranch, and two marine teleosts. Other investigators have recorded resting potentials of about 70 mV in atrial strips of the skate heart, in the goldfish atrium and ventricle, and in the trout ventricle. Jaeger recorded somewhat larger resting potentials from the atrium and ventricle of the roach, *Leuciscus rutilus L.*

Initiation of the heartbeat usually occurs in the sinoatrial node, but the actual site and extent of the pacemaker region varies from fish to fish. Jensen was able to record pacemaker potentials from a large number of fibers in the branchial (systemic) and portal heart of the hagfish. Those fibers having a pacemaker potential were situated in both the atrium and the ventricle of the branchial heart. Mc-William and Kisch have shown that in fish many regions of the heart are capable of pacemaker activity.

The conduction velocities of the wave of excitation spreading out from the pacemaker region are slower than in mammalian hearts. The rate of spread of excitation varies in different parts of the heart, and there are fast conducting pathways in the trout ventricle, *Salmo gairdneri*, that transmit the wave of excitation rapidly to the apex and cause it to contract before other parts of the ventricle closer to the atrium and bulbus are excited.

The electrocardiogram has been recorded from a number of fish and consists of a P wave followed by a QRS complex and a T wave. Oets recorded a V wave which preceded the P wave and was associated with a contraction of the sinus venosus in the eel. Chapman *et al.* recorded a typical electrocardiogram from the hagfish; however, another diphasic deflection resulting from activity of the portal heart appeared on the record. As might be expected it had no phase relationship with the electrocardiogram of the systemic (branchial) heart. Tebfcis recorded a wave complex (termed the "Bd complex") in the electrocardiogram of the Port Jackson shark. The Bd complex normally occurs between the QRS complex and the T wave and is associated with the depolarisation of the conus. The exact position of the Bd complex in the ST segment is variable and may even coincide with the T wave. In a few preparations, Teb°cis recorded a Br complex between the T wave of one cardiac cycle and the P wave of the next cycle. The Br complex is associated with repolarisation of the conus. Teb7cis calculated the conduction velocity of the tonal wave of excitation from the time delay between two tonal electrograms recorded from different points along the length of the conus and found that the velocity of conduction was very slow, of the order of 24 cm/sec. The conduction velocity of the excitation wave passing over the surface of the ventricle of the Port Jackson shark is much faster, in the range of 40100 em/sec.

Mechanical Properties

There are differences in function between the teleost and elasmobranch heart. The elasmobranch heart is contained in a noncompliant pericardium and has a contractile conus. The teleost heart is within a thin and compliant pericardium and has a noncontractile but very elastic bulbus.

In teleosts, the rate of atrial filling is determined by venous pressure. The sinus venosus appears to play little or no role in actively moving blood through the heart, but it forms part of an extensive venous reservoir serving the atrium. The volume of the atrium prior to systole is adequate to fill the ventricle, and the atrioventricular valves do not open until atrial systole. Therefore, contractions of the atrium in the presence of sinoatrial valves to prevent the reflux of blood into the sinus venosus serve to fill the ventricle, and, unlike the situation in mammals, there is no direct inflow of blood into the ventricle from the venous system during ventricular diastole.

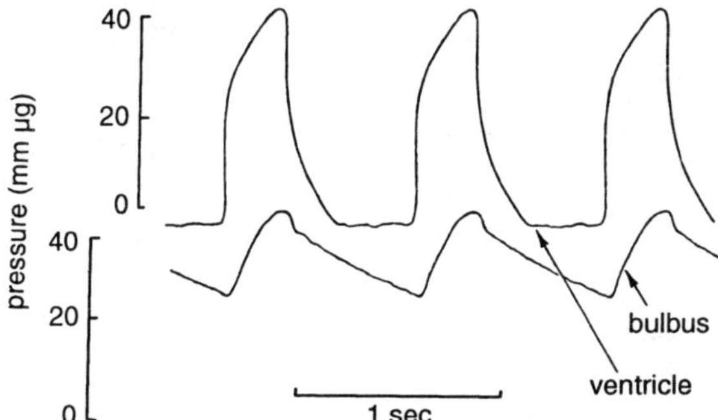

Figure 3.2 : Pressures recorded from the ventricle and bulbus of the lingcod, Ophiodon elongatus.

Contractions of the ventricle result in large increases in intraventricular pressure, the atrioventricular valves close and, in teleosts at least, the ventricle contracts isovolumetrically until pressures rise above that in the bulbus, when the valves at the exit of the ventricle open and blood leaves the ventricle. Pressures in the ventricle, bulbus, and ventral aorta rise simultaneously and do not differ by more than 1 mm Hg, indicating a low resistance to flow between the ventricle and ventral aorta. Intraventricular pressure falls as the ventricle relaxes; the exit valves close and relaxation is isovolumetric until atrial systole opens the atrioventricular valves and fills the ventricle once more.

Blood flow in the ventral aorta and pressures in the ventricle and bulbus have been recorded in the lingcod, *Ophidon elongatus*. Flow in the ventral aorta can be related to three phases of the cardiac cycle. There is a rapid increase in ventral aortic blood flow as the ventricle contracts, a rapid decrease in flow as the ventricle relaxes, and, finally, a slow decline in the rate of flow as the volume of the bulbus decreases. Closure of the valves between the ventricle and bulbus occurs at the transition from a rapid to a slow decline in

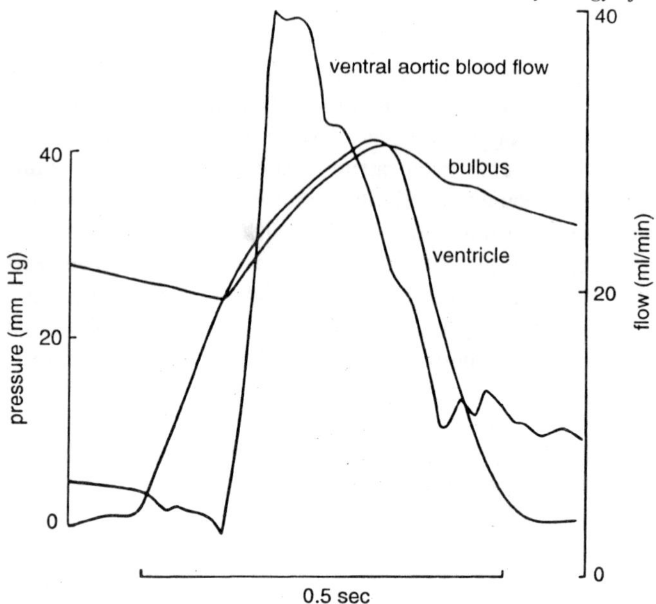

Figure 3.3 : Pressures in the ventricle and bulbus and blood flow in the ventral aorta replotted and superimposed upon one another to illustrate the relationships of the pressure and flow in the ventricle, bulbus, and ventral aorta of the lingcod, Ophiodon elongatus.

ventral aortic blood flow rate. Flow in the ventral aorta only falls to zero if the heart rate is very low. Usually the minimum flow rate of blood is between 9 and 12 ml/min in a 2-3-kg lingcod, *Ophiodon elongatus*. The mean blood flow is 15 ml/min, the period when blood flow is due to the elastic rebound of the bulbus is 44% of each cardiac cycle, and mean blood flow in the ventral aorta during this period is 10 ml/min. Blood flow due to the elastic rebound of the bulbus therefore represents about 29% of the total cardiac output. At lower rates the proportion of blood flowing during this period is undoubtedly larger. Thus the bulbus plays a significant role in maintaining blood flow in the ventral aorta during ventricular diastole. If the bulbus were absent, and the ventricle connected directly to a short ventral aorta leading to the afferent branchial arteries, pressures developed by the ventricle would need to be much larger than those in the presence of a bulbus if cardiac output were to be maintained at the same level. The presence of a bulbus therefore dampens the oscillations in pressure and flow imposed upon the ventral aorta by contractions of the ventricle and also diminishes the work done by the ventricle in maintaining a given cardiac output.

In elasmobranchs, contractions of the heart within a noncompliant pericardium produce subatmospheric pressures within the pericardial cavity. These subatmospheric intrapericardial pressures have been recorded in lungfish as well as in a number of elasmobranchs. The reduction in intrapericardial pressure produces aspiratory forces which increase venous return to the heart. This increase in venous return presumably augments cardiac output, and Hanson found that opening the pericardial cavity of the ratfish, *Hydrolagus colliei*, decreased cardiac output. However, Satchell and Jones were unable to detect any change in cardiac output when the pericardium of the Port Jackson shark, *Heterodontus portus jacksoni*, was opened.

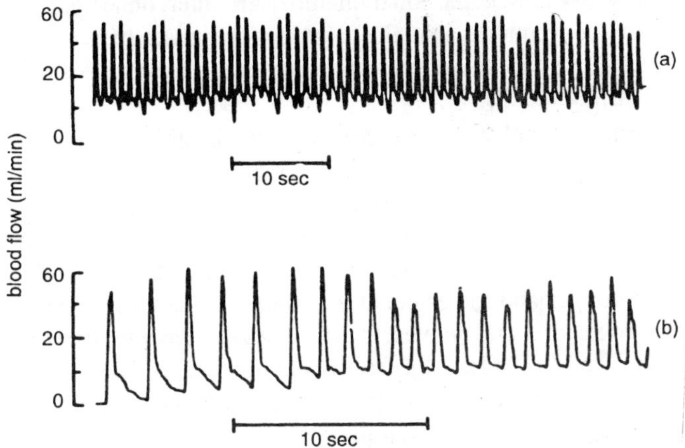

Figure 3.4 : The effect of changing heart rate on the stroke volume and velocity of blood flow in the ventral aorta of a 2.7-kg lingcod, O hiodon elongates. At low heart rates the blood flow falls to zero during diastole. At high heart rates the elastic rebound of the bulbus maintains blood flow in the ventral aorta during diastole. (A) Fish immobile in tank, heart rate normal. (B) Changes in blood flow associated with recovery from bradycardia induced by a short burst of muscular activity.

The magnitude of the intrapericardial subatmospheric pressure depends on the rigidity of the pericardium and the magnitude and rate of change in volume of the heart. The respiratory movements have also been shown to exert an effect on the magnitude of the intrapericardial subatmospheric pressure. Ventricular ejection leads to a rapid decrease in the volume of blood contained within the heart; hence, the major reduction in intrapericardial pressure occurs during ventricular systole. Conal systole causes a less marked reduction in intrapericardial pressure. The aspiratory forces produced by contractions of the ventricle and conus within a noncompliant pericardium increases venous return to the heart and augments atrial filling.

There are several sets of valves in the conus of the elasmobranch heart. Only the most distal valves are closed during ventricular diastole, maintaining the pressure difference between the ventral aorta and the conus and ventricle. The valves in the conus nearer the ventricle are patent, and the ventricular and tonal chambers are interconnected. Pressures rise in both the ventricular and conal chambers during atrial systole, indicating that contractions of the atrium move blood into the conus as well as the ventricle. As in the teleost, ventricular filling during diastole is solely dependent upon atrial systole.

As the proximal conal valves are open prior to ventricular systole and the ventricular lumen and tonal chamber are interconnected, blood moves out of the ventricle into the conus at the onset of ventricular systole. Thus there can be no truly isovolumetric phase to ventricular systole in elasmobranchs.

The number of valves in the conus varies in different fish. The Port Jackson shark has three sets of valves with three semilunar valves in each set. Satchell and Jones refer to the valves as lower (proximal), middle, and upper (distal) tonal valves. The lower valves are those near the junction of the conus and ventricle. The middle valves divide the conus into upper and lower conal chambers. A diagram by Hanson shows four rows of tonal valves in the dogfish, ***Squalus acanthias***, whereas a drawing by Sudak of a freshly obtained heart from a smooth dogfish, *Mustelus canis*, shows only proximal and distal tonal valves.

During ventricular diastole the proximal (lower and middle) conal valves of the Port Jackson shark are open, the ventricular and conal chambers are interconnected, and the pressure difference between ventricle and conus and ventral aorta is maintained by the closed distal (upper) tonal valves. The proximal and middle valves are not large in the Port Jackson shark, and when the conus is relaxed their free edges do not meet and the valves are incompetent. Conal systole narrows the tonal lumen, and under these conditions the valves are competent. The conus swells at the onset of ventricular systole as blood moves from the ventricle into the conus. Pressures rise simultaneously in the ventricular and tonal chambers and eventually exceed that in the ventral aorta; the distal (upper) conal valves open and blood flows into the ventral aorta. Conal systole begins before the end of ventricular systole. The contraction starts at the junction of the ventricle and conus and passes forward toward the ventral aorta. Conal systole aids the closure of first the proximal (lower) and then the middle tonal valves as the ventricle relaxes. As the conus relaxes the proximal (lower) and middle valves become incompetent, there is a small backflow, and the distal (upper) tonal valve closes. This valve, unlike the proximal and middle valves, is competent in the absence of tonal systole; hence, it remains closed after the conus relaxes. Satchell and Jones recorded blood flow in the ventral aorta of the Port Jackson shark and found that flow occurs only during the period between the opening of the distal conal valves and the closing of the proximal tonal valves. During this period the changes in flow and pressure in the ventral aorta can be ascribed to ventricular systole. The period between closure of the proximal and distal valves corresponds to tonal systole, and during this period, Satchell and Jones did not record any blood flow in the ventral aorta. There is hardly any backflow in the ventral aorta as the valves close, but if tonal systole is prevented, the proximal (lower) and middle valves are incompetent throughout the cardiac cycle and there is a large backflow of blood in the ventral aorta as the distal valves close. The conus is within a rigid pericardium, and the subatmospheric intrapericardial pressures tend to dilate the conus and increase the incompetence of the tonal valves. If the intrapericardial pressure is experimentally reduced, the backflow in the ventral aorta increases as the distal valve closes. Thus, Satchell and Jones suggest that tonal systole plays no part in ejecting blood into the ventral aorta but maintains the competence of the proximal and middle tonal valves for a short period of time subsequent to the ventricular ejection of blood. The suggestions made by Satchell and Jones are not mutually exclusive. Conal systole could play a part in ejecting blood into the ventral aorta and also maintain the competence of the conal valves. Sudak, Johansen *et al.*, and Hanson have reported data that either indicate or demonstrate enhanced flow in the ventral aorta that can be ascribed to conal systole. Thus, in most elasmobranchs studied, tonal systole maintains flow in the ventral aorta during ventricular relaxation.

A series of tonal valves are required because the conduction velocity of the wave of excitation over the conus is relatively slow. The number of sets of valves present is presumably related to the length of the conus and the velocity of conduction. A series of valves is required if proximal portions of the conus relax, and the valves become patent, while systole is still occurring in more distal portions of the conus. Thus a series of valves permits the progressive contraction and relaxation of the conus and reduces any backflow resulting from conal systole.

Work done by the contracting ventricle and the efficiency of

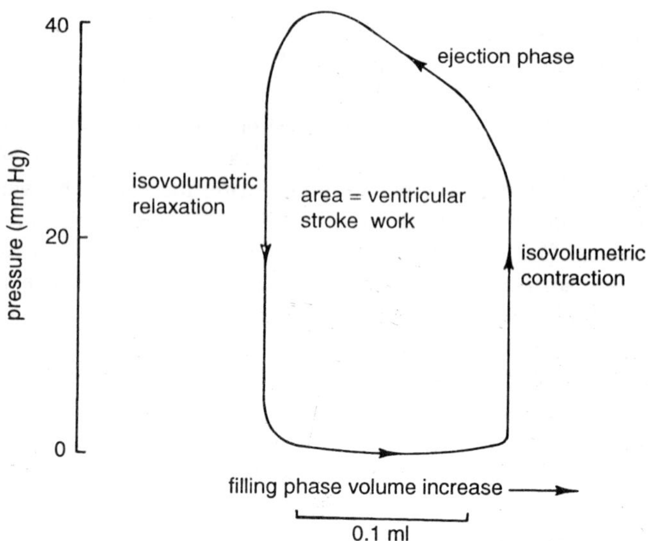

Figure 3.5 : Pressure and volume relationships in the ventricle of the lingcod, Ophiodon elongates.

contraction have not been calculated for any fish heart. Work done is force times distance moved, which, in terms of the heart, is pressure times volume ejected. Pressure and flow continually change during ventricular systole. Work is done only during periods when blood is ejected from the heart. A measure of ventricular stroke work can be obtained by plotting ventricular pressure against ventricular volume. The area of the enclosed loop is a measure of the stroke work of the ventricle. Although ventricular pressures have been measured, there are no reported data in the literature of associated volume changes in the ventricle during systole, or indeed in any other part of the heart. In Figure 6, changes in ventricular volume were calculated from increments of ventral aortic flow. This is not an accurate assessment of volume change in the ventricle because, during systole, the volume of of blood flow in the ventral aorta. The bulbus is increased, and the volume of blood flowing into the ventral aorta is less than that leaving the ventricle. During ventricular diastole, when the valves between the bulbus and ventricle close, the volume of the bulbus decreases and the elastic rebound of the bulbus maintains flow in the ventral aorta. Thus the estimates of ventricular volume changes reported in Figure elsewhere in the chapter are on the low side.

The construction of pressure-volume loops for the ventricle and bulbus or conus, under a variety of conditions, would enable one to assess the influence of the bulbus or conus on the pattern of blood flow in the ventral aorta.

Regulation of Cardiac Activity

Nervous Regulation

All fish hearts, with the exception of that of the hagfish, are innervated by a branch of the vagus nerve. Stimulation of the vagus in teleosts and elasmobranchs causes a marked bradycardia which can be blocked by atropine. The application of acetylcholine has been shown to slow the heart, and it has been assumed that the cardiac vagus contains cholinergic fibers. Some investigators, however, have been able to produce cardioacceleration by vagal stimulation, and recently adrenergic fibers have also been found to innervate the trout heart.

The lamprey heart is innervated by a branch of the vagus nerve. This nerve has not been stimulated directly, but stimulation of the medulla oblongata causes cardioacceleration. The effects of nervous stimulation are similar to those produced by the application of acetylcholine on the heart. Acetylcholine has a positive chronotropic (increased rate of contraction) but a negative inotropic (decreased force of contraction) effect on the lamprey heart. The effects of acetylcholine cannot be explained in terms of the release of catecholamines from stores within the heart or of the action of these compounds on β-adrenergic receptors known to be present in the lamprey heart. The effects of acetylcholine are not blocked by atropine, but they are blocked by curare. The lamprey heart, like most other fish, appears to receive a cholinergic nervous supply; the receptor sites, however, are different. The lamprey cholinergic receptors are blocked by curare, whereas those in teleosts and elasmobranchs are blocked by atropine.

A rise in dorsal aortic blood pressure in teleosts causes a radycardia which can be blocked by atropine. Irving *et al.* have shown that increases in blood pressure increase the level of afferent activity in nerves innervating the branchial arches of the dogfish. Recently, Laurent has recorded chemoreceptor and baroreceptor discharges from nerves innervating the pseudobranch in teleosts. Dorsal aortic pressure varies little in the intact animal, and the pseudobranch may play a role in regulating heart rate, via the level of vagal tone, to maintain a constant dorsal aortic pressure.

The level of vagal tone to the heart varies between species and is altered, within a single species to a variable extent, by a large number of parameters. Bradycardia has been observed in response to light flashes, mechanical vibrations, salinity changes, atmospheric pressure changes, anoxia, removal from water, and touch. The exact nature of the response varies from fish to fish, some fish slow their heart in response to almost any stimulus, whereas others are more selective in terms of the stimuli required to evoke bradycardia. Most of the above parameters probably slow the heart by increasing the level of vagal tone. The functional significance of many of the heart rate changes is not clear. Usually only charges have been measured. A decrease in heart rate does not necessarily indicate a decreased cardiac output. In the trout, bradycardia which develops during hypoxia is offset by an increase in stroke volume in such a way that only minor changes in cardiac output are observed.

The level of vagal tone to the heart oscillates in phase with the breathing cycle in many fish and tends to inhibit the heart as the mouth opens. Satchell has suggested that this sinus arrhythmia serves to correlate maximum flows of blood and water at the respiratory surface.

Swimming in freshwater teleosts is associated with a decrease in vagal tone and an increase in heart rate. The magnitude of the increase is related to the level of vagal tone prior to activity. Large heart rate changes indicate a high level of vagal tone before exercise, and *vice versa*. When Chondrichthyes or some marine teleosts are forced to swim, exercise is associated with an increase in vagal tone and a bradycardia, which disappears if the activity is prolonged. The bradycardia may not occur during spontaneous swimming.

Aneural Regulation

Starling's law of the heart is a basic aneural cardiac control mechanism. This law states that the energy of contraction is a function of the initial length of the muscle fiber. The greater the end diastolic volume, the greater will be the force of contraction and therefore the stroke volume. The venous input pressure to the heart usually determines end diastolic volume.

Catecholamines are known to increase both the rate (positive chronotropic effect) and myocardial contractility (positive inotropic effect) of mammalian hearts. The increased myocardial contractility results in a larger stroke volume being ejected from an unchanged end diastolic volume, i.e., there is greater systolic emptying of the heart. Both the positive inotropic and chronotropic responses result from the action of catecholamines on β-adrenergic receptor sites in the heart. Catecholamines, therefore, alter the nature of the Starling relationship between end diastolic volume and stroke volume by their action on β-adrenergic receptors in the heart.

The fish heart appears to obey Starling's law. The *in situ* aneural hagfish heart responds to increased filling by increasing its force of contraction. The isolated *in vitro* trout heart, *Salmo gairdneri*, also obeys Sartling's law. As the input pressure is raised both stroke volume and apparent stroke work (calculated by multiplying stroke volume by the difference between the mean output and input pressures) increase. Increasing input pressures also cause the rate to increase in the isolated heart. In all probability, this results from a direct effect of stretch on the pacemaker fibers causing changes in ionic conductance. Both the increase in rate and stroke volume contribute to a large increase in cardiac output as the input pressure is raised.

A rise in temperature increases heart rate both in the isolated preparation and in the intact animal. In the *in vitro* heart, the rate increases are offset by decreases in stroke volume as temperature increases, and cardiac output remains constant. Decreases in temperature, therefore, appear to have a positive inotropic effect on the heart.

Epinephrine alters the relationship between end diastolic volume and stroke volume in the isolated trout heart. In all cases, epinephrine increases cardiac output and apparent heart work (calculated by multiplying cardiac output by the difference between mean output and input pressures to the heart; in all of these experiments the heart had to pump saline through a fixed resistance to a height above the input level). The effect of epinephrine on stroke volume and heart rate is temperature dependent. At low temperature (6°C), increased epinephrine levels raise heart rate but have little effect on stroke volume, whereas at high temperatures (15°C), stroke volume is increased but heart rate is decreased.

In the *in vitro* trout heart, stroke volume and heart rate are affected by the input pressure, temperature, and epinephrine. At low temperatures (6°C), the positive inotropic effect of a decrease in temperature masks the positive inotropic effect of epinephrine, and only rate increases are observed when epinephrine levels are raised. At high temperatures (15°C), however, the increased inotropic effect of epinephrine is seen and stroke volume increases when epinephrine levels are raised. At this temperature the initial rate is high. Systolic emptying is increased so greatly when epinephrine levels are raised that a longer interval is needed to fill the heart again. Heart rate appears to be sensitive to stretch of the heart as well as to temperature and epinephrine; hence, as the heart requires more time to fill at 15°C when epinephrine levels are increased, the stretch component determining rate is reduced and heart rate falls. Thus epinephrine exerts both a chronotropic and inotropic effect on the *in vitro* trout heart; the positive chronotropic response predominates at low temperatures, whereas the positive inotropic response predominates at high temperatures. Epinephrine is known to have several metabolic effects on the mammalian heart. It promotes glycogen breakdown by increasing the level of cyclic AMP, which leads to an increase in the active form of the enzyme glycogen phosphorylase. Cyclic AMP is known to sensitise actomyosin toward changes in calcium concentration. Thus cyclic AMP plays a role in both the inotropic and glycogenolytic effects of epinephrine. Both the glycogenolytic and inotropic effects are blocked by β-adrenergic blocking agents in mammals. The β-adrenergic blocking agent, Inderal, blocks the response to epinephrine in the isolated trout heart and causes a general myocardial depression similar to that seen in mammals. The a-adrenergic blocking agent, phenoxybenzamine, had no effect on the response of the isolated trout heart.

High levels of catecholamines have been found in cyclostome hearts. Much lower levels of adrenaline and noradrenaline have been demonstrated in the hearts of other fishes, and adrenaline usually represents more than 50% of the total catecholamines present. Catecholamines have been shown to increase the rate and force of contraction of both teleost and elasmobranch hearts, and β-adrenergic receptors have been demonstrated in the isolated plaice and trout heart. Surprisingly, catecholamines have little effect on the hagfish heart; however, Chapman *et al.* did find that adrenaline restored loss of vigor caused by the application of reserpine in the hagfish heart.

In the intact animal the activity of the heart is determined by the sum total of many factors that influence the heart. Johansen observed an increase in stroke volume when venous return was experimentally increased in the cod, *Gadus morhua;* no rate changes were observed. Labat *et al.* observed an increase in heart rate in the catfish when the temperature was increased or if saline was injected into the hepatic vein, increasing venous return to the heart. Laffont and Labat found that below 8°C epinephrine injections caused a decrease in heart rate. The bradycardia recorded by Laffont and Labat may have been the result of an increase in vagal tone, resulting from a change in blood pressure caused by the action of epinephrine on the circulatory system.

When trout swim there are large changes in cardiac output which are the result of small increases in heart rate and large increases in stroke volume; also the total peripheral resistance to flow decreases. There is no vagal tone to the heart of trout during rest or exercise as long as the fish is in well-aerated water; therefore, any changes that occur during activity cannot be explained in terms of a decrease in vagal tone to the heart. The level of circulating catecholamines increases during exercise in trout. There are β-adrenergic receptors in the heart and adrenergic receptors in the gills and elsewhere in the circulation. The effect of epinephrine is to decrease the resistance to flow in the gill circulation. Thus the changes in cardiac output occurring during exercise may be caused, at least in part, by the action of catecholamines on p receptors in the heart and on other adrenergic receptors in the peripheral circulation. One would suspect, from data on the mammalian circulatory system, that other factors, such as the production of metabolites, play a role in decreasing peripheral resistance during exercise in fish.

Exercise in the lingcod, *Ophiodon elongatus*, is accompanied by stroke volume and cardiac output increases only when the fish is atropinised. Exercise in the nonatropinised fish is accompanied by an increase in vagal tone and a bradycardia, which decreases cardiac output. Possibly this is a response to the stimulus which provokes exercise rather than to exercise itself. Lingcod usually-remain motionless at the bottom of the tank and must be prodded or startled to produce short bursts of swimming. An increase in vagal tone may not be a physiological response to spontaneous swimming in this animal. Hanson found that spontaneous activity in a number of Chondrichthyes was accompanied by an increase in cardiac output; if, however, the animal was disturbed and then swam away an initial bradycardia was observed.

An increase in temperature acts directly on the heart, increasing the intrinsic rate of the pacemaker cells. In the intact lingcod, *Ophiodon elongatus*, stroke volume remains constant over a wide range of temperatures, and a rise in temperature causes an increase in cardiac output owing to an increase in heart rate. The *in vitro* heart of the trout behaves in a different way; a rise in temperature increases heart rate but decreases stroke volume, and cardiac output remains constant over a wide range of temperatures. In the intact animal, temperature changes affect the whole animal, whereas in the *inn vitro* preparation only the response of the heart is recorded. Many factors, including the magnitude of the peripheral resistance to blood flow, can affect stroke volume. There are some indications that the peripheral resistance to blood flow decreases as temperature increases, and this would tend to maintain stroke volume in the face of a negative inotropic effect on the heart caused by a temperature increase.

The balance of evidence indicates that changes in cardiac output in fish are associated with changes in stroke volume with some adjustment of heart rate. The rate changes are mediated via cholinergic nerve i nervating the heart, changes in temperature, and changes in the level catecholamines. Stroke volume changes occur in response

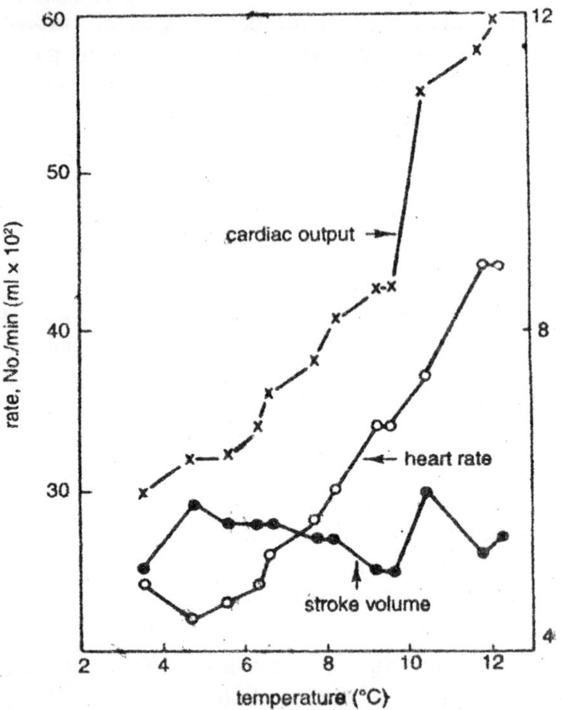

Figure 3.6 : The effect of temperature on heart rate, stroke volume, and cardiac output in the lingcod, Ophiodon elongatu.

temperature to increased levels of catecholamines and increases in venous return to the heart

The Regulation of Venous Return

An increase in venous return will increase end diastolic volume and increase stroke volume as described by Starling's law. An increase in venous return can be achieved in a number of ways. A decrease in total peripheral resistance will increase the vis a tergo effect of the heart on venous return. In Chondrichthyes, increased cardiac activity will lower the negative pressure within the rigid pericardium and raise the aspiratory forces moving venous blood back to the heart. Mobilisation of blood from storage organs and venous reservoirs will also aid venous return. The pumping activity of skeletal muscle in the presence of suitably placed valves is another powerful mechanism for increasing venous return during exercise.

Hagfish possess several accessory hearts, namely, the portal, cardinal, and caudal hearts. The action of these hearts plus the pumping action of contractions of the gill musculature aid the circulation of the blood and regulate venous return to the systemic (branchial) heart.

The effectiveness of skeletal muscle contractions in increasing venous return is dependent on the presence of valves directing blood flow toward the heart. Valves have been demonstrated in the segmental arteries and veins in the caudal regions of the Port Jackson shark, *Heterodontus portus jacksoni*. These valves are at the entrance and exit of the myotomal muscles and are so arranged that the contracting muscles pump blood from the arterial to the venous side of the circulation, thus increasing venous return to the heart. Venous valves have been found in teleosts. Blood flow in the dorsal aorta of some teleosts may be augmented by the presence of a stiff ligament extending into and partially dividing the lumen of the dorsal aorta. The ligament extends from the basioccipital to the caudal region and undulations of the body, pressing the walls of the dorsal aorta against this stiff ligament, may propel blood in a caudal direction.

Mobilisation of blood from storage organs or venous reservoirs may aid venous return and increase cardiac output. Exercise in trout is accompanied by a decrease in splenic volume. Hepatic vein sphincters in elasmobranchs regulate the volume and transit time for blood in the liver, and their relaxation would permit the mobilisation of blood stored in the liver at times of increased demand on cardiac output. The hepatic vein sphincters relax in response to the topical application of adrenaline and contract in response to acetylcholine. Lowering the dorsal fin in the ratfish decreases. the amount of blood in the dorsal fin sinus and may be used to increase venous return in this fish.

GENERAL PROPERTIES OF THE CARDIOVASCULAR SYSTEM

Cardiac Output

Three methods have been used to determine cardiac output in

fish. These are the application of the Fick principle, the dye dilution method, and the direct measurement of the velocity of blood flow in the ventral aorta using either an electromagnetic or a Doppler ultrasonic blood flowmeter system.

Recorded values of cardiac output in a number of elasmobranchs are in the region of 25 ml/kg/min. The fish investigated include dogfish, *Squalas acanthias* and *Scyliorhinus stellaris*. Hanson reported cardiac outputs for the dogfish, *Squalus suckleyi*, of between 9 and 23 ml/kg/min. The cardiac output of the ratfish, *Hydrolagus colliei*, is about 21 ml/kg/min. Cardiac output per kilogram falls with increasing weight in the skate, *Raja binoculata*.

By comparison, measured values for cardiac output in teleosts, are variable, ranging from 5 to 100 ml/kg/min. Most values, however, fall within the range of 15-30 ml/kg/min.

Peak ejection velocities of blood flow in the ventral aorta of the Californian Horn shark (*Heterodontus francisci*), skate (*Raja binoculata*), and dogfish (*Squalus suckleyi*) are between 8 and 20 cm/sec which occur 120 msec after the aortic valves open in the Horn shark and 450 msec in the skate; acceleration of the blood reaches a maximum of 120 cm/sect in the Horn shark.

Blood Volume

The blood volume of a fish consists of that portion of the extracellular space contained within the cardiovascular system, plus the volume of erythrocytes, leukocytes, and platelets in the blood. It consists of a plasma volume and the volume of cells within the blood, which when expressed as a percentage of the total blood volume is the hematocrit. Both plasma volume and hematocrit vary considerably in fish.

Mott has reviewed the earlier literature on blood volume measurements in fish. Holmes and Donaldson have recently completed an exhaustive review of the subject which will only be discussed briefly here. Thorson has estimated blood volume from measurements of plasma volume and hematocrit in a large number of fish. Agnatha have very large blood volumes in excess of any other fish. The blood volume of Chondrichthyes is about 6.6% body weight, whereas that of Chondrostci, Holostei, and teleosts (both marine and freshwater species) is approximately 3% body weight. Thorson pointed out that the evolutionary trend in fish appeared to be toward smaller blood volumes, the significance of which is not clear. Schiffman and Fromm, Conte *et al.*, and Ronald *et al.* also reported blood volumes of approximately 3% body weight in trout and cod. Most investigators have used methods developed for determining blood volume in mammals. Many of these methods depend on the injection of a dye and the measurement of its subsequent dilution in the blood. It is important that the dye be evenly distributed throughout the vascular compartment before blood volume is determined. Most investigators, in accord with mammalian practices, have allowed 30-40 min for equilibration of the dye. Smith found that Evans blue (T-1824) was not evenly distributed in the blood of salmon until at least an hour had elapsed, and reported blood volumes of 6% body weight for. a number of salmonids, approximately twice that previously reported for any other teleost. This very slow mixing of the dye indicates the possibility of a slow turnover of blood in some parts of the vascular system.

Violent bursts of swimming result in small increases in weight in the trout. The concentrations of hemoglobin and plasma proteins oscillate in phase with one another, indicating changes in the blood" water concentration during exercise. There is an initial hemodilution followed by a hemoconcentration as exercise continues. The weight gain and initial hemodilution could be the result of a net influx of water into the blood across the gills, followed by an osmotic shift of water into the muscles, causing the hemoconcentration. Exercise, or any form of disturbance, increases blood catecholamine levels in salmonids, which in turn alter the pattern of blood flow through the gills and increase the rate of ion, water, and gas exchange. Artificially raising the blood adrenaline levels in trout results in the fish gaining weight. Thus any form of disturbance, which causes a rise in blood catecholamine levels in freshwater fish, may cause a hemodilution resulting from a net influx of water across the gills.

Blood Distribution

Circulation time can be calculated by dividing blood volume by cardiac output per minute. If cardiac output is 25 ml/kg/min and blood volume is 5% body weight, the calculated circulation time will be 2 min. Mott reported that blood took 5.6 $\pm$ 1.9 sec to pass through the gills and appeared in the visceral veins 20 sec after leaving the heart. The observed circulation time is much more rapid than the calculated value. The estimated time however is only an average value, and the circulation time through different pathways must vary.

Stevens found that the blood was unevenly distributed to different parts of the body of the rainbow trout, *Salmo gairdneri*. White myotomal muscle, constituting 66% of the total weight of the fish, contains only 15.8% of the total blood volume (assuming a blood volume of 5% body weight). Red myotomal muscle contains between two and three times as much blood per unit weight as white muscle, which is in approximate agreement with the ratio of the number of capillaries in the two types of muscle. Combined, the red and white myotomal muscle contains about 20% of the blood and represents two-thirds of the body weight. The remaining one-third of the body therefore contains 80% of the total blood volume. If one assumes the blood volume of the trout studied by Stevens to be 5% body weight, 4 ml of blood are contained in 33% of the body weight in a 100-g fish. The blood volume in parts of the body other than the myotomes is therefore about 12% tissue weight.

The apparent distribution of blood does not change as a result of exercise in the trout, except for a slight decrease in spleen blood volume and a rather surprising increase in gut blood volume. Small changes in the diameter of vessels cause large changes in vascular

Table 3.1 : The Distribution of Blood to Various Tissues in Rainbow Trout, *Salmo gairdneri*, Assuming a Blood Volume of 5% *Body Weight.*

Tissue	A (Body weight, %)	B (Blood volume in tissue, %)	A/B
White muscle	66	15.8	0.24
Red muscle	1.0	6	6.0
Heart	0.2	2	10.0
Gills	3.9	7.6	19.0
Gut	5.1	2.4	0.47
Liver	1.4	4.0	2.9
Spleen	0.3	1.4	4.7
Blood in arteries, veins, heart, and kidney	3.0	60.0	20.0
Remainder	19.1	0.9	0.005
Total	100.0	100.0	

resistance and blood flow but only small changes in the volume of blood in the vessels. Stevens measured only the instantaneous blood volume in various tissues. The procedure he adopted did not enable him to detect changes in blood flow. Very little is known about the relative changes in flow to various organs during exercise in fish.

Skeletal Muscle Circulation

Exercise

There are two types of muscle fiber in the myotomes of fishes which can be recognised by their color. Red or dark fibers form thin, lateral superficial sheets just under the skin, white fibers make up the rest of the underlying muscle mass. The two types of muscle fiber form discrete motor systems, receiving separate nervous innervation and having different enzymic distributions, mitochondrial content, and fiber diameter. White muscle is used in burst swimming of short duration, and energy is supplied by anaerobic glycolysis. During burst swimming there is a rapid depletion of muscle glycogen and a large lactate production. The lactate diffuses slowly into the blood and up to 12 hr postexercise recovery is necessary before the low preexercise lactate levels are restored. White muscle fatigues rapidly, and the effects of burst activity on glycogen depletion and lactate production are cumulative during the recovery period.

White myotomal muscle, constituting the major portion of the body, is only used in burst activity of short duration. Fish, therefore, drag twothirds of their body around simply to effect escape reactions and various other burst responses. The fish is streamlined and neutrally bouyant and the cost of moving this volume of muscle is probably small, especially if the maintainance and circulation of the muscle is minimised. The sparse vascularisation limits the oxygen supply and during activity the muscle must operate anaerobically. The poor vascularisation will also contribute to the slow release of lactate from the muscle and so minimise the effects of a massive lactate production on the rest of the body.

Red myotomal muscle is used in sustained swimming and is almost impossible to fatigue. There are no changes in red muscle glycogen content during activity, and energy is supplied by the oxidation of fats. The blood volume and number of capillaries per unit weight are three times that in white muscle. Scombroids and certain elasmobranchs, in addition to the superficial red muscles, have a deeper lying red portion of the myotome, correlated with an ability to sustain high cruising speeds and muscle temperatures that are above ambient.

Nothing is known about the regulation of blood flow to the red and white muscle fibers during exercise. Contractions of skeletal muscle probably create a vis *a tergo* force driving blood back to the heart. Satchell demonstrated that an isolated trunk preparation of the Port Jackson shark (*Heterodontus portusjacksoni*) was able to pump and increase the flow of a perfusate during flection of the trunk. Arteries in the postpelvic region of the trunk arising from the dorsal aorta have valves which prevent the reflux of blood into the dorsal aorta during muscle contraction. Veins entering the caudal vein in the postpelvic region have valves which prevent reflux of blood into the segmental vessels. The caudal vein is encased in bone, and the contracting muscles squeeze all vessels in that region except the caudal vein propelling blood from the arterial to venous side of the circulation. Flow through the trunk preparation was somewhat higher after electrical stimulation than before, indicating that some other factors also play a role in increasing muscle blood flow during exercise.

Temperature Regulation

The muscles of fish are cooled by the blood, loss of heat through the general body surface is insignificant. Metabolic heat, which warms the blood, is quickly lost at the gill surface, where blood comes into close proximity to the water. Water has a high specific heat, and thermal diffusion is ten times more rapid than gaseous diffusion; thus, thermal equilibrium between blood and water must occur very rapidly in the gills. Tuna and lamnid sharks, however, can maintain muscle temperatures well above ambient. This is particularly true of the bluefin tuna, *Thunnus thynnus*, which can control its muscle temperature so that the warmest part of the muscle mass varies only 6°C over a 20°C range of water temperatures. The red muscle, used in sustained swimming, is the warmest part of the whole muscle mass. This ability to thermoregulate is related to the presence of a rete mirabile formed by arteries and veins supplying the red muscles and to the encapsulation of the red muscle by the white muscle. The retia of bluefin tuna are formed from arteries and veins arising from a pair of cutaneous vessels. The segmental arteries arising from the dorsal aorta are of secondary importance in the bluefin tuna, unlike skipjack and yellowfin tuna, which have central as well as lateral rctia. The retia act as countercurrent heat exchangers in such a way that venous blood warms arterial blood entering the muscle and reduces heat loss. The retia, therefore, form a thermal barrier retaining metabolic heat in the tissues and preventing its loss in the gills.

The size of the rete must be regulated to conserve heat but allow the free passage of gases in and out of the muscle. The red muscles operate aerobically, and the design of the rote must take advantage of the differences between thermal and gaseous diffusion. Because thermal diffusion is rapid, a relatively small rete is required to allow thermal

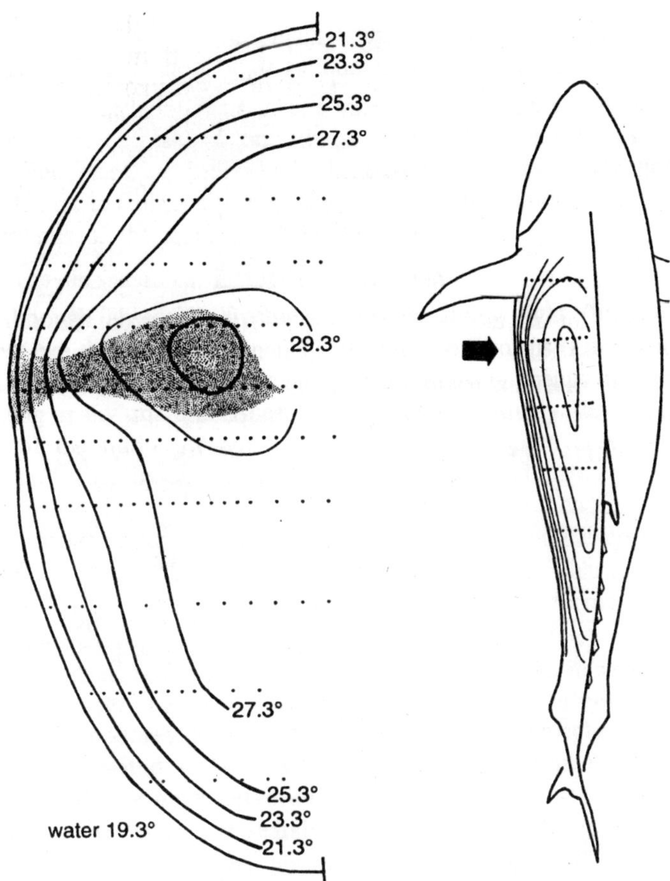

Figure 3.7 : Temperature distribution in cross section and frontal section of a bluefin tuna. Temperatures were measured with thermistor probes at positions indicated by the dots. Heavy curves are isotherms plotted at 2°C intervals. Dark (red) muscle is indicated by stippling.

equilibration of arterial and venous blood; gases diffuse much more slowly, and so there is probably only an insignificant exchange of gases across this rete. Thermal and gaseous diffusion coefficients across the rete, however, have not been measured.

The ability to thermoregulate varies. Large bluefin tuna appear to have a greater capacity to thermoregulate than small fish. Skipjack and yellowfin tuna-can maintain an elevated muscle temperature, but it is only a fixed value above ambient over a wide range of temperatures. Large bluefin tuna can maintain a red muscle temperature of between 26° and 32°C, while the ambient temperature varies between 6° and 30°C. At low water temperatures the muscle is 20°C above ambient, whereas at water temperatures of 30°C the muscle is only 2°C above ambient. This implies regulation of muscle temperature, which could be achieved by altering the functional size of the rete and/or by altering blood flow through the rete.

Lamnid sharks have a similar rete structure to bluefin tuna. Both groups can thermoregulate and are able to swim swiftly, at a sustained speed, as a result of the extra power available from warm muscles. The bluefin tuna exists over a wide range of temperatures and migrates long distances across several temperature zones. Carey and Teal report an instance of a bluefin tuna traveling from the Bahamas to Norway, a distance of 4200 nautical miles, in less than 50 days. This is an excellent testimonial of the ability of warm red muscles to maintain sustained swimming at high speed.

THE EFFECTS OF SOME SUBSTANCES ON THE CIRCULATORY SYSTEM

When injected into or otherwise applied to, the circulatory system of fish, a large number of compounds are known to cause changes in blood pressure and flow. The-author has attempted to list the effects of a few compounds on the cardiovascular system of fish; some of these responses have already been discussed in preceding sections.

Acetylcholine

Acetylcholine increases the vascular resistance to flow through the gills of teleosts directing blood away from the secondary lamellae. The resistance to flow in the systemic circulation of elasmobranchs is decreased. Acetylcholine has a negative chronotropic effect on teleost and elasmobranch hearts but acts as a cardiac stimulant in the lamprey (Falck -*et al* Methacholine, succinylcholine, pilocarpine, neostigmine, and *d*-tubocurarine all have similar but less marked effects than acetylcholine, causing bradycardia and fall in blood pressure in the eel.

Atropine

Atropinisation increases heart rate in both elasmobranchs and teleosts, blocking the effect of efferent activity in cholinergic fibers innervating the heart. Atropine injected into the vascular system increases heart rate in both the resting and active dogfish causing an increase in ventral aortic systolic blood pressure but a decrease in peak ejection velocity and stroke volume of the heart.

Catecholamines

Both adrenaline and noradrenaline increase ventral and dorsal aortic blood pressure and may either increase, decrease, or have no effect on heart rate when injected into the intact fish. A- decrease in heart rate is the result of reflex vagal inhibition on the heart, probably resulting from a rise in blood pressure, and can be blocked by atropine. The direct effect of catecholamines on the heart is stimulatory. Adrenaline has a positive chronotropic and inotropic effect on the lamprey, teleost, and elasmobranch heart; but it has little or no effect on the hagfish heart. Adrenaline and noradrenaline dilate the vessels in the teleost and clasmobranch gill decrease the resistance to flow through the gills, and increase blood flow in the secondary lamellae. Phenoxybenzamine, an a-adrenergic receptor blocking agent, diminishes the effect of catecholamines on the circulation. Isoprenaline, a b-adrenergic stimulant, decreases the resistance to blood flow through the gills suggested that isoprenaline dilated vessels in the systemic circulation of the eel. Hanson concluded that catecholamines increase systemic resistance to flow in elasmobranchs. Keys and Bateman found that adrenaline increased the resistance to flow through the systemic vessels in the eel's tail.

Tyramine and ephedrine, substances that cause the release of catecholamines from bound sites in mammals, caused a bradycardia and a transient fall, followed by a rise, in blood pressure in the eel, similar to that produced by noradrenaline. Cocain, which prevents the release and reduces the storage of noradrenaline in mammalian adrenergic nerves, causes a bradycardia and hypertension when injected into the eel. The bradycardia can be abolished by atropine.

Reserpine

This substance depletes the stores and limits the uptake of catecholamines in mammalian tissues. Injections in the eel cause the chromatophores to dilate and the swim bladder to swell, changes consistent with catecholamine depletion. Reserpine depletes the catecholamine stores in both the lamprey and hagfish heart and causes a loss in vigor in the hagfish heart which can be restored by adrenaline.

Eptatretin

Eptatretin is an unstable amine or amide but, although it has similar effects, it is not a catecholamine. It has been found only in the hagfish heart, but has a marked positive chronotropic and inotropic effect on the heart of many vertebrates, including the hagfish.

Histamine

Histamine causes vasoconstriction in the perfused gills of pike and a fall in blood pressure in the ventral aorta of the eel. Chan found that histamine had little or no effect on blood pressure in the dorsal and ventral aorta of the eel. Chan used injected doses of 200 μg, some 100 times greater than that used by Mott.

5-Hydroxytryptamine (Serotonin)

Chan found that injections of 25 μg of 5-hydroxytryptamine caused a slight increase, followed by a decline, in arterial blood pressure in the eel. No changes in heart rate or pulse pressure were recorded.

Renin and Angiotensin II

Angiotensin II, formed by the action of renin on angiotensin I, is a potent vasoconstrictor in mammals. Angiotension II, injected into eels, causes a marked increase in both dorsal and ventral aortic blood pressure, but it has no effect on heart rate or pulse pressure. The corpuscles of Stannius and kidney extracts, possible sources of renin in fish, also cause a rise in blood pressure.

Oxytocin and Vasopressin

The neurohypophysial hormones oxytocin and vasopressin cause a prolonged and marked rise in ventral aortic blood pressure, but they have little effect on the dorsal aortic blood pressure in the eel. The magnitude of the response varies with diverent hormones, the order of potency is 4-serine 8-isoleucine oxytocin > oxytocin > 8-arginine oxytocin > arginine vasopressin > lysine vasopressin, with lysine vasopressin having almost no effect on arterial blood pressure. The action of these hormones on the circulation appears to be related to interaction between the hormones, catecholamines, and a-adrenergic receptor sites.

Urophysial Extracts

Mugil uophysial extracts increase ventral and dorsal aortic blood pressure when injected into the eel. The eel has a caudal lymph heart, consisting of two chambers, an atrium, and a ventricle. The lymph heart collects fluid from the lateral cutaneous and hemal lymphatic vessels and pumps it into the caudal vein. Extracts from the eel urophysis increase the rate of beat of the caudal lymph heart.

4

EXCRETION

Vertebrates eliminate some metabolic wastes through the gut and the skin, but most are eliminated through special excretory organs, t-e kidneys. In elimination fishes and other aquatic animals have a particular problem in that their gills and oral membranes are permeable both to water and salts. In the ocean the salinity of the water is more concentrated than that of the body fluids of the fish, and water is drawn out, but salts tend to diffuse inward; Fence marine fishes drink sea water. In contrast, in fresh water, fishes lose salts and take up water through the gills because their internal salt concentration is greater than that of their surroundings.

Many nitrogenous wastes of fishes pass through the kidneys that also assist in water-salt balance (homeostasis) by the excretion or retention of certain minerals. The gills also take a prominent part in waste excretion, eliminating mainly ammonia.

The typical fish kidney is made up of many individual units or nephrons, each consisting of a renal corpuscle (Malpighian body) and a kidney tubule. The tubules join in collecting ducts that finally lead to the outside through the mesonephric duct with its various terminal modifications. The Malpighian body is made up of a glomerulus, a blood vessel tightly coiled with afferent nd efferent arterioles and encapsulated by thin kidney cells (Bowman's capsule). Some teleosts have kidneys with frequent small glomeruli or a few medium-sized glomeruli; others show a tendency for the glomeruli to become smaller, less frequent and poorly vascularised as in the horned sculpins (*Myoxocephalus*), whereas still others have very few glomeruli, and these are nonfunctional in their adult stages; for example, the goosefish

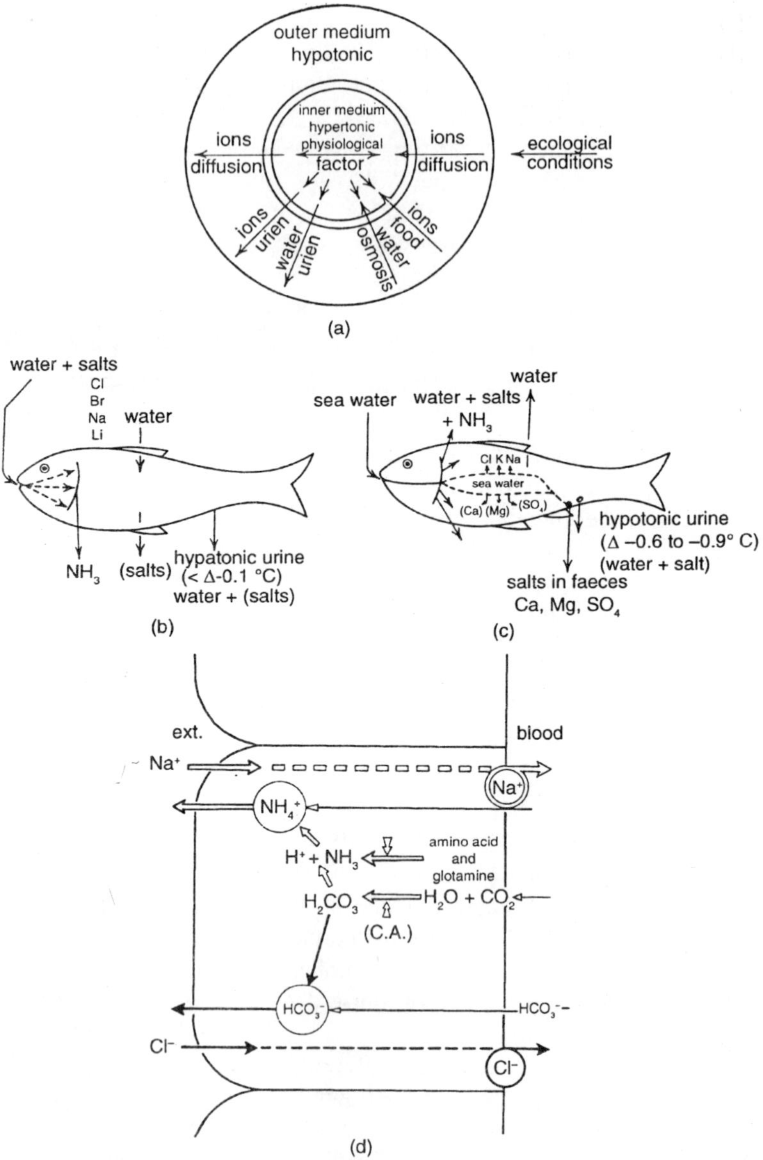

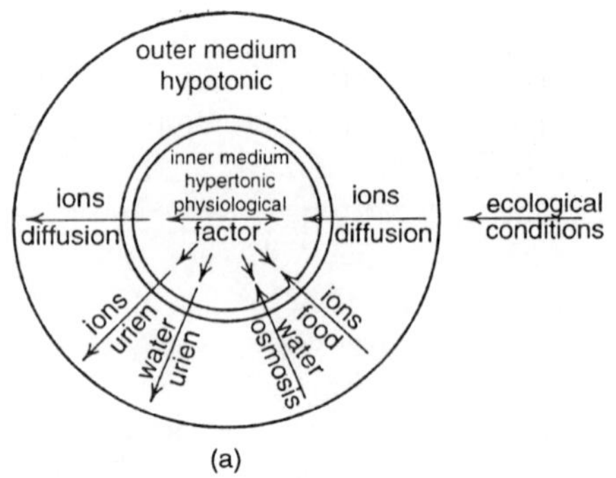

(a)

(b)

(c)

(d)

Figure 4.1 Principal processes involved in osmoregulation in (a) freshwater fishes, (b) inflow and outflow of salts and water in a freshwater fish, and (c) osmotic regulation in marine bony fishes. (d) Schematic representation of ionic exchanges in the branchial cell of the goldfish, Carassius auratus. aSe, deamidation and deamination enzymes; (C. A.) carbonic anhydrase; Lr active transp ,rt of ions requiring energy.

(*Lophius*) loses the connection between glomeruli and kidney tubules when the fish becomes mature. Finally there are a fair number of marine fishes, including Antarctic bony fish, which are lacking glomeruli entirely (for instance, the toadfishes, *Opsanus*). The glomerulus and the capsule together act as ultrafilters where blood pressure brings about nonselective dialysis of molecules up to a molecular weight of about 70,000; thus, serum proteins are retained in the blood. The excretory fluid undergoes alteration on its way through the tubules where glucose, various minerals, other solutes, and in some cases (sculpins, Cottidae) water are reabsorbed into the blood by an energy-requiring process. Presence of antifreeze glycoproteins and aglo'merulism of the kidneys probably played important roles in permitting Antarctic fish to adapt to near freezing water temperatures. Since these small molecular weight glycoproteins are not filtered into the formative urine, the fish avoid the energy expenditures that would be necessary for their tubular reabsorption. Control of filtration and reabsorption takes place through hormone action; in fishes the adrenal cortex, thyroid, suprarenal bodies, gonads, hypothalamus, and perhaps the pituitary are involved in this regulation.

Emphasis here will be placed on functional aspects of nitrogen excretion and related processes; freshwater, marine, and diadromous, (migrating to and from the sea) fishes are treated separately-as are sharks, skates, and rays (Elasmobranchii)-because they have developed a high urea tolerance through which they achieve osmotic homeostasis differently from other fishes.

Freshwater Lampreys and Bony Fishes

The osmotic pressure of body fluids depends on thei. mineral and organic compound content. In all freshwater fishes this pressure is higher than that of the surrounding water and there is a tendency for water to diffuse into the animal wherever there is a water-permeable membrane, that is, the gills, the oral membranes, and the intestinal surface. The skin and its mucus greatly reduce water permeability although small amounts of water do enter a fish trough the skin. Freezing point depression, Δ, of the blood of 'freshwater lampreys (some Petromyzonidae) is –0.38° to –0.46°C, and that of freshwater teleosts varies around -0.57°C when o of the outside water is virtually zero.

To cope with the steady inflow of water resulting from the differences in tonicity between the internal and external media, freshwater fishes produce a copious and h ghly dilute urine that is hypotonic, with regard to the fish. Lampreys in fresh water have a urine flow that normally varies between 15 and 36 percent of the body weight per day. The comparable measurements for freshwater bony fishes (Osteichthyes) range between 5 and 12 percent. The urine has a freezing point depression (Δ) of around -0.025°C for freshwater fishes; higher values of Δ are rare, although -0.09°C has been recorded for an African lungfish (*Protopterus*). The main work of the kidney in freshwater fishes, therefore, lies in water excretion; certain nitrogenous compounds, usually amounting to only a fraction of the total excreted nitrogen, also pass to the exterior by way of the kidney and its associated ducts. The kidney further functions in the retention of sugars and other vital solutes. Some salt losses occur since fishes in fresh water are hypertonic to the outside. These losses are made up by selective absorption of salt through the gills against the natural diffusion gradient.

Kidneys of Freshwater Fishes

Larvae of the lampreys (Petromyzonidae) have functional pronephric kidneys until they are 12 to 15 millimeters long. The nephrostomes have funnels that open into the pericardial coelom. The composition of body fluids is determined by permeability of the body surface and by the developing gill mechanism for the regulation of ionic equilibrium. Coelomic fluids are swept into the nephrostome, at the site of the glomus, the circulatory component of the pronephric tubule; substances to be conserved are reabsorbed into the blood and the remaining aqueous filtrate passes by duct to the urogenital sinus. In metamorphosis of the larvae, the mesonephric kidney takes over and the pronephric openings into the coelom disappear. Arterial blood is filtered through the glomerulus witliout selective retention of solutes, while the convoluted tubule is instrumental in reabsorbing salts and other substances.

With some exceptions, such as the bowfin (*Arnia*) and the gars (*Lepisosteus*) which have nephrostomes opening into the anterior

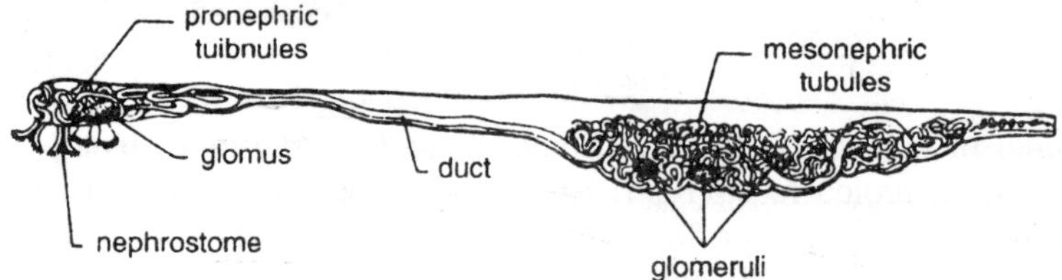

Figure 4.2 : Kidney of a lamprey (Petromyzon) *larva of 22 millimeters showing relationship of pronephric and mesonephric portions.*

coelom, adult freshwater rayfin fishes (Actinopterygii) have a mesonephric kidney closed off from the body cavity. Here again excess water is filtered from the blood through glomeruli, and salt and sugars are reabsorbed into the bloodstream through the epithelium of the kidney tubules and returned into the surrounding capillaries.

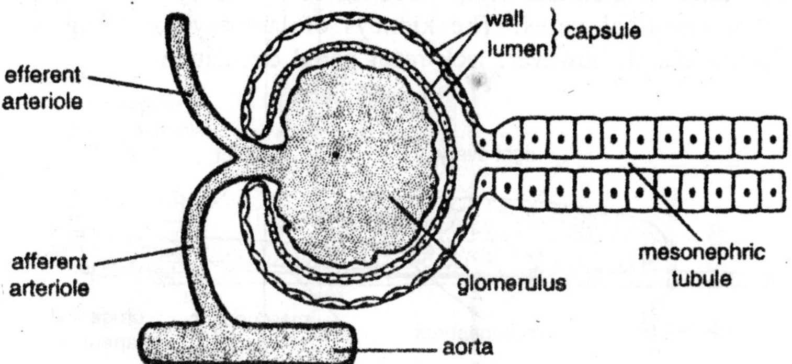

Figure 4.3 : Diagram of a renal corpuscle (glomerulus plus capsule) at the end of a uriniferous tubule as in a fish mesonephros.

The typical kidney of freshwater rayfin fishes (Actinopterygii) has many glomeruli that are larger than those of the typical marine rayfins with glomerular kidneys; the blood supply is also well developed. The kidney of freshwater fishes is often larger, in relation to the body weight, than that of marine fishes and measurements testify to the passage of great amounts of water through the organ. The number of glomeruli in one kidney from a number of freshwater species often exceeds 10,000 whereas many truly marine species have far fewer of them. These characteristics are related to salt content of the environment and therefore to the volume of urine excreted, as shown by a comparison of kidneys of salmons (*Oncorhynchus*) retained in freshwater to those of the same age in the sea; the former have significantly more glomeruli than the latter. The most telling comparison can be made, however, with the diameter of the glomerulus itself; in several freshwater species this measure ranges from 48 to 104 microns (mean 71), whereas in several marine species it ranges from 27 to 94 microns (mean 48). Some extreme glomerular variations exist; a tropical river-dwelling pipefish (*Microphis boaja*) has none and neither do any of the goosefishes (Lophiidae) that also invade fresh water. The tubules that drain the Bowman's capsules of freshwater fishes have an open neck segment, proximal and distal convoluted portions and, in some forms, an initial collecting region.

The fish kidney has a dual blood supply—the renal artery and the renal portal veins. The renal artery sends blood to the glomeruli where the high arterial blood pressure helps to produce the glomerular filtrate. The renal portal veins are connected to the capillary network around the kidney tubules and their blood joins that coming from the capillary bed within the glomerulus; all blood leaves the kidney region through the postcardinal veins. The kidneys of lampreys and hagfishes (Cyclostomata), however, lack renal portal circulation.

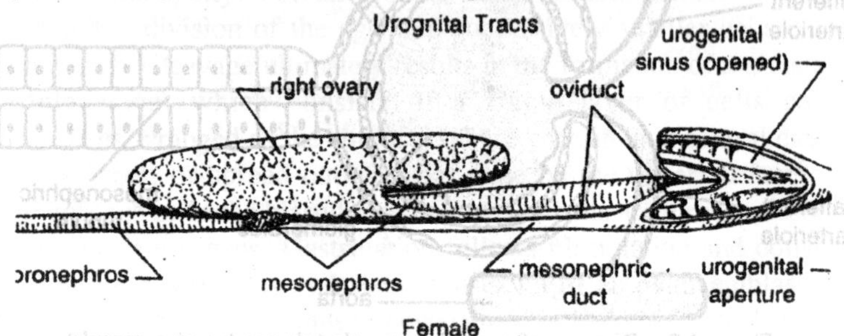

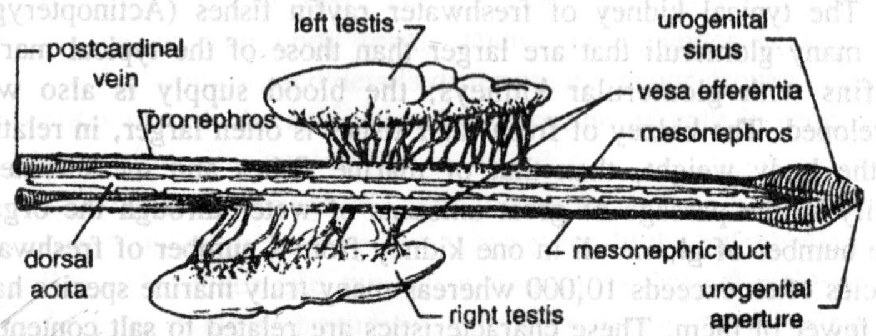

Figure 4.4 : Diagram of relationships of kidney, gonads, urinary ducts, urogenital sinus, and blood vessels in a male and female holostean fish, gar (Lepisosteus).

The urine of freshwater fishes contains creatine, creatinine, unidentified nitrogenous compounds some of which may be amino acids, and a little urea and ammonia. Urinary nitrogen amounts to between 2 and 25 percent of the total nitrogen excreted by freshwater fishes. The bulk passes out through the gills as ammonia (56 percent or more in the carp, *Cyprinus*). The remainder is probably urea and other simple compounds of nitrogen that also leave through the gills. The urine volume, as was already mentioned, depends largely on the water permeability of body surface, gills, and oral lining. Not all fishes have an equally impermeable mucus; water will diffuse twenty times as fast into a lamprey (*Lampetra*) as it does into an eel (*Anguilla*), but only twice as fast into a goldfish (*Carassius*); the urine volumes vary accordingly.

Most important for the outward diffusion of substances from the body of a fish is the ratio of gill to body surface; small fishes, with a high metabolic demand, have relatively larger gill surfaces than their bigger species mates. Pronounced differences in gill-surface to body-surface ratios also exist among species and are reflected in the urine volume produced per unit body weight and unit time. The most unfavorable gill to body surface ratio in any of the fishes is that of the lampreys (Petromyzonidae).

The passive intake of water through gills and mouth and consequently also urine production are temperature dependent; lampreys (*Petromyzon*) kept at 2° to 3°C "bailed out" surplus water through urine production at a rate of 60 milliliters per kilogram of body weight per day, whereas, at 18°C, they passed 500 milliliters per kilogram per day. The extreme range of these values is between 6 and 50 percent of the body weight, a copious urine flow indeed if we consider that freshwater bony fishes (Osteichthyes) rarely accumulate in a day more than 20 percent of their body weight as urine and terrestrial mammals, 1.5 percent.

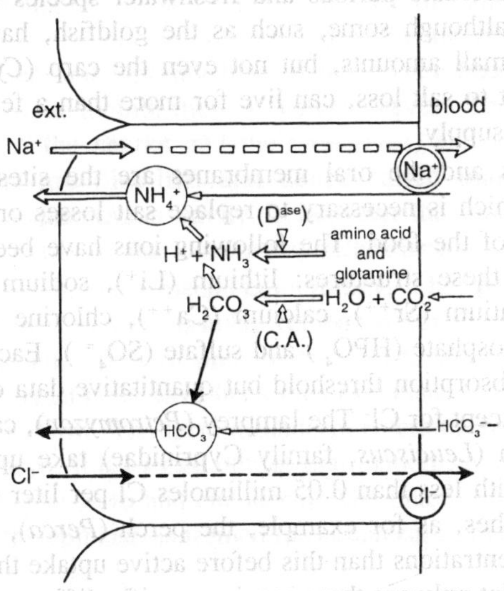

Figure 4.5 : Schematic representation of the kidney unit (nephon) of different fish groups.

In connection with copious urine secretion, several freshwater bony fishes have developed a sometimes pronounced posterior evagination of the mesonephric ducts that serves for urine storage; the structure is frequently referred to as a urinary bladder.

Some salts, especially chlorides, are also lost in the urine and through the mucus by fishes. Gills and mouth are yet another site of chloride loss, estimated to equal all others of the body. The ratio of urinary to total chloride losses varies with different species, being 1:100 in a lamprey (*Petromyzon*) and certain salmons (Salmoninae), and 1:35 in the carp (*Cyprinus*), pointing to reabsorption of filtered chlorine in the kidney tubules. Total chloride losses also vary considerably; the lamprey (*Petromyzon*) loses about seven times as much as the carp (*Cyprinus*) when chloride passage out of these fishes is compared per *100* grams of tissue per hour.

The Role of the Gills of Freshwater Fishes in Replacing Salt Losses

Osmotic regulation in freshwater animals is necessary for the maintenance of a fixed internal salt concentration, an important component of homeostasis. Salt loses vary according to species; the Atlantic salmon (*Salmo salar*) for instance, may, in fresh water, lose up to 17 percent of its body chlorides in a day by diffusion whereas the goldfish (*Carassius auratus*) loses only *5* percent. Fishes can also fast for considerable periods and freshwater species rarely, if ever, drink water although some, such as the goldfish, have been shown to swallow small amounts, but not even the carp (*Cyprinus*), a fish most resistant to salt loss, can live for more than a few days without external salt supply.

The gills and the oral membranes are the sites of active ion absorption which is necessary to replace salt losses or to supplement the minerals of the food. The following ions have been shown to be absorbed by these structures: lithium (Li^+), sodium (Na^+), cobalt (Co^{++}), strontium (Sr^{++}), calcium (Ca^{++}), chlorine (Cl^-), bromine (Br^-), acid phosphate (HPO_4^-) and sulfate ($SO_4^=$). Each of these ions has its own absorption threshold but quantitative data on their uptake are meager except for Cl^- The lamprey (*Petromyzon*), carp (*Cyprinus*), and the roach (*Leuciscus*, family Cyprinidae) take up chloride ions from water with less than 0.05 millimoles Cl per liter (2 ppm); other freshwater fishes, as for example, the perch (*Perca*), require higher chloride concentrations than this before active uptake through the gills takes place. Not only are there species-specific differences with regard to the uptake of chloride and other ions, but it is also highly probable that such differences extend to physiological races, depending on the chemical characteristics of the water in which they occur. The rate of absorption corresponds to the rate of losses by diffusion; the lamprey

(*Petromyzon*) absorbs 90 micromoles of Cl per 100 grams of body tissue per hour, the Atlantic salmon (*Salmo salar*), 30, and several minnows (Cyprinidae), between 4 and 30. These values show that active replacement of chlorides by absorption through the gills can considerably exceed passive losses. Many minerals are undoubtedly replaced through the food in sufficient amounts, but some may have replacement mechanisms similar to that ascertained for chlorides.

A mechanism for the uptake of Na^+ and Cl^- by gills of a freshwater fish. Hydrogen ions (H^+), produced from the reaction catalysed by carbonic anhydrase, react with ammonia (NH_3^+) to produce ammonium ($NH4^+$). Most ammonia apparently results from deamidation and oxidative deamination in the gills but a trace of ammonia will be in the blood as a result of these reactions in the liver. Ammonium ions diffuse passively out of the gill cells. In exchange, an equivalent amount of positively charged sodium (Na^+) is carried through the gills by active transport, an energy requiring process. Bicarbonate (HCO_3^-) from the carbonic anhydrase reaction diffuses out of the gill cells and chloride ions (Cl^-) are transported actively through the gills. The net result is a sodium-ammonium exchange and a chloride-bicarbonate exchange. The trademarks of chloride cells involved in active transport of salts are an abundance of mitochondria and a highly developed labyrinth of, agranular membranes (endoplasmic reticulum). The autonomic centers of the brain in the brainstem and thalamus exert a powerful influence on salt uptake because fishes with cut spinal cords show a severe reduction in salt- and water-regulatory capacity. Endocrine control of salt and water balance in fishes has also been shown. The gill epithelium and that of the proximal convoluted kidney tubules show certain physiological similarities inasmuch as such agents as the mercury (Hg) ion upset the Na and Cl uptake in both organs, probably through impairment of the enzyme succinic dehydrogenase.

There are great differences among freshwater fishes with regard to salt tolerance so that one may arrange them into groups as stenohaline (relatively intolerant to salinity changes), the freshwater fishes generally, or euryhaline (salt-change tolerant). Among the latter can be ranged the many diadromous species, such as the eel (*Anguilla*), the Atlantic salmon (*Salmo salar*), and also certain sticklebacks (*Gasterosteus*) and killifishes (*Fundulus*). Among stenohaline fishes there is variation in the capacity to adjust to a higher than normal salinity; the carp (*Cyprinus*) and the goldfish (Carassius) can tolerate

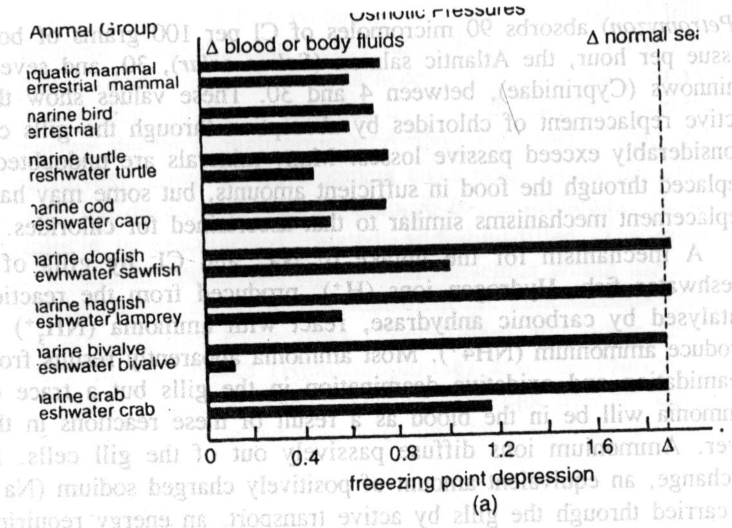

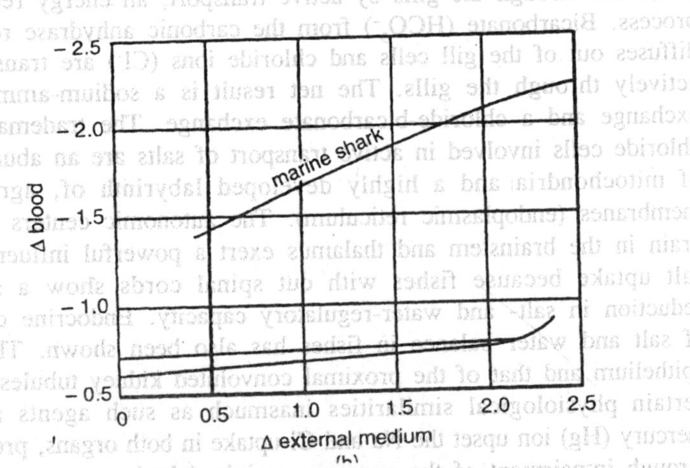

Figure 4.6 : Osmotic pressures of blood and body fluids (a) and variation in blood concentration of three fishes in waters of different concentrations (b). The marine shark is a cat shark (Scyliorhinus), the marine eel is a conger (Conger), and the freshwater (euryhaline) eel is the European species of Anguilla.

salinities up to 17 parts per thousand, equivalent to a freezing-point depression (o) of -0.9°C and also equivalent to a concentration of salts considerably higher than that in their own bodies. Concentrations of both tissue salt and urinary salt rise under such conditions and urine flow is reduced accordingly. The slower such stressful external conditions are imposed on the fish, the greater its tolerance of change.

Capabilities of freshwater fishes to adjust to abnormally high concentrations of nontoxic salts also depend on species-specific factors such as gill to body surface ratio, gill histology, neurosecretory and/or hormonal control of membrane permeability as well as oxygen and and temperature levels. Racially determined physiological variations within species also exist in this regulatory function.

Marine Fishes

In contrast to the freshwater environment, marine fishes live in a medium that is hypertonic to the body fluids and tissues and thus they tend to lose water and gain salts through their osmotic membranes. To counteract the water losses, marine fishes drink sea water and thus increase the salt content of the body fluids. Whereas dehydration is prevented by this process, excess salts must be eliminated. This homeostatic mechanism is peculiarly circuitous and energy consuming.

Kidney Function and Water Balance in Marine Fishes

Since marine fishes are forced by osmotic conditions to conserve water, urine volume is greatly reduced compared to that of freshwater species. As little as 3 milliliters of urine per kilogram of body weight per 24 hours have been measured, with a tonicity comparable to that of the blood. The kidney tubules are apparently capable of water retention, as known for marine sculpins (Cottidae), because the glomerular filtrate is commonly five times as voluminous as the amount of urine finally excreted. Structurally the kidneys of marine fishes vary greatly; the hagfishes (*Myxine*) have a primitive mesonephros with the pronephros persisting in the adult. The function of the pronephros is not clear. It may serve in some ionic regulatory capacity, but critical experiments have not been performed to examine this possibility. As in freshwater species of lampreys (Petromyzonidae), the circulation to the glomerular and tubular portion of the nephron (kidney unit) is solely arterial, no venous renal portal circulation exists. Some killifishes (*Fundulus*), sculpins (*Cottus*), and the mudskipper (*Periophthalmus*) have been reported as having functional pronephroi but the kidney of rayfin fishes (Actinopterygii) in general is truly mesonephric.

Up to 90 percent of the nitrogenous wastes of marine fishes may be eliminated through the gills mostly as ammonia and as small amounts of urea, but the urine also contains traces of the above compounds. Ammonia appears to be formed from nontoxic precursors at the site of excretion because it is toxic to tissues and only a trace of it occurs in the blood. The urine of marine bony fishes (Osteichthyes) contains creatine, creatinine, some unidentified N compounds, and trimethylamine oxide (TMO). The problem of the metabolic origin of TMO in fishes is unresolved. The osmoregulatory advantage of its presence in the body fluids of marine forms is clear however. Possible explanation for the much higher levels of TMO in marine as compared to freshwater species could include the higher TMO content in the diets of marine fishes, the synthesis of TMO by intestinal microorganisms, and differences in retention capabilities of gills and kidney as well as the possibility of some endogenous synthesis of TMO.

Salt Balance in Marine Fishes

Among marine cyclostomes, the hagfishes (Myxinidae) occupy a peculiar position inasmuch as their blood has an osmotic concentration that approaches that of sea water, though with a lower chlorine content; urea concentrations in the blood have also been found to vary, most likely depending on whether the hagfish had fed on a bony fish (Osteichthyes) or a member of the shark group (Elasmobranchii) prior to analysis. Being nearly or quite iso-osmotic to their surroundings, the hagfishes do not have to drink water and use whatever water they derive from their food for the formation of their own urine. The slime produced by hagfish skin is high in Mg^{++}, Ca^{++}, and K^{+} and may serve as a primitive mechanism for secretion of these cations. Marine bony fishes eliminate their surplus salts, which come mainly from the food and sea water swallowed, through the gills and the gut; only traces leave through the urine. Univalent ions, especially the chloride surplus, pass from intestine to blood stream and leave through the gills whereas divalent ions, for example, magnesium (Mg^{++}) and calcium (Ca^{+}), remain in the intestine where the alkaline nature of the fluids promotes their combination with oxides and hydroxides to form insoluble compounds which are then passed out with the feces. Chloride or salt cells in *marine fishes eliminate excess* chlorine ions, whereas they absorb them in freshwater fishes. These cells have been found on the bases of the gill lamellae as in an eel (*Anguilla*) as well as on the oral membranes, as in certain killifishes (*Fundulus*). There is uncertainty about the participation in ion excretion and absorption of the entire oral epithelial surface as contrasted with the work of special secreting cells.

Among marine as among freshwater bony fishes, there are some species that are more, and others which are less stenohaline. Adjustment to higherand lower-than-normal salt concentrations can be made. Mudskippers (*Periophthalmus*) often meet with 45 parts per thousand salinity in isolated tidepools and many marine species can gradually adjust to living in brackish or even almost fresh water; direct transfer over the entire salinity range from 35 parts per thousand to *0* parts per thousand often fails even in otherwise tolerant species such as the milkfish (*Chanos*). An important factor in adjustment to lowered salinity by marine fishes is the presence of a high concentration of calcium ions. Since calcium lowers cell permeability to both salts, and water, the slippery dick (*Halichoeres bivittatus*), a reef-dwelling wrasse (Labridae), can be kept healthy and will feed in a mixture of less than ^{1}/lo salt to %o fresh water filtered through calciferous coral sand.

Many experiments on osmoregulation and excretion have been complicated by temporary laboratory diuresis. The condition is caused by handling or otherwise upsetting the fish and manifests itself by impaired gill and kidney function, a temporarily increased urine flow, and loss through the kidneys of such ions as magnesium.

Diadromous Fishes

When an eel (*Anguilla*) reaches the freshwater habitat on its upstream migration it faces problems of salt depletion and overhydration as compared to the tendency toward dehydration and salt excess in the ocean. A young salmon (*Oncorhynchus*) migrating downstream to its ocean habitat comes up against these conditions in reverse. It follows then, that both anadromous and catadromous fishes must be versatile in their osmotic adjustment, have glomerular kidneys that can adjust to differences in urine volumes due to different salinities, and possess gills and oral membranes capable to cope both with the uptake and the secretion of certain ions against existing diffusion gradients. Experiments with the eel (*Anguilla*) suggested that the chloride cells can function both as secreting and absorbing units. It should be noted that diadromous fishes are adjusted either to the freshwater or to the saltwater phase of the life cycles and that such adjustments are primarily dependent on genetically determined physiological changes. The sea lamprey (*Petromyzon*), for example, after having entered fresh water in order tc spawn cannot adjust to salinities higher than 17 parts per thousand, and young salmons (*Oncorhynchus; Salmo salar*) cannot successfully enter the sea before their chloride cells are well developed. The ease of transition depends on anatomical peculiarities such as the gill-to-body surface ratio, well illustrated in different races of stickleback (*Gasterosteus*), which show differential rates of adjustment tc fresh water on their spawning migrations along the coasts of western Europe.

Changes in endocrine activity are usually simultaneous with, or precede, changes in salt- and water-balance mechanisms; the pituitary, the thyroid, and the gonads are primarily concerned with changes in physiological adjustment prior to and during migration. An increase in thyroid activity has been reported for salmon (*Oncorhynchus*) on downstream migration, possibly to facilitate the energy-demanding process of salt excretion in sea water, among other functions. Pituitary and gonadal changes often lead to appetitive behavior, such as the preference for fresh water by maturing sticklebacks (*Gasterosteus*) and are therefore important in the timing of migrations.

Sharks, Rays, and Skates

Sharks, rays, and skates, the Elasmobranchii, so adjust their internal osmotic pressure that no or little water passes through their permeable membranes; they can thus avoid some of the osmotic stresses of other fishes. Blood and tissue fluids of marine elasmobranchs have a freezing point depression, o, of more than -2.0°C, a value slightly higher than that for sea water. Elasmobranch blood, is thus hypertonic with regard to the outside water. This hypertonicity is in part due to a high chloride-ion content, higher than is found in the blood of rayfin bony fishes (Actinopterygii). However between 42 and 55 percent of the osmotically active blood solutes are nitrogenous compounds, mostly urea. The tissues of marine elasmobranchs have a high content of urea, as much as 2 to 2.5 percent. Little or no urea is lost in shark gills, a feature paralleled only by the kidney tubules of mammals among all other vertebrate organs. In the elasmobranchs there is a slight water intake through the gill and mouth membranes but the rest of the body surface is relatively impermeable to water. Consequently only a scant flow of hypotonic, urea-containing urine is produced. Representative urine-flow measurements for sharks range from 2 to 24 milliliters per kilogram of body weight per 24 hours, but the volume of glomerular filtrate is about 80 milliliters per kilogram per day; water reabsorption takes place in the proximal convoluted tubules. Trimethylamine oxide (TMO) makes up from 7 to 12 percent or more of that part of osmotic pressure in sharks, rays, and skates which is caused by organic compounds. TMO appears to be important in elasmobranch osmoregulation because, like urea, it is reabsorbed from the glomerular filtrate in the renal tubules.

The water regimen of marine elasmobranchs is regulated first by water intake through the gills so that the blood becomes diluted and urine flow increases. Consequently, blood and tissue urea become lowered and urine flow slows down. This negative feedback can always provide the animals with sufficient fresh water to meet their urinary and other metabolic water requirements. Salts are excreted primarily by the kidneys and the rectal gland. The volume of urine and of rectal-gland fluid is nearly equal; however, the composition of the two fluids differs in that the latter is essentially a hypertonic sodium chloride solution with little Mg^{++} compared to urine. Minor amounts of Na' are excreted by the gills of elasmobranchs. Some reabsorption of salts occurs in the kidney tubules. The young and the few freshwater species of elasmobranchs have special osmotic problems.

Sharks, skates, and rays either bear live young or enclose their eggs in horny urea-containing, impermeable egg cases. In either situation, development takes place in a high-urea environment and the young are born with the mechanisms of urea retention. The serum has a 0 of about -1.0°C, and the salts resemble in percentage composition those of freshwater bony fishes (Actinopterygii) with a blood o of about -0.57°C. The differences of -0.43°C or more result from urea in blood and tissues and it follows that freshwater elasmobranchs must have highly glomerular kidneys and a copious flow of hypotonic urine. Nitrogen is mainly excreted as ammonia through the gills. Urea is also an excretory product of freshwater elasmobranchs; about a third of it passes through the kidneys and its ducts but the bulk leaves through the gills. Chloride losses through diffusion and urine are relatively high but it is not known if freshwater members of the shark group can take up this ion from the water by gill absorption.

ENDOCRINE CONTROL OF EXCRETION AND OSMOREGULATION

The volume of urine output and salt balance is regulated in fishes, as well as in other vertebrates, by endocrine secretions.

Hormones may influence the kidney either by an increase or decrease of blood pressure that alters the filtering rate into the capsule of the renal corpuscle, and thus the amount of fluid excretion. Hormones may also affect renal excretion by specific action on tubule cells to change permeability and reabsorption rates of specific substances. In fishes, where gills and kidneys share the osmoregulatory process, hormones also influence filtration or absorption processes at the gills.

In man and in amphibians, posterior pituitary extracts contain fractions that directly influence urine output by constricting the afferent glomerular arterioles. Experiments with eels (*Anguilla*) show no such effect on the water balance although hypophysectomised killifishes (*Fundulus*) can only be kept healthy in fresh water by injection of pituitary extracts. It is not known to what extent this treatment influences water- as opposed to electrolyte-balance.

Information on the effects of fish endocrine organs and substances on mineral balance is somewhat more extensive than that on urine output, although still very far from adequate. Adrenal cortical hormones that influence sodium and chlorine equilibria in higher vertebrates also play a role in the regulation of gill and kidney functions of fishes. Blood of trouts (*Salvelinus; Salmo*) contains hydrocortisone and corticosterone as well as other cortical hormones and related compounds. Corticosteroids administered to trout lessen sodium elimination through the kidneys and increase its excretion through the gills either by a change in the permeability of the gill epithelium or by affecting the rate of its uptake. In marine fishes, which live in a sodium-rich medium, such a mechanism of sodium conservation in the kidneys is not needed, although a control of gill permeability to the movement of sodium ions would be beneficial. Evidence of such action in marine species is, so far, scanty and not conclusive.

Other endocrine tissues in the fish body, implicated to function in mineral balance include: hypothalamus, thyroid, suprarenal bodies, gonads, and urohypophysis.

Neurosecretory cells in the preoptic nucleus of the hypothalamus lose their secretions rapidly in several marine fishes, including dragonets (*Callionymus*), sand lances (*Ammodytes*), and mullets (Mugilidae) when the fishes are temporarily placed in hypertonic sea water. An hour after return to normal sea water, hypothalamic neurosecretory cells are refilled; there follows also a recovery of secretory cells in the neurohypophysis.

The strictly aquatic larvae of the mudskipper (*Periophthalmus*) have a less active thyroid than the adults which lead semi-aquatic lives. Thyroid injections of this fish may induce it to leave the water for lengthened periods. It appears that the thyroid hormone, apart from influencing other changes during metamorphosis, affects the water and mineral balance of fishes, not only in adaptation to temporary life outside the water but also in changes from fresh to salt water such as occur in many migratory fish species. Salmon smolt (downstream migrant juveniles of *Salmo salar*) develop chloride-secreting cells coincident with or perhaps even as a consequence of thyroid hyperactivity that occurs in their development and at a time when they begin their seaward migrations. Detailed investigations on the mode of action of thyroid secretions on osmoregulatory tissues in fishes have not yet been done.

Experiments with perfused gills of eels (*Anguilla*) reveal that adrenalin, the hormone produced in the suprarenal gland(s), has a strong vasodilatory effect on gill vessels and reduces or stops chloride secretion, which normally takes place there. However, it is not known whether adrenalin also influences chloride exchange in the normal, intact animal.

There are differences between male and female fishes of one species in blood calcium and chloride levels and in the water content of their tissues as in the cods (*Gadus*), the puffers (*Tetraodon*), and salmon (*Salmo*). These may be due to gonadal hormones since it is known from mammalian examples that female sex hormones especially exert a hydrating effect by means of sodium retention.

Finally, acetylcholinestarase, the enzyme that destroys the neurohumoral substance acetylcholine, influences the sodium uptake in frog skin and also has the same function in the anal papillae of dipterous insect larvae. In view of the wide distribution of the substance and its comparable effects in widely differing animal groups one may expect that acetylcholinesterase could also play a role in the sodium balance of fishes.

5

RESPIRATION

Studies of gas exchange in fish have concentrated on the movement of oxygen and carbon dioxide across the gills. In comparison the transfer of gases between blood and tissues in fish has received very little attention except for a few special exchange sites that have been studied in detail. These are the rete mirabile and gas gland of the teleost swim bladder and, to a lesser extent, the rete in the choroid layer of the eye of teleosts.

Fish breathe either seawater, freshwater, or air or some combination of these media. There are a number of surfaces in fish for the exchange of gases between blood and the medium. The primary site, the gills, is designed for gas exchange between blood and water. There are many accessory respiratory surfaces, which in general are associated with the movement of fish from an hypoxic aquatic to an aerial environment and are designed for gas exchange between blood and air. The problems an animal has to face when breathing seawater or freshwater are similar, although the differences in ionic content, osmolarity, and gas solubility require small changes in the functioning of the gas exchange system. The problems associated with aquatic gas exchange, however, are different from those associated with aerial gas exchange. This chapter is confined to a discussion of gas transfer between water and blood and between blood and the tissues. Air breathing in fishes is discussed in the chapter by Johansen, this volume.

Oxygen is brought into close contact with the gills by the bulk flow of water. Oxygen diffuses across the gills into the blood down a gradient of between 40 and 100 mm Hg P_{O_2}. Oxygen is transported in the blood from the gills to the capillaries; it then diffuses across the capillary walls into the tissues. Gas gradients between blood and tissues in fish have not been measured, but tissue gas tensions are probably in the range of 1-15 mm Hg P_{O_2} and 3-15 mm Hg P_{CO_2}. The partial pressure of oxygen in the environment is dissipated as oxygen molecules move from the water into the tissues via the blood. The largest pressure drop, and hence the major resistance to the movement of oxygen, occurs at the gills between water and blood.

GAS EXCHANGE BETWEEN BLOOD AND WATER

Gas exchange between blood and water occurs across the gills and the general body surface. The surface area of the gills is between 10 and 60 times that of the rest of the body and is the more important site for gas exchange. Much more is known about the movement of oxygen and carbon dioxide across the gills, and therefore gas exchange across the gills is discussed separately from and in greater detail than gas exchange across the skin.

Gas Exchange Across the Gills

Gas exchange across the gills of fishes has been reviewed by Krogh, Black, and Fry. More recently, Hughes and Shelton, Hughes , Lenfant and Johansen, Rahn, Garey, Kylstra *et al.*, Hanson, Randall *et al.*, Robin and Murdaugh, Dejours *et al.*, and Piiper and Baumgarten-Schumann have discussed aquatic gas exchange in detail; they have shown how the design of the exchanger can be related to the properties of the media on either side of the respiratory epithelium, in this case blood and water. The diffusion of gases in both tissues and water is extremely slow, and the design of the gas exchange system is such that diffusion is kept to a minimum and largely restricted to the movement of oxygen and carbon dioxide across the gill epithelium. Gas molecules are delivered to or removed from the gill epithelium by the bulk flow of water and blood.

Except for some larval forms, gills are ubiquitous in fish. The fine structure of the gills of teleosts has been described by Hughes and Grimstone, Rhodin, Newstead, and Hughes and Datta Munshi. Other groups of fish have received less attention, references and a general description of the anatomy of the gills of elasmobranchs, cyclostomes, and Dipnoi can be found in Daniel, Fry, Chapman *et al.*, and Johansen and Strahan.

The gills form a sievelike structure placed in the path of the respiratory water flow. The secondary lamellae form the side walls of this sieve and probably represent the major respiratory portion of the gill structure. The total surface area of the secondary lamellae is about 5 cm^2/g body weight. There is a countercurrent or multicapillary arrangement of the flows of blood and water on eitherside of the gill epithelium; the epithelium is usually between 1 and 5 μ in thickness. The ratio of the flows of blood and water is somewhere between 1: 10 and 1:80. The ratios of the content per mm Hg partial pressure of both oxygen and carbon dioxide in water and blood are between 1:10 and 1:20.

Dimensions of the Gills

A number of workers have measured the dimensions of the gills of teleost fish and estimated the total surface area of the secondary lamellae, which is generally considered to represent the anatomical respiratory surface area. The surface area of the gills discussed below refers to the total surface area of the secondary lamellae.

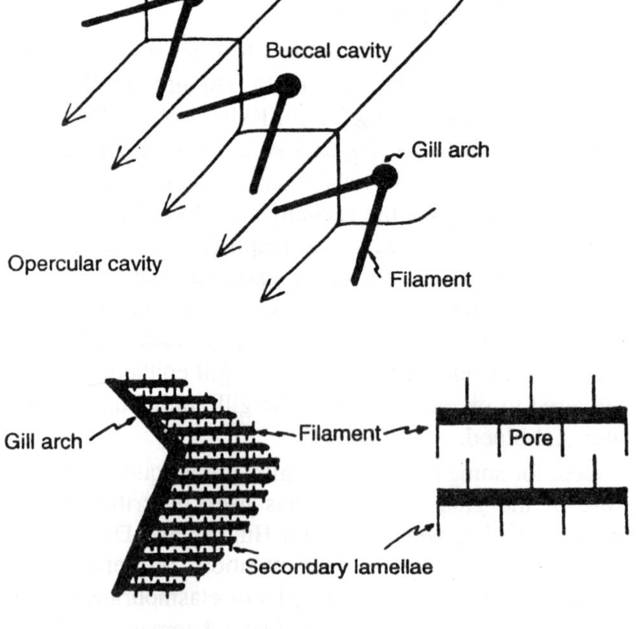

Figure 5.1 : Diagram illustrating the general structure of the teleost gill arch, possible arrangement of the secondary lamellae in forming pores through which water flows, and the general pattern of water flow over the gills.

Table 5.1 : Various Gas Coefficients and Some Other Constants

Temp (°C)	Gas solubility coefficients (ml/liter/mm Hg) Freshwater			Seawater (salinity 34-35‰)			Water vapor pressure (mm Hg)	Viscosity of water (poise, dynes/ em²/sec)
	O_2	CO_2	$CO_2//O_2$	O_2	$0O_2$[b]	CO_2/O_2		
0	0.0647	2.254	35	0.0497	1.90	38	4.58	0.0175
5	0.0567	1.871	33	0.0433	1.57	36	6.54	0.0149
10	0.0505	1.571	31	0.0390	1.34	35	9.21	0.0128
15	0.0455	1.341	30	0.0353	1.16	33	12.79	0.0112
20	0.0414	1.155	28	0.0324	1.00	31	17.54	0.0098
25	0.0381	0.999	26	0.0300	0.88	29	23.76	0.0087
30	0.0351	0.875	25	0.0281	0.77	28	31.82	0.0078
37	0.0322	0.683	21	0.0256	—	—	47.07	—

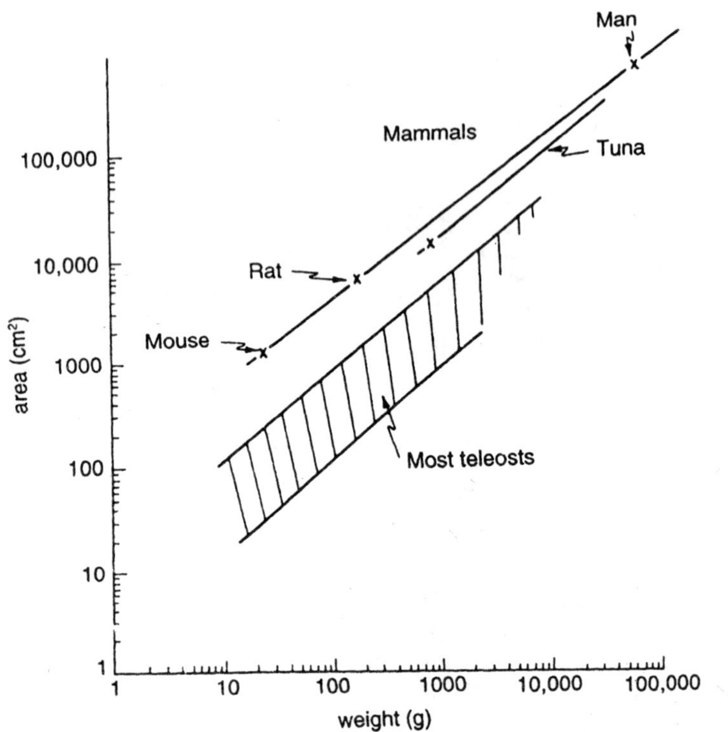

Figure 5.2 : The relationship between respiratory surface area and body weight for teleosts and a few mammals. The shaded area covers most of the recorded values for gill area in teleost.

The average surface area of the gills of teleosts, compiled from the data of Hughes and Gray on 31 species of teleost fish, is 4.9 cm^2/g body weight. There is considerable variability around this average value. Gray recorded a gill area of 11.58 cm^2/g for the mackerel, whereas that of the goosefish is only 1.43 cm^2/g body weight. Smaller fish have a larger gill area per unit weight than heavier fish of the same species, the area increasing with weight to a power of about 0.85. Most active fish have a larger gill area than sluggish forms. The surface area of the gills in tuna is much larger than normally encountered in fish and approaches the lung surface area of a terrestrial mammal. This is undoubtedly related to the high levels of sustained activity that can be maintained by tuna. In general, however, the area of the respiratory surface is smaller in fish than in terrestrial mammals. This may be related to either the lower rates of oxygen consumption in fish or the dual function of the gills. The gills are an important site for ion and water exchange as well as gas exchange. An increase in the size of the gills increases ion and water exchange as well as gas exchange. Fish in freshwater may have to restrict the surface area of their gills in order to reduce ion and water exchange, and it is possible that the large surface area of the gills of tuna could only have evolved under the reduced osmotic load of seawater.

The surface area of the gills tends to be reduced in air-breathing fish. In these fish the secondary lamellae are often far apart, and unlike ion-air-breathing fish, the lamellae do not collapse or stick together when the fish is in air. Many air-breathing fish cannot live by aquatic respiration even in well-oxygenated water. The climbing perch, *Anabas,* an air-breathing fish, has a gill area of about 1.44 em^2/g and the respiratory epithelium is about 20 μ thick. Gas exchange across the gills of this fish cannot be very significant, and the gills probably function primarily in ion exchange. The area is presumably reduced to prevent loss of oxygen to the water and to restrict ion and water movement.

The values presented and discussed above are anatomical areas of the gills. The functional area need not, and probably does not, equal the anatomical area. The anatomical area is a measure of the maximum possible functional area. Hughes estimated that the respiratory surface was between 60 and 70% of the total lamellar surface. This value was based on the area of blood channels in the secondary lamellae. However not all the blood pathways may be utilised at any instant in time, and the functional area may be smaller than 60-70% of the lamellar surface area. There are alternate nonrespiratory pathways that shunt blood past the secondary lamellae which can be used to decrease lamellar blood flow. The following equation from Hughes relates the functional area A′ to the oxygen uptake $\dot{V}_{O_2}$ across the gills, d' the diffusion distance between blood and water, D' the oxygen permeation coefficient, and ΔP_{O_2}, the oxygen gradient between blood and water across the gill epithelium:

$$A' = \frac{\dot{V}_{O_2} \cdot d' \cdot 760}{D' \cdot \Delta P_{O_2}}$$

If one assumes that D' is the same as that for frog connective tissue and d' is 2 μ, then the oxygen uptake and gas gradient data of Stevens and Randall, when applied to the above equation, indicate that the functional surface area of a resting trout is only 20% of the measured lamellar area reported for this fish by Hughes.

The thickness of the gill epithelium is usually between 1 and 5 μ, but under certain circumstances the liberation of mucus from cells in the gills may increase the diffusion distance between blood and water. The production of mucus cells is probably regulated by the hormone prolactin, and mucous is released under a variety of conditions including the movement of fish from seawater to freshwater. The release of mucus may increase the diffusion distance for both ions and water as well as gases. The discussion between Hughes, Strang, Reid, and Pattle is germane to this topic. Pattle points out that "One of the functions of mucus at respiratory surfaces may be to prevent excessive diffusion of water whilst allowing adequate diffusion of gases". Nothing is known about the rates of diffusion of water, ions, or gases through the mucus produced by the skin and gills of fishes.

Water Flow Over the Gills

A unidirectional flow of water over the gills is produced by the action of muscular pumps. The pioneering work of van Dam indicated that between 60 and 80% (percent utilisation) of the oxygen in the water passing over the gills is utilised by the fish. The more recent studies of Saunders, Garey, Holeton and Randall, Piiper and Schumann, and Stevens and Randall have shown that percent utilisation is variable between fish and in the same fish under different conditions.

The cost of ventilating the gills with such a dense medium is generally considered to be high and may be as much as 30% of the total oxygen uptake. Garey, however, has pointed out that the high density of water is offset by a low resistance to flow through the gills and has calculated that the cost of breathing in carp is only 2% of the total oxygen uptake, a value similar to that for man.

There are probably differences in the cost of breathing among fish. Fish, like carp and tench, in still water, must pump water over their gills. These fish probably have the highest cost of ventilation, and so they tend to have low ventilation volumes (ventilation volume is the volume of water passing over the gills per unit of time) and high percent utilisation of oxygen in the water. Fish swimming rapidly or fish that can maintain position in fast flowing water need only open their mouths to ventilate their gills. Gill ventilation in this case is related to swimming speed, water flow rate, resistance to water flow through the gills, and the gape of the mouth and operculum. At high swimming speeds the problem may be to reduce water flow over the gills rather than maintain an adequate rate of ventilation. In these fish the cost of ventilating the gills is probably small and related to the increased resistance to forward motion produced by passing water over the gills. Salmonids can maintain position in fast flowing water using only frictional forces between the fish and the bottom. As water flow is increased the amplitude of breathing decreases; residual breathing movements are always present, however, even in high flows of water supersaturated with oxygen. In this instance, ventilation of the gills is largely a byproduct of maintaining position in a stream; hence, the cost of breathing is probably not an important factor in the total energy budget of the fish, ventilation volumes are high, and the percent utilisation of oxygen is low. The remora stops breathing in water velocities greater than about 60 cm/ sec, ventilation volume is probably regulated at high velocities by altering the gape of the mouth. At low velocities the fish actively ventilates its gills. This animal rides on the body of sharks and uses the swimming efforts of its host to ventilate its gills. There are no breathing movements in mackeral and tuna and forward motion of the fish through the water acts to ventilate the gills. Muir and Kendall refer to this as "ram" ventilation. Thus some fish can probably maintain a high ventilation volume at low cost while others must continually pump water over their gills. Those animals that can maintain a high ventilation volume will have a low percent utilisation of oxygen from the water. The advantage of a low percent utilisation is that the water P_{O_2} at the respiratory surface remains high along the whole length of the secondary lamellae.

If the respiratory surface area is infinitely large, the diffusion distance between blood and water infinitely small, and gas exchange across the gills passive, then, as there is a countercurrent arrangement of the flows of blood and water in teleosts it is theoretically possible for the P_{O_2} in water leaving the gills to be in equilibrium with that of venous blood. In practice, however, the P_{O_2} in water leaving the gills is not in equilibrium with that of venous blood in either teleosts or elasmobranchs. Thus not all the oxygen that can be removed from the water is utilised by the animal. There are two possible explanations for this: first, there could be a large diffusion resistance across the gills, and, second, some of the water may not come into close contact with the gills and be shunted past the gills. This volume of water can be considered as a water shunt (V_D shunt) and expressed as a percentage of the total ventilation volume.

If the diffusion resistance across the gills is negligible, then the magnitude of the water shunt can be calculated from the following equation:

$$V_D \text{ shunt} = \dot{V}_G \frac{\left(P\text{E}_{O_2} - P\text{veq}_{O_2}\right)}{P\text{I}_{O_2}}$$

where $\dot{V}_G$ is the total gill ventilation, $P\text{I}_{O_2}$ the partial pressure of oxygen in inspired water, $P\text{E}_{O_2}$ that in expired water, and $P\text{veq}_{O_2}$ that in water having the same P_{O_2} as venous blood entering the gills. If the expired water has the same P_{O_2} as venous blood then both the diffusion resistance and the water shunt are zero. In practice, however, $P\text{E}_{O_2}$, is always greater $P\text{veq}_{O_2}$ than and there is probably both a diffusion resistance and a water shunt. The magnitude of the actual water shunt can be calculated from the following equation:

$$V_D \text{ shunt} = \dot{V}_G \frac{\left(P\text{E}_{O_2} - P\text{veq}_{O_2} - \Delta P_{O_2}\right)}{P\text{I}_{O_2}}$$

where ΔP_{O_2} is the oxygen gradient between blood and water across the gill epithelium. Using measurements of the dimensions of the respiratory surface, ΔP_{O_2} can be calculated by rearranging an equation from Hughes:

$$\Delta P_{O_2} = \dot{V}_{O_2} \cdot \frac{d' \cdot 760}{D' A}$$

where $\dot{V}_{O_2}$ is the oxygen uptake, d′ is the thickness of the gill epithelium, A is the area of the secondary lamellae, and D' is the Krogh permeation coefficient for oxygen in the gill epithelium [assumed to be the same as that for frog connective tissue]. The calculated oxygen gradient across the gill epithelium (2 μ thick) is between 2 and 8 mm Hg at standard rates of oxygen consumption. This gradient will obviously be larger if the functional area of the gills is decreased or the diffusion distance is increased.

Direct measurements of ΔP_{O_2} have not been made, and the actual gradient that exists across the gill epithelium will depend on the functional rather than the anatomical dimensions of the respiratory surface. The functional area of the gills will depend on the extent of gill vascularisation and on the number of capillaries open to blood flow. There are no adequate estimates of the functional area of the gills but it will be less than the anatomical area. In the absence of any measurements, if one assumes that ΔP_{O_2} is 20 mm Hg, then the water shunt in the trout based on the data of Stevens and Randall is 60% of the total ventilation volume.

Water flow over the gills can therefore be divided into a series of separate volumes or flows. First, there is that portion of the water flow which contains oxygen that passes into the blood, which is the respiratory water flow or volume. Second, there is some water that is not brought into close contact with the respiratory epithelium, which is the water shunt. Finally, there is the remainder of the water flow, which can be considered as a residual volume, the magnitude of which will be determined by $P\text{V}_{O_2}$.

The water shunt can be further divided into component volumes. These are a diffusion dead space volume V_D diff$_{O_2}$ an anatomical dead space volume V_D anat$_{O_2}$ and finally a distribution dead space volume V_D dist$_{O_2}$. They are described below.

Diffusion Dead Space

Water is probably in contact with the respiratory surface for about 1 or 2 sec, and since the rate of diffusion of gases in water is slow water must be brought into close contact with the gills if it is to exchange gases with the blood. Distances between lamellae may be so large and/or flow through the lamellae so rapid that there is not sufficient time for all water to reach equilibration with the blood. In this case there will be persistent gradients in the water, and these may be considered as representative of a diffusion dead space volume.

Distribution Dead Space

If ventilation of the pores formed by the secondary lamellae is high, or if the pores are ventilated unequally, more oxygen may be delivered to all or part of the respiratory surface than is required to saturate the blood; hence, there may be a distribution dead space associated with unequal ventilation and perfusion of the gills.

Anatomical Dead Space

Water flowing over the gills of teleosts may pass through the pores formed by the secondary lamellae or spill between the edges of the filaments. Only water passing through the pores will be involved in gas exchanged, and water passing between the ends of the filaments can be considered as part of the water shunt. Hughes referred to this portion of the water flow as the anatomical dead space volume.

The percent utilisation of oxygen is a measure of the size of the respiratory volume and is an inverse measure of the size of the combined residual volume and water shunt.

The size of the water shunt varies in fish, as mentioned above, and may be as high as 60% of the total ventilation volume in the trout. The relative contribution of the anatomical, diffusion, and distribution dead space volume to the water shunt is difficult to assess. The size of the anatomical dead space will be related to the arrangement of the filaments to each other and to the ventilation volume. The anatomical dead space is probably larger at high water flow rates. The position of the filaments is controlled by muscles at their base which could play a role in regulating the size of the anatomical dead space. The author has observed movements of trout gills during normal breathing movements, and these may be caused by changing water velocities or by the action of the muscles at the base of the filaments. Whatever the cause, the effect of altering the relative position of filaments to each other must be to change the size of the anatomical dead space, which is probably very variable with time in a single fish, as well as between different species of fish. There appears to be a tendency to enlarge supporting structures and fuse parts of the gill in fish exposed to high ventilation volumes. In tuna, the secondary lamellae of adjacent filaments are used to form a compact sievelike gill structure. This presumably helps to maintain flow between the secondary lamellae and reduces the size of the anatomical dead space, but it must also increase the resistance to flow through the gills. Fusion of gill parts, however, need not always be associated with the maintenance of flow between secondary lamellae. Hughes has suggested that fusion in *Amia* serves to prevent collapse of the gill sieve when the animal is in air.

Thus water flow over the gills can be divided into a series of component volumes as follows:

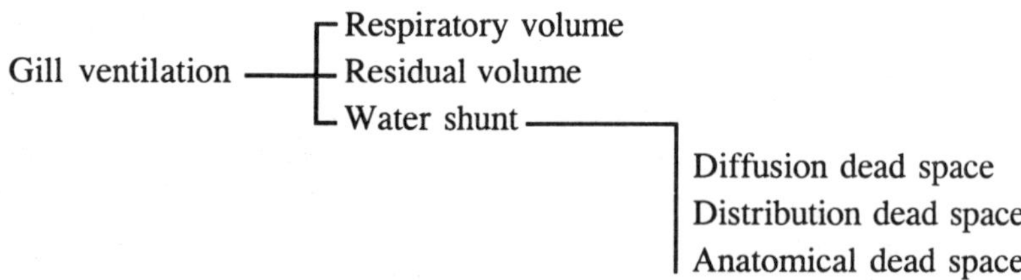

Kylstra *et al.* have analysed gas transfer in water-breathing dogs and in the gills of fish. They related the size of the diffusion dead space $\left(V_{D\,\mathrm{diff}_{O_2}}\right)$ in the gills of fishes to the rate of diffusion (D), the time for diffusion (t), and the distance over which diffusion takes place (a) and have derived the following equation which permits the evaluation of the diffusion dead space for oxygen at the gills:

$$V_{D'\,\mathrm{diff}_{O_2}} = \frac{1}{1+3D\cdot t/a^2} \times \dot{V}_G$$

The volume of water contained within the pores of the gills can be calculated by multiplying the surface area of the lamellae (A) by half the distance *(d)* between 'successive secondary lamellae. The time *(t)* that a particular volume of water is in contact with the respiratory surface can be determined from the equation

$$t = 2\dot{V}_{G\ \text{pore}} / Ad$$

where $\dot{V}_{G\ \text{pore}}$ is the water flow between the secondary lamellae. The time *(t)* can be determined for a variety of values of $\dot{V}_G$ using the anatomical data of Hughes. Values for *t* in the resting trout are of the order of 0.4-2 sec depending on ventilation volume. The diffusion dead space can be calculated from the equation of Kylstra *et al.*; *D* is 1.43 $\times 10^{-5}$ em^2/sec at 5°C (assuming a temperature coefficient at 2% °C). The diffusion distance (a) is taken to be half the distance between

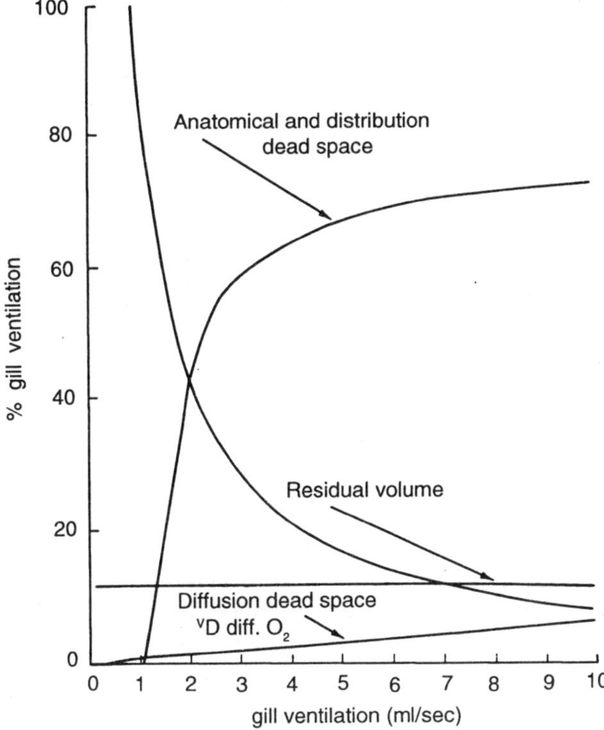

Figure 5.3 : The effect of ventilation volume on the size of the anatomical and distribution dead space, the residual dead space, and the diffusion dead space assuming an oxygen uptake of 0.47 ml/min and a venous oxygen tension of 20 mm Hg.

successive secondary lamellae on a gill filament. The calculated values of diffusion dead space are small in trout and carp, largely because of the very small distances between secondary lamellae. Large water shunts past the gills cannot therefore be explained in terms of a large diffusion dead space. In the trout the diffusion dead space is only 7% $\dot{V}_G$ even when $\dot{V}_G$ is 10.3 ml/sec (t = 0.4 sec) and all the water passes between the secondary lamellae. A large water shunt can only be explained in terms of a large anatomical or distribution dead space.

A low percent utilisation of oxygen from the water may therefore be associated with either a high venous oxygen tension $\left(P_{V_{O_2}}\right)$, a large anatomical dead space, or a large distribution dead space. Large differences between $P_{V_{O_2}}$ and $P_{E_{O_2}}$, (expired water oxygen tension) may be associated with either a large anatomical or distribution dead space. The size of the anatomical dead space can be reduced by maintaining contact between the tips of adjacent filaments and lamellae, thus forcing water between the secondary lamallae. The distribution dead space could be reduced by ensuring maximum blood flow to those portions of the gill receiving maximum water flow. There is no evidence at present that this occurs in fish. Measured values for $P_{E_{O_2}} - P_{V_{O_2}}$ are 100 mm Hg in the trout 40 mm Hg in the carp, and 46 mm Hg in the dogfish, indicating a large water shunt in these fish.

It is important to note that, although there may be large differences between the mean P_{O_2} values for blood and water across the gills and arterial P_{O_2} values may be very different from those in inspired water, this does not indicate that the gills are not effective in transferring oxygen between water and blood. In fact, blood leaving the gills is usually between 85 and 95% saturated with oxygen.

Blood Flow Through the Gills

In fish, unlike mammals, the respiratory and systemic circulations are in series rather than in parallel. Blood ejected from the heart is conveyed by the ventral aorta to the afferent branchial arteries supplying each gill arch. Blood flows from the afferent to the efferent branchial arteries through the capillary bed of the gills. These efferent branchial vessels join to form the dorsal aorta through which blood passes to the general body circulation. Figure 4 illustrates the vascularisation of the gills of a salmonid. There are a number of alternate pathways of varying distance from the water interface, for the passage of blood through the gills. Blood may flow through either a few or all of the secondary lamaellae on each filament. Some blood may bypass the lamellae, flowing through capillaries joining afferent and efferent vessels within the filaments. Alterations in the distribution and volume of blood flow through these channels will change the functional surface area of the gills and the diffusion distance between blood and water, thus affecting the capacity of the gills to transfer gases. At low rates of oxygen uptake in the eel some blood is probably shunted past the secondary lamellae, and blood leaving the gills is not fully saturated.

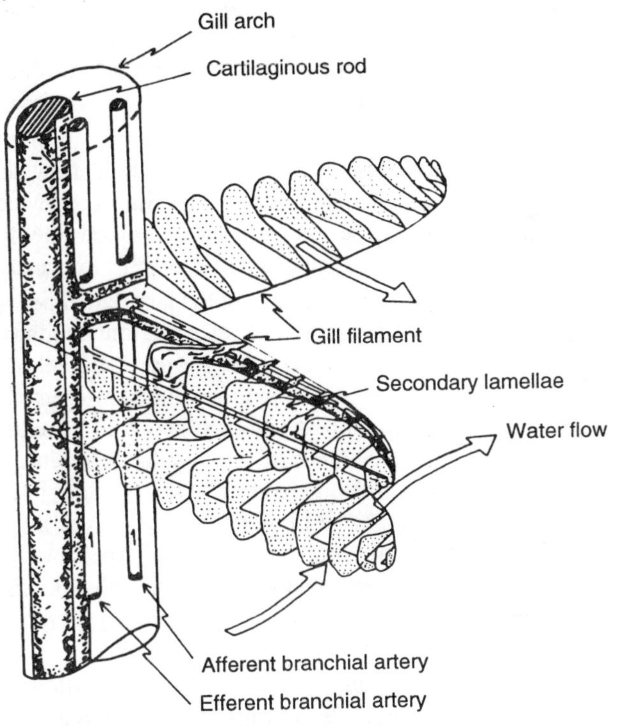

Figure 5.4 : A diagram to illustrate the pattern of blood flow through and water flow over the gills of a teleost fish.

Thus the eel can increase the arterial-venous oxygen difference by raising arterial saturation as well as by lowering venous saturation. There are probably gill blood shunts in elasmobranchs and Dipnoi as well as in teleosts.

Capillaries in the secondary lamellae are close to the water interface and are undoubtedly more involved in gas transfer than those in the gill filament. Steen and Kryusse have shown that adrenaline increases lamellar blood flow and the percent saturation of the blood leaving the gills of the eel. Adrenergic receptors are present in fish gills, and catecholamines are known to cause a marked vasodilation of the gills. Exercise in salmonids is associated with an increase in the level of circulating catecholamines and a rise in dorsal aortic blood pressure. This rise in dorsal aortic blood pressure is not seen after a-adrenergic receptor blockade with phenoxybenzamine. There is a decrease in the resistance to blood flow in both the respiratory and systemic circulations during exercise and an increase in the transfer factor of the gills for oxygen. The circumstantial evidence cited above indicates that increased levels of catecholamines alter blood flow through the gills in a way that either increases the functional surface area of the gills or decreases the diffusion distance in order to increase the rate of gas exchange across the gills.

Acetylcholine increases the resistance to blood flow through the gills and decreases lamellar blood flow. Hypoxia in the water flowing over the gills of trout increases the resistance to blood flow through the gills but does not impair the capacity of the gills to transfer gases.

The pattern of capillaries in the secondary lamellae of the tuna appear to be different from the rest of teleosts. The capillaries are not divided into respiratory and nonrespiratory vessels, but all blood passes through the secondary lamellae. Tuna have typical afferent and efferent vessels, but each filament afferent gives off about 20 lamellar afferents to each secondary lamellae. Each of the lamellar afferents subdivides forming a large number of blood channels in the secondary lamellae, all of which are respiratory. The respiratory surface area of the gills of the tuna may be regulated by altering the number of lamellar afferents open to blood flow. In some tuna species there are valvelike flaps in the filament afferent that may play a role in regulating blood flow to the lamellae.

The capillaries in teleost gills are wide enough to allow the passage of nucleated red blood cells. The erythrocytes are about 11 μ long and 6.5 μ wide in tuna, similar in size to those of *Scomberomorus*. The diameter of the capillaries appears to be less than that of the erythrocyte, which becomes sausage-shaped as it is forced through the gill capillaries. Red blood cells in fish swell markedly if blood CO_2 levels increase.

The importance of countercurrent exchange in fishes has been stressed by a number of investigators. Higher oxygen levels in arterial blood than efferent water have been recorded in both teleosts and elasmobranchs indicating the presence of a functional countercurrent arrangement of the flows of blood and water. However, in a number of instances, the recorded oxygen level in arterial blood was below that of the efferent water. These observations led Lenfant and Johansen to question the presence of a countercurrent system in elasmobranchs. Robin and Murdaugh discussed the effects of various arrangements of the flows of blood and water on gas exchange and concluded that a countercurrent arrangement of the flow was not present in elasmobranchs. The morphology of the elasmobranch gill is in close agreement with the multicapillary arrangement of the circulation and Piiper and Schumann used a serial multicapillary arrangement of the flows of blood and water to explain

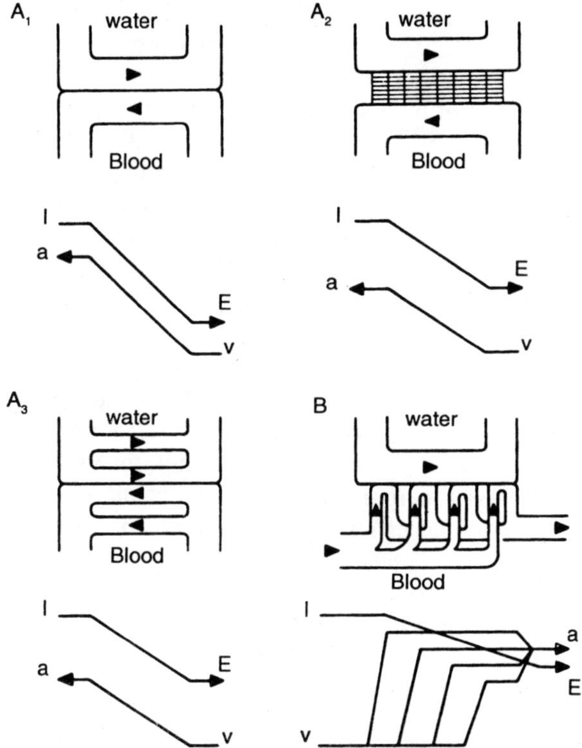

Figure 7.5 : Diagrams to illustrate possible arrangements of water and blood flows in the gills.

the phenomenon of P_{O_2} levels in arterial blood of the dogfish that were higher than in expired water. Thus, both teleosts and elasmobranchs can have arterial oxygen tensions higher than that in expired water; however, in teleosts, these are the result of a countercurrent arrangement of flows, whereas in elasmobranchs a serial multicapillary system may predominate. These high arterial oxygen tensions are also explicable in terms of the active transport or exchange diffusion of oxygen across the gill epithelium. Although active transport or exchange diffusion of oxygen cannot be excluded, there is no evidence for the involvement of these processes in oxygen transport across the gills.

Gas Content in Blood and Water

The concentration of CO_2 and O_2 in both seawater and freshwater is extremely variable. The amount of oxygen dissolved in water per mm Hg varies with the temperature and ionic content of the water. All the oxygen in water is in physical solution; CO_2 in blood and water and O_2 in blood is not simply in physical solution, and the gas content per mm Hg varies with the partial pressure of the gas. The relationship between gas content and the partial pressure of gas in solution is described by a series of CO_2 and O_2 dissociation curves of water and blood. Carbon dioxide is between 20 and 30 times more soluble in water than oxygen. Fully saturated blood contains between 10 and 20 times more oxygen, than water at the same partial pressure. Arterial blood contains twice as much CO_2 as O_2 even though the partial pressure of O_2 may be 40 times that of CO_2.

The rate of gas exchange across a respiratory epithelium depends on the dimensions of the epithelium, the concentration gradient, and the diffusion coefficient of the gas. The diffusion coefficient of carbon dioxide is only slightly less than that for oxygen, and the concentration gradient for free carbon dioxide is similar to that for physically dissolved oxygen in the reverse direction. Consequently, the exchange of oxygen and carbon dioxide across the gills occurs at more or less the same rate. Although the concentration gradients for carbon dioxide and oxygen across the gills are similar, the tension gradient for carbon dioxide across the gills is much less than that for oxygen in the reverse direction, because the solubilities of oxygen and carbon dioxide in blood and water are different.

The oxygen capacity of fish blood is variable between species. The antarctic icefish have no erythrocytes or hemoglobin in their blood and an instance of a carp without hemoglobin has been reported by Schlicher. Using carbon monoxide poisoning of the hemoglobin, it has been shown that trout, *Salmo gairdneri,* can exist without functional hemoglobin as long as the temperature is below 5°C. At these temperatures the oxygen demands of the tissues can be met by that in physical solution in the blood; at higher temperatures hemoglobin is required to increase the oxygen carrying capacity of the blood to meet the increased oxygen requirements of the fish. The oxygen capacity of most fish blood is between 4 and 10 vol %. Tuna have blood oxygen capacities that are

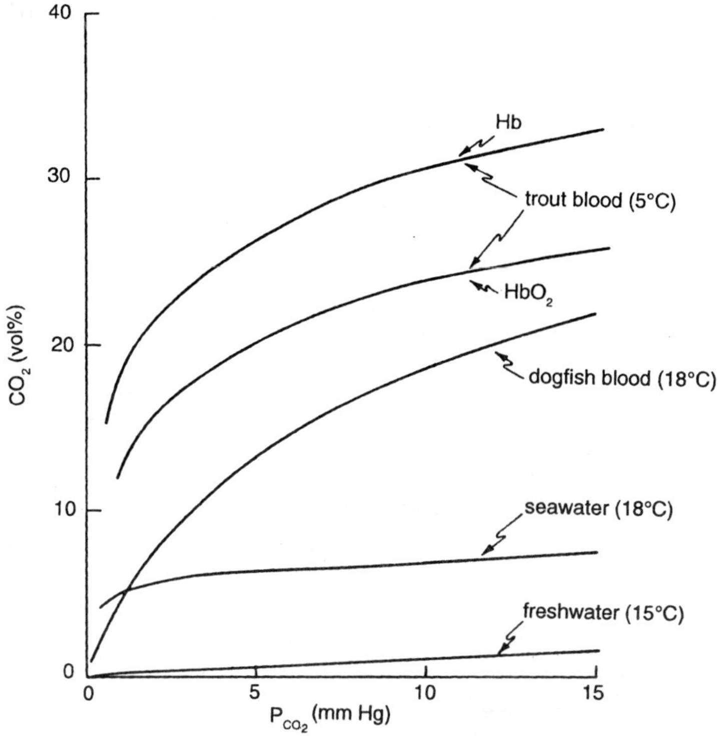

Figure 5.6 : A variety of CO_2 dissociation curves.

much higher than most fish and hemoglobin levels of up to 20 g/100 ml of blood. Hemoglobin content of the blood and hematocrit increase in response to hypoxia in fish as in mammals. Hemoglobin, hematocrit, and the number of red blood cells increases, but the volume of each erythrocyte decreases with an increase in temperatures in salmonids. Thus hemoglobin levels and the red blood cell surface area available for gas exchange increase in the summer, correlated with a rise in temperature and activity and hence the oxygen requirements of the fish.

If the solubility of a gas in solution is high, or if many of the gas molecules in solution are in a bound form, large volumes of the gas can enter or leave the solution without causing large changes in partial pressure. Therefore, the form of the oxygen or carbon dioxide dissociation curve has an effect on the rate of change of the gradient across the gills when gases are diffusing from one solution to another. The steeper the slope of the dissociation curve, the smaller the change in partial pressure per unit volume of gas exchanged.

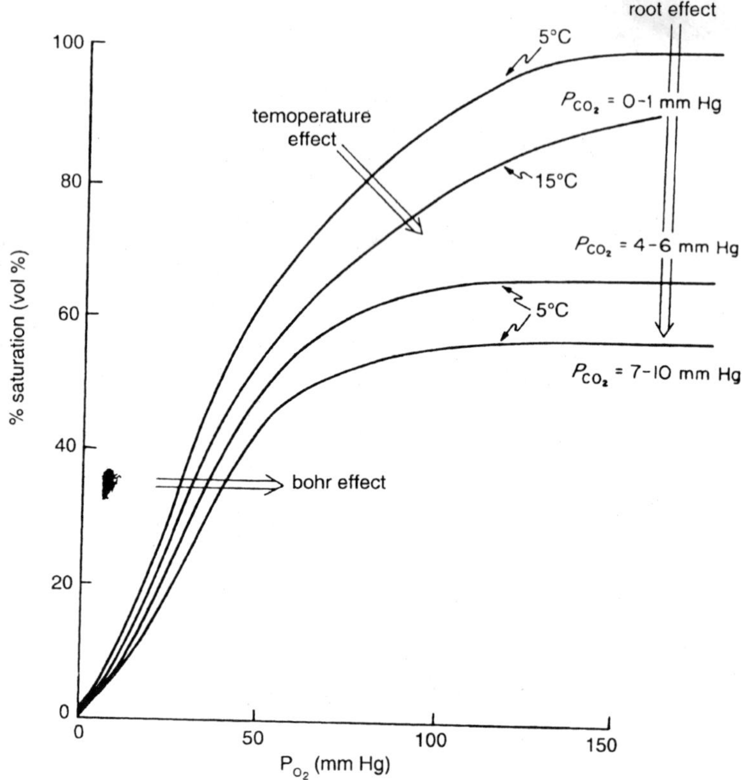

Figure 5.7 : Oxygen dissociation curves for the blood of trout (Salmo gairdneri) at different CO_2 levels and different temperatures.

Fish blood has a high affinity for oxygen and is 95% saturated at relatively low partial pressures of oxygen compared with pressures required to saturate mammalian blood. Decreases in temperature increase the oxygen affinity of the blood, except in tuna, where the oxygen affinity of hemoglobin is largely independent of temperature. The P_{50} (partial pressure for 50% O_2 saturation) is 17 mm Hg (10°C) in the dogfish, 38 mm Hg (5°C) in the trout, and 4 mm Hg (10°C) in the carp. At low P_{O_2} levels (2 mm Hg) carp blood is 95% saturated at P_{O_2} tensions as low as 25 mm Hg. This means that the P_{O_2} of carp blood will not rise above 25-30 mm Hg until the hemoglobin is fully loaded, provided the rate of reaction of oxygen with hemoglobin is equal to or more rapid than the rate of entry of oxygen into the blood. The amount of oxygen that enters the blood will be increased if the respiratory area is increased, the diffusion distance decreased, or the oxygen activity gradient across the gills increased. The oxygen gradient can be increased either by lowering the blood P_{O_2} or by increasing the water P_{O_2}. A hemoglobin with a high affinity for oxygen will maintain low P_{O_2} levels in the blood until the hemoglobin is fully saturated. High ventilation volumes with a low percent utilisation of oxygen will maintain a $_{P02}$ in water which is close to ambient along the whole length of the secondary lamellae. A countercurrent arrangement of the flows of blood and water in teleosts also enables the fish to maintain a large O_2 gradient between blood and water along the whole length of the secondary lamella. The mean oxygen pressure gradient across the gills is similar in carp and trout. The carp, however, has a high utilisation of oxygen from the water and a hemoglobin with a very high affinity for oxygen, whereas the trout has a low utilisation of oxygen from water passing over the gills and a hemoglobin with a lower affinity for oxygen. The maintenance of large oxygen gradients enables the fish to utilise a smaller respiratory area for a given oxygen uptake and so reduce ion and water exchange across the gills.

Blood oxygen dissociation curves have been described for a number of fish. Recently, Forster and Steen have measured the half-time for some oxygen hemoglobin reactions in the blood of eels, but apart from these data very little is known about oxygen hemoglobin reaction times and these may play a role in limiting the rate of gas exchange across the gills. In teleost fish an increase in CO_2 content or a decrease in pH of the blood causes not only a reduction in the affinity of hemoglobin for oxygen (Bohr effect) but also a reduction in the oxygen carrying capacity of the blood (Root effect). Unlike the blood of terrestrial mammals, teleost hemoglobin never becomes fully saturated once the P_{O_2} is more than 1-2 mm Hg. The magnitude of this "Root effect" is largest at CO_2 tensions between 1 and 5 mm Hg and decreases with an increase in P_{CO_2} above this range. If the CO_2 tension in blood fully saturated with oxygen is increased, oxygen is released from the hemoglobin (Root-off shift), and if the CO_2 tension is lowered the reverse occurs. The half-time for the eel Root-off shift at 23°-25°C, produced by an increase in P_{CO_2}, is 87 msec, whereas the half-time for the reverse Rooton shift is 9 sec. The asymmetrical reaction velocities of the Root shift are important in unloading gases in the swim bladder, but a slow Root-on shift will impair oxygen transfer across the gills.

The time course of the Root-on shift is related to the reaction velocity of the uncatalysed CO_2 hydrationdehydration reaction in the plasma. The presence of a CO_2 sink, for example, water passing over the gills, will cause a more rapid reduction of plasma and intracellular P_{CO_2} than recorded under the conditions of Forster and Steen's experiments, and hence the rate of the Root-on shift will be more rapid as blood passes through the gills. Blood leaving the gills is usually 95% saturated with oxygen in trout. Incomplete saturation of arterial blood has been recorded in the eel but this was probably related to the utilisation of venous shunts in the gills rather than the result of a slow Root-on shift.

Acclimation to high CO_2 levels in the water has been reported to depress the magnitude of Root effect. This acclimation process is associated with an increase in hemoglobin levels with little change in blood oxygen capacity. These observations require further investigation.

Teleosts generally have a marked Bohr and Root effect, both of which are absent or small in elasmobranchs. There is no Root effect in Dipnoi, but there is a pronounced Bohr effect. The reader is referred to the chapter by Riggs, this volume, for further information on this subject.

The much higher CO_2 than O_2 solubility in water means that if the respiratory quotient is around unity, the changes in CO_2, tension in the water as it passes over the gills are small and of the order of a few mm Hg whereas the changes in P_{O_2} are large. Under these circumstances the ratio of the P_{O_2} to P_{CO_2} changes will be equal to the ratio of the solubilities of O_2 and CO_2 in water, that is, between 20 and 30:1. This effect of solubility on the changes in partial pressure is described by the O_2 : CO_2 diagram of Rahn.

The relationship between the amount of CO_2 in solution and the partial pressure of the gas is not linear in either blood or water. The CO_2 dissociation curves are curvilinear in water below 2 mm Hg Pc02, particularly in seawater or highly carbonated freshwater. Carbon dioxide entering water is buffered by a carbonate-bicarbonate system reducing the magnitude of the P_{CO_2} increase. As water passes over the gills of teleosts the excretion of ammonium ions increase the CO_2 buffering capacity of water, further reducing the magnitude of the rise in P_{CO_2} in efferent water. This latter effect is probably not seen in elasmobranchs because they are ureotelic and not ammonotelic.

Carbonic anhydrase has been located in fish red blood cells and the gill epithelium and bicarbonate is exchanged for chloride by an exchange diffusion mechanism across the gill epithelium of freshwater teleosts.

Thus because of the presence of carbonic anhydrase and an exchange diffusion mechanism, bicarbonate as well as CO_2 enters the water passing over the gills. If this is so then the slow formation of CO_2 from bicarbonate in water will delay the rise in P_{CO_2} until the water has left the respiratory surface. A body of water is in contact with the gill epithelium for about a second, the time required for the formation of CO_2 from bicarbonate in water is of the order of several seconds; hence, the major portion of the rise in P_{CO_2} will occur after the water has left the respiratory surface. The relative importance of the buffering system, the excretion of ammonium ions, and the slow formation of CO_2 from bicarbonate, in water, in maintaining the CO_2 gradient across the gills will depend on the water flow rate and the ambient temperature, but all tend to reduce or retard the rise in water P_{CO_2} and maintain the P_{CO_2} gradient between blood and water.

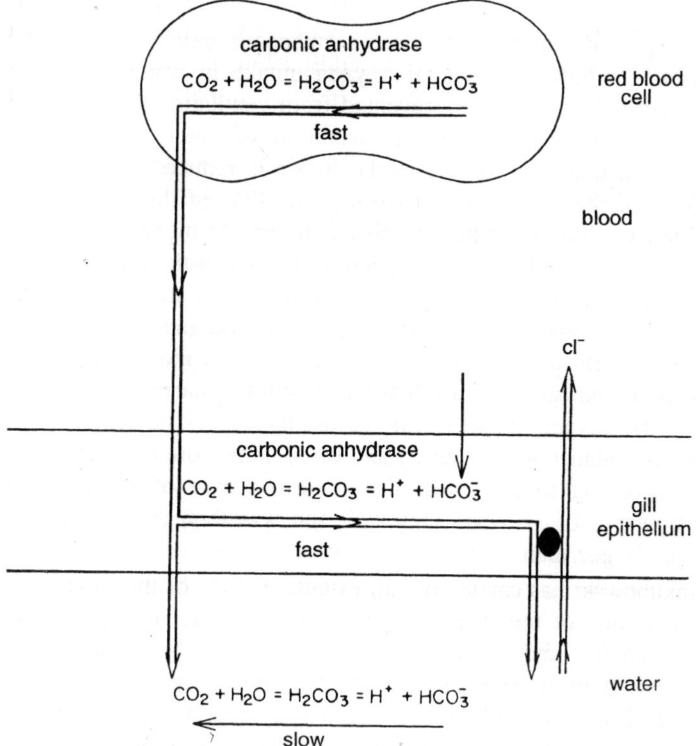

Figure 5.8 : Possible pathway for the elimination of CO_2 across the gills of freshwater teleost fish.

The rate of formation of CO_2 in the red blood cell, and hence the rate of excretion of CO_2, will depend upon the presence of carbonic anhydrase. The activity of carbonic anhydrase in fish blood is only about 2% of that in mammals, and as tentatively suggested by Maren the CO_2 gradient that does exist across fish gills may result from a limitation in the rate of formation of CO_2 in the blood. Inhibition of carbonic anhydrase with acetozolamide causes an increase in P_{CO2} in the blood. Seasonal changes in the level of carbonic anhydrase have been observed in the plaice, *Hippoglossoides platessoides*, and these changes may be important in regulating CO_2 excretion across the gills on a long-term basis.

Gas Exchange across the Skin

The amount of gas transfer between blood and water occurring across surfaces other than the gills varies considerably from fish to fish.

Krogh showed that eels have a cutaneous oxygen uptake in air that is up to 60% of the total oxygen uptake in water at the same temperature. His eels were prevented from ventilating their gills. Eels normally pump air over their gills when out of water. Berg and Steen have shown that the oxygen uptake in air is reduced to between 40 and 63% of that in water and from 31 to 47% of the oxygen enters the blood via the gills when the fish is in air. At higher temperatures there is a larger reduction in oxygen uptake when the animal is in air compared with the oxygen uptake in water and a smaller percentage of the total oxygen uptake enters the body via the gills. Berg and Steen observed that the longitudinal fin becomes more red when the eel is in air indicating a vasodilation of skin capillaries. In water, 85-90% of the oxygen uptake occurs across the gills. The eel has a well-developed cutaneous circulation and often comes out on land, to steal peas, according to Day who also reports a case of an eel "which lived upwards of 31 years in a well and was then choked by a frog that was larger than it could swallow".

Jakubowski has carried out an extensive study of the structure and vascularisation of the skin of a variety of teleosts. He reports values of between 0.5 and 1.5 cm^2/g body weight for the surface area of blood vessels in the skin of a variety of teleosts, as well as ratios for the surface area of gill to skin capillaries, ranging from 3:1 in the pond loath to 10:1 in the carp and flounder. Although the area of capillaries in the skin is large, it is important to note that the capillaries are covered by an epidermis, ranging in thickness from 31-38 μ in the flounder to 263 and 338 μ in the eel and pond loath, respectively. The differences in thickness largely result from differences in size of mucous and goblet cells within the epidermis. The number of mucous cells and the thickness of the skin are probably controlled by prolactin. By comparison the gill epithelium is usually between 1 and 5 μ in thickness.

Jakubowski could not find any relationship between the extent of the skin vascularisation, thickness of the epidermis, and the amount of oxygen uptake occurring through the skin. Gas exchange across the skin will not only depend on the extent of skin vascularisation but also on the gas gradients, diffusion distances, and the blood flow and number of capillaries open in the skin. Nothing is known about many of the parameters which affect gas exchange across the skin.

There appears to be considerable variability in the extent to which fish respire through their skin. In many fish with thick, poorly vascularised skins, cutaneous respiration is probably not important. Obvious exceptions to this are larval fishes in which the gills have not yet developed. Here cutaneous respiration must be very significant.

GAS EXCHANGE BETWEEN BLOOD AND TISSUES

There are very few direct measurements of tissue gas tensions in fish. Haning and Thompson have recorded tissue P_{CO_2} levels of 9.2 mm Hg in the catfish. Gas tensions in venous blood are an indication of tissue gas partial pressures; however, there is a gas gradient between the tissues and blood, the magnitude of which is unknown. Oxygen and carbon dioxide levels in mixed venous blood appear to be variable both between species and within a single species. Carbon dioxide tensions in trout venous blood are about 5 mm Hg, doubling during activity. Values are generally lower for carp and dogfish. The P_{O_2} in mixed venous blood is 3.2 mm Hg in the carp, 10 mm Hg in the dogfish, and 19 mm Hg in the trout. In the absence of direct measurements one might assume that tissue P_{CO_2} is between 3 and 15 mm Hg and P_{O_2} is between 1 and 15 mm Hg.

Almost nothing is known of the relative size of, and blood flow to, the various capillary beds in the systemic circulation of fish. Red myotomal muscle contains about 3 times as much blood and has 3 times the number of capillaries per unit weight as white myotomal muscle. White myotomal muscle, used during violent burst responses, operates anaerobically. P_{O_2} levels are presumably much lower, and P_{CO_2} levels much higher, than those in red myotomal muscle, which operates aerobically. Stevens has estimated the volume distribution of blood to parts of the body of rainbow trout. He demonstrated that, except for a diminished spleen blood volume, the distribution of blood was unaffected by violent swimming.

The retention of lactic acid in white myotomal muscle for a considerable period after violent exercise or hypoxia is enhanced by the small number of capillaries and may be aided by an ischemia during and after exercise or hypoxia.

Gas exchange across the rete mirabile and gas gland of the swim bladder has been studied in detail. These studies are reviewed in the chapter by Steen, this volume. Studies of other systemic capillary beds are sparse and poorly documentated.

Some teleosts have a rete mirabile in the choroid layer of the eye. High oxygen levels in the vitreous humor are associated with the presence of a rete, and in some cases the oxygen tensions are in excess of one atmosphere. Carbonic anhydrase and bicarbonate levels in fish eyes are in excess of those in the plasma.

The concentration of bicarbonate is possibly associated with the formation of the aqueous humor and the maintainence of intraocular pressure in fish as well as mammals. Bicarbonate accumulation in the aqueous humor of the rabbit is owing to formation from CO_2 and not the transport of the ion. In fish the rete could maintain CO_2 levels in the eye that are higher than those in the plasma, which are only of the order of 2-5 mm Hg, and therefore assist in bicarbonate concentration and the maintainence of intraocular pressure. Therefore, high oxygen levels in the eye may not be functionally significant but only a byproduct of the presence of a rete.

Some fish are viviparous, and in a few instances a complex system exists for exchanging material between adult and young. The anatomy of circulatory structures associated with development of viviparity in fish have been described, but nothing is known of the exchange processes in these maternal fetal connections.

METHODS OF ANALYSIS OF THE TRANSFER OF GASES

The $O_2 : CO_2$ diagram has been used extensively to analyze gas exchange in mammals. In aquatic respiration the relationship between P_{CO_2} and P_{O_2} in expired water is not linear, as in air, for a given respiratory quotient and a series of ventilation volumes. This nonlinearity results from the CO_2 buffering capacity of the water and the slow rate of formation of CO_2 from bicarbonate, both of which tend to reduce the P_{CO_2} in water in contact with the respiratory surface. In fish, oxygen tensions on either side of the gills do not approach an equilibrium point, and some of the factors that determine the rate of CO_2 removal appear to be different from those that govern O_2 uptake. This, plus the general paucity of relevant data, makes an extensive analysis of gas exchange in fish using the $O_2 : CO_2$ diagram difficult. No system for the graphical analysis of gas transfer in fish, similar to the O_2: CO_2 diagrams, has been developed. Although there is no adequate method for the extensive analysis of the interaction of several parameters, a few investigators have attempted to analyze certain aspects of gas exchange in fish.

Transfer Factor

The transfer factor of the gills for a particular gas is a measure of the ability of the gills to transfer that gas per unit gradient. The transfer factor of the gills for oxygen T_{O_2} is defined as

$$T_{O_2} = \frac{\dot{V}_{O_2} \text{ (oxygen uptake)}}{\Delta P_{O_2} \text{ (mean oxygen gradient across gill epithelium)}}$$

Randall *et al.* calculated the mean gradient using the following equation

$$P_{O_2} = \frac{1}{2}\left(P_{I_{O_2}} + P_{E_{O_2}}\right) = \frac{1}{2}\left(P_{a_{O_2}} + P_{V_{O_2}}\right)$$

where $P_{I_{O_2}}$ is the inspired oxygen level, $P_{E_{O_2}}$ the expired oxygen level, $P_{a_{O_2}}$ the arterial oxygen tension, and $P_{V_{O_2}}$ the venous oxygen level. Piiper and Baumgarten-Schumann, criticizing this method on the grounds that the mean gradient was affected by the slope of the blood oxygen dissociation curve, used a modified Bohr integration technique to determine T_{O_2}. Uptake was divided into a series of steps, and then the T_{O_2} for each step calculated. The mean partial pressures for each step were obtained using an oxygen dissociation curve. The total T_{O_2} value was obtained by summation of the individual T_{O_2} values for each of the steps of V_{O_2}. They found that the differences were small in the calculated values of T_{O_2} and T_{CO_2}, obtained for dogfish using the method of Randall *et al.* and the Bohr integration technique.

The transfer factor of the gills for oxygen $\left(T_{O_2}\right)$ in the resting trout is 0.0056 ml/min/mm Hg/kg, increasing by a factor of five during moderate swimming. The ratio of T_{O_2} (the transfer factor of the gills for carbon dioxide) to T_{O_2} is about 20:1 decreasing to 14:1 during activity. That is, the transfer factor of the gills for oxygen increases more than the transfer factor for CO_2 during swimming in the trout. The reported T_{O_2} for the dogfish is 0.0080 ml/min/mm Hg/kg, a value very similar to that for the resting trout. The ratio of T_{O_2} to T_{CO_2} is 56:1 in the dogfish.

The transfer factor may be altered by changing the functional area of the gills or diffusion distance between blood and water by changing the pattern of blood or water flow over the respiratory surface. The thickness of the boundary layer of water at the gill surface may also be altered and affect the transfer factor, but this effect is probably not important. The velocity of the various hemoglobin reactions and the rate of bicarbonate formation will also affect the respective transfer factor of the gills for oxygen and carbon dioxide. The relative importance of these parameters in determining the size of the transfer factor has yet to be investigated.

Hughes and Shelton and Hughes introduced a term analogous to the transfer factor. This term "the number of transfer units" is determined by the ratio between the capacity of the gills for transfer of gases and the load imposed on the system by the blood flow. In practical terms it is difficult to assign a numerical value to this term, and the transfer factor appears to be a more suitable estimate of the transfer capacity of the gills.

The Effectiveness of Gas Transfer

This term introduced by Hughes and Shelton is a measure of the relative ability of the system to transfer a particular gas. The effectiveness of gas transfer is the ratio of the actual rate of gas exchange to the maximum rate of gas exchange possible. The maximum rate of oxygen removal from water occurs (assuming that no active processes are involved) when water leaving the gills is in equilibrium with venous blood entering the gills.

The effectiveness of gas exchange can be applied to the removal of oxygen from water, the uptake of oxygen by the blood, the removal of CO_2 from the blood and the uptake of CO_2 by water, this has been done for trout and dogfish. In both trout and dogfish the effectiveness of oxygenating the blood is high, ranging from 79% in the unanesthetised dogfish to 95-100% in the trout. The effectiveness of CO_2 removal from the blood is not so high (35-60%) in trout or dogfish, but in both cases it is considerably more effective than CO_2 removal from human blood. The effectiveness of CO_2 uptake by the water is between 38 and 43% in dogfish and 4-6% in trout. The low effectiveness of oxygen removal from water (11-30%) passing over the trout gills is related to the very high ventilation volumes and the low utilisation of oxygen recorded in this animal. The difference between the effectiveness of oxygen removal from the water and the percent utilisation of oxygen is that effectiveness takes into consideration the venous P_{O_2}. Effectiveness, therefore, is a more accurate estimate of the efficiency of oxygen removal from the water. Comparison of the effectiveness of gas transfer between fish and man indicate that the CO_2 uptake by and O_2 removal from the medium and O_2 uptake by the blood is similar in man and fish.

Capacity Rate Ratio

Hughes and Shelton compared the capacity rates of water $\left(\dot{V}_G \cdot \alpha w_{O_2}\right)$ and blood $\left(\dot{Q}_G \cdot \alpha b_{O_2}\right)$ and discussed the relevance of a changing capacity rate ratio of water and blood on gas exchange across the gills. ($\dot{Q}_G$ = gill blood flow; $\dot{V}_G$ = gill water flow; αb_{O_2} = oxygen content of blood per mm Hg; αw_{O_2} = oxygen solubility coefficient in water.) The oxygen content of blood is not linearly related to P_{O_2} but changes in proportion to the slope of the oxygen dissociation curve. The shape of the dissociation curve is affected *by* P_{CO_2}, pH, and temperature of the blood. The solubility of oxygen in water is affected by temperature and ionic content of the water. The value of using capacity rate ratio in an analysis of gas exchange in fish is limited by the large number of variables that must be measured before capacity rate ratio can be determined. The capacity rate ratio of water to blood in the resting and active trout is about 5.

Ventilation-Perfusion Ratio

Although the effects of changes in the ventilation-perfusion ratio on gas exchange have been extensively studied in mammals, very little attention has been paid to the effects of variations in this ratio in fish. The $\dot{V}_G / \dot{Q}$ ratio is extremely variable in fish and ranges from 10:1 in the carp and dogfish to 80:1 in the trout. The ratio increases in the trout during hypoxia. There has been no extensive analysis to date of the effect of changing $\dot{V}_G / \dot{Q}$ ratios on gas exchange in fish.

Although there is little concrete data on changes in the ratio of blood to water flow at the gills, there are a number of observations on the relationship between heartbeat and breathing movements in fish. The heart beats during a particular phase of the breathing cycle in many fish, and there may be a 1:1, 1:2, or 1:3 ratio between heart and breathing rates. In the trout, synchrony between a phase of the breathing cycle and the heart beat occurs during hypoxia.

The eel often stops breathing for several seconds or breathes through only one side of the gills; during these periods of reduced or no breathing the heart rate slows. A type of Cheyne-Stokes breathing often occurs in quiet, inactive teleosts. As the breathing rate oscillates heart rate also changes; the rate slows as breathing is reduced or absent. Thus there appears to be a correlation between heart and breathing rate in many fish under a variety of conditions, indicating some overall regulation of the ventilation-perfusion ratio.

THE EFFECTS OF VARIOUS PARAMETERS ON GAS EXCHANGE

Temperature

The major effect of temperature on gas exchange is to change the oxygen requirements of the fish. A change from 10° to 20°C increases the standard oxygen uptake in the goldfish by 254%; the amount of oxygen in the water decreases by only 18% and is offset by a decrease in viscosity of water and a more rapid rate of diffusion of gases. Krogh's permeation coefficients increase by 1%/°C. The fish must therefore adapt its respiratory system to meet the increased oxygen demand as temperature rises.

There are a large number of studies demonstrating the effect of temperature on the oxygen uptake of both resting and active fish. The increase in oxygen uptake with temperature is associated with an increase in cardiac output and ventilation volume. The peripheral resistance to blood flow decreases with increasing temperature, reducing the amount of work required of the heart to maintain a given cardiac output as the temperature is raised.

The rate of oxygen transfer across the gills is limited either by the cost of moving blood and water past the respiratory surface or by the magnitude of the transfer factor of the gills for oxygen. Brett showed that maximum levels of sustained activity of salmon at temperatures above 15°C could be raised by supersaturating the water with oxygen. These observations indicate that, under the conditions of his experiments at least, either the cost of delivering oxygen to the gills or the transfer factor of the gills for oxygen is limiting oxygen uptake.

Hypoxia

The effects of hypoxia on O_2 uptake and CO_2, removal have been studied in the trout and the carp. A reduction of the oxygen level in the water causes an increase in breathing rate and amplitude, a decrease in heart rate, but an increase in stroke volume of the heart in the trout. Cardiac output varies little but the ventilation-perfusion ratio increases because of a large increase in water flow over the gills. The transfer factor of the gills for oxygen is increased, resulting in a decrease in the mean pressure gradient for oxygen between blood and water across the gills. The partial pressure of oxygen in venous blood falls as the animal utilises the venous oxygen reserve. The end result of the increase in transfer factor and ventilationperfusion ratio and the decrease in venous oxygen content is that oxygen uptake is maintained in the face of a decreased amount of oxygen in the water. Oxygen uptake falls at low oxygen levels in the water, and there is an increased production of lactate by the fish. The exact level at which oxygen uptake falls varies between fish. Teleosts appear to be able to regulate oxygen uptake over a wide range of oxygen levels in the water; elasmobranchs appear to be less able in this respect. Both groups become more active in hypoxic conditions and attempt to leave the oxygen-depleted environment.

An increase in ventilation volume associated with a decreased heart rate appears to be a general response of fish to hypoxia. The increase in ventilation volume maintains the delivery of oxygen to the respiratory surface as the oxygen content falls. The decreased heart rate is offset by an increase in stroke volume in the trout such that there are only small changes in cardiac output. During hypoxia, therefore, the pattern of blood flow, rather than the cardiac output, changes. The changing pattern of blood flow through the gills may, as suggested by Satchell, augment gas exchange.

Exercise

There have been a large number of studies on the changes in oxygen uptake with exercise in fish. Stevens and Randall and Randall *et al.* have investigated gas exchange across the gills of trout during swimming. Moderate exercise in the trout is associated with an increase in cardiac output, ventilation volume, and the gill transfer factor for oxygen. The increase in cardiac output is largely the result of an increase in stroke volume.

Although both water and blood flow increases, the ventilation-perfusion ratio remains constant. The transfer factor of the gills for oxygen increases in phase with the changes in oxygen uptake in such a way that the gas gradient across the gills remains constant. Hence, the oxygen partial pressures in afferent and efferent blood and water vary little throughout the period of exercise.

The maximum oxygen uptake is 4-8 times the resting consumption.

Maximum oxygen uptake increases in salmon as the temperature rises to 15°C and then varies little for further increases of temperature. Supersaturation of the water increases the maximum oxygen uptake, indicating that at temperature above 15°C either the cost of ventilating the gills or the transfer factor of the gills for oxygen is limiting oxygen uptake.

An increase in the oxygen transfer factor in the trout is associated with an increased rate of sodium exchange and also possibly a net influx of water into the body across the gills.

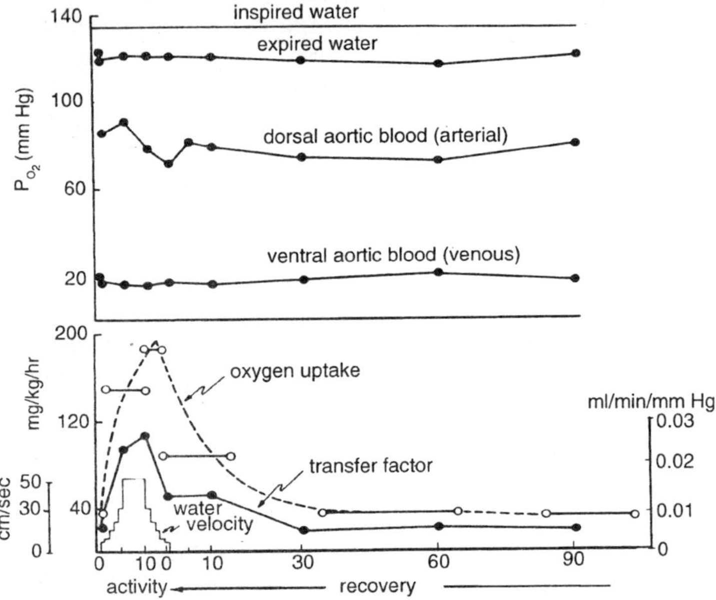

Figure 5.9 : Change in oxygen uptake, transfer factor for oxygen, and afferent and efferent blood and water oxygen tensions in the trout.

The changes in transfer factor are probably in part related to changes in the pattern of blood flow in the gills caused by increased levels of circulating catecholamines. The increased flux of sodium across the gills of trout during swimming in freshwater is probably alleviated by a rise in blood cortisol levels which causes a reduction in sodium excretion in the kidney.

THE REGULATION OF BREATHING

In this chapter performance of the systems involved in ventilation of gills with water will be considered, especially in relation to the way in which the basic patterns of activity are produced and coordinated. The adaptive modification of these patterns in response to changes in internal and external environment will also be discussed. To a lesser extent, consideration will be given to perfusion of the gills with blood in relation to ventilation and gas exchange. Events at the level of the cell are excluded even though they set the level of oxygen consumption and carbon dioxide production and thus create the demands to be met by the ventilatory system in its widest sense. Nor is this chapter concerned with details of the gas exchange process since this, too, is discussed elsewhere. The broader problem to be discussed here is essentially that of the regulation of ventilation and perfusion so that the gas exchanger is able to function and, as far as possible, meet the needs of the cells. Previous articles in which these aspects of fish respiration have been reviewed are those of Black, Fry, Hughes and Shelton, and Hughes.

If a multistage system such as that involved in the transfer of respiratory gases between environment and cell enzymes is to operate satisfactorily, then control becomes a matter of some complexity. Prosser has pointed out that an organism can survive in an altered environment in different ways. At one extreme it may change its internal state with the environment (physiological adjustment), or alternatively it may regulate the internal state at a constant level over wide environmental fluctuations (physiological regulation). The control systems responsible for physiological regulation seldom result in complete homeostasis so that adjustment and regulation invariably occur together, particularly at the extremes of the regulated range. This is true of the fish respiratory system, and although environmental changes result in adaptive regulation of an immediate or longer term type these are frequently accompanied by adjustments in metabolism.

THE RESPIRATORY PUMP

In the gill breathing vertebrates there is a surprising diversity of mechanisms whose function is one of maintaining a flow of water over the respiratory surface. Water is usually taken in through the mouth and, in those fish possessing it, the spiracle, is passed over the gills, and leaves through separate branchial or common opercular openings on either side of the body. This unidirectional flow pattern is thought to be the most effective way of bringing the environment into intimate contact with the gill epithelium since anatomical dead space is not necessarily a characteristic of such a system, although it may in fact exist. It has also been claimed that unidirectional flow reduces the work of ventilation as compared to a tidal flow system. This generalisation may be open to question since an elastic recovery system can prevent undue energy loss in reciprocating or tidal systems. In fact, the adult lamprey possesses a tidal ventilation system, the water being taken in and ejected through the external openings of the gills.

Many of the differences in pumping mechanism such as the presence of velum, spiracle, and operculum depend in the first place on the phylogenetic position of the fish. Other differences of a less obvious type are related to habitats in which the fish live. Details of many of these differences will be given in the following account, although the teleosts will be treated in greatest detail because they are the most fully documented group of fish.

Teleosts

The Pattern of Water Flow

General Features of the Breathing Movements and Pressures

It is probably true to say that the recent developments in the study of fish breathing movements began with the work of Woskoboinikoff and Balabai and van Dam. These workers introduced the concept of a continuous gill curtain separating buccal and opercular cavities. They also suggested that water flow over the gills was essentially a continuous process even though the water entered and left the overall pumping system in a discontinuous manner. Woskoboinikoff and Balabai concluded that two pumps, one in front and one behind the gill curtain, were responsible for maintaining the flow. Attempts to substantiate these claims were made by measuring pressures in the two parts of the system by means of simple water manometers. The views of these three workers were summarised and extended by Henschel in a generalised scheme of teleost breathing mechanisms.

Modern manometric methods were applied to the problem by Hughes and Shelton working on trout, *Salmo,* roach, *Rutilus,* and tench, *Tinca.* Pressure measurements in buccal and opercular cavities showed that the gills offered appreciable resistance to water flow so that a differential pressure was always found, usually with the gradient from buccal to opercular cavity. Together with records of the breathing movements obtained from tine films, the pressure records offered good evidence of a dual mechanism responsible for maintaining water flow. The breathing cycle was accordingly divided into two major phases as the opercular suction pump (phase 1) or the buccal pressure pump (phase 3) predominated, with periods of transition when the differential pressure was

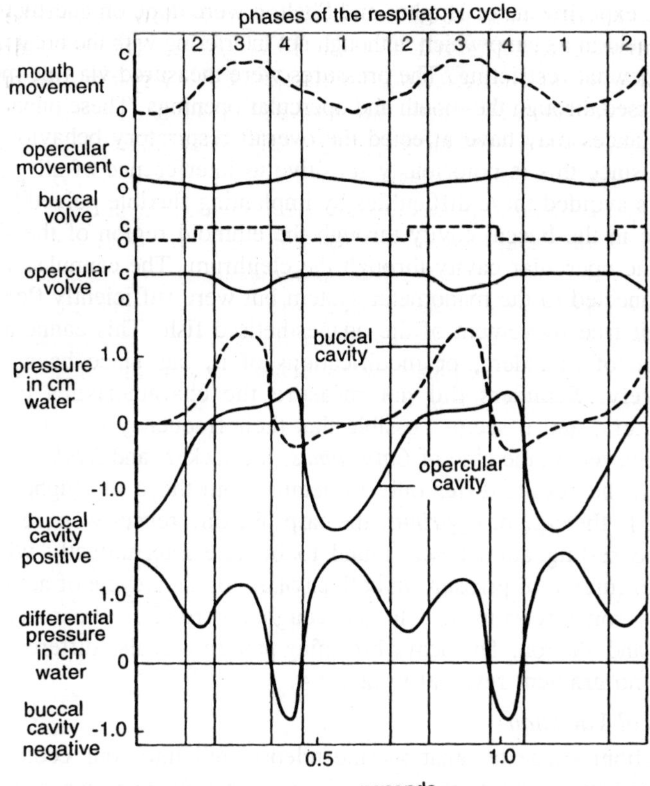

Figure 5.10 : Trout (70 g). The breathing movements of the mouth and operculum, together with associated pressure changes in the buccal and opercular cavities. The differential pressure between these cavities is shown in the bottom trace.

reduced (phase 2) or reversed (phase 4). During phase 1, the mouth was open, water was entering the buccal cavity and passing over the gills to the opercular cavities whose valves were closed. In phase 3, the mouth and buccal valves were closed, water was passing through the gills from the contracting buccal cavity and leaving the system through the opercular openings.

The concept of a dual pump, although a useful one, arises from pressure measurements made on either side of a gill resistance, relates primarily to water flow through the gills, and has little anatomical basis since there is mechanical interaction throughout the system. This point will be returned to later. However, the concept is a useful one provided that the full extent of interaction between pumps in producing flow into the system, through the gills, and out to the environment is appreciated.

The experiments of Hughes and Shelton were done on anesthetised animals held in a clamp which, although not interfering with the breathing, was somewhat restricting. The pressures were measured via fine metal tubes passed through the mouth and opercular openings. These unnatural circumstances may have affected the overall respiratory behaviour of the fish since this is notoriously sensitive to interference of all types. Saunders avoided these difficulties by implanting flexible polyethylene cannulae in the buccal cavity through the ethmoid region of the skull and in the opercular cavity through the cleithrum. The cannulae were then connected to the manometer system but were sufficiently flexible to permit free movement of the unanesthetised fish. This cannulation technique of Saunders, or modifications of it, has since been used extensively. Saunders did not measure the characteristics of his manometers, and it seems possible that their frequency response was low. However, in the case of *Catostomus,* the sucker, and *Ictalurus,* the bullhead, the results differ only in details from those of Hughes and Shelton. In the case of *Cyprinus,* the carp, the differences were marked since the resting animal was found to breathe intermittently with a double excursion of pressure in both cavities for each cycle of activity. There is some resemblance to the cough recorded in the roach by Hughes and Shelton. The heavy breathing pattern was also different and showed no gradient reversal in phase 4.

Ecological Variation

Baglioni suggested that marine teleosts fell into four ecological categories between which there were fairly clear differences in respiratory mechanism. The variation was mainly in the degree of development of the branchiostegal apparatus below the opercular flap. Hughes recorded breathing movements and pressure in a range of marine teleosts and, in general, confirmed the validity of Baglioni's classification.

In his first group are pelagic fish which never rest on the bottom and in which the operculum is well developed with a small branchiostegal apparatus. Hughes found a great deal of variation in the pressure characteristics of fish in this category. In wrasse, *Crenilabrus,* and herring, *Clupea,* the two pumps were balanced, but the buccal pump predominated in the horse mackerel, *Trachurus,* and the opercular pump in the whiting, *Gadus.* There has been very little work done on the mechanics of ventilation in pelagic fish which are moving, although clearly this is of the greatest significance. Devices in which water currents can be maintained, such as the Brett respirometer or the Bainbridge fish wheel, make this type of investigation possible.

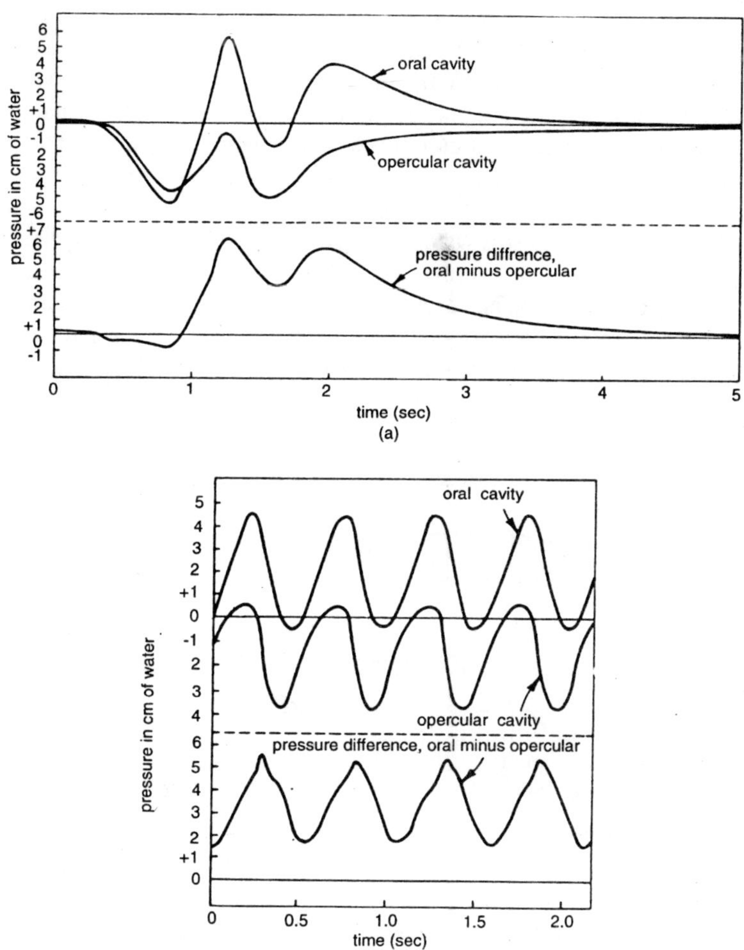

Figure 5.11 : Carp. Pressure changes in the buccal and opercular cavities, together with differential pressure curves. The records were taken from (a) quietly breathing fish and (h) fish stimulated to breathe heavily by increasing the carbon dioxide content of the water.

It has been known for some time that adequate oxygenation of the blood in mackerel depends on forward swimming, and it has been assumed that the gills could not be ventilated in stationary fish. Many other fish have been observed to reduce their breathing movements as they move forward. For example, Brett found that breathing stopped in sockeye salmon during violent exercise. It seems reasonable that animals living in a dense fluid should exploit this method of ventilation which may have little or no effect on the total drag of the body. The tuna ventilates its gills entirely by means of its movement through the water and regulates the stream by the degree of mouth opening. Brown and Muir suggest that the drag of the gill system is some 5-10% of the total drag in tuna swimming at 66 cm/sec. An ingenious examination of the problem was carried out by Muir and Buckley in *Remora,* a fish which is a poor swimmer but which moves rapidly through the water when riding on sharks. In still water this fish ventilated its gills by balanced buccal and opercular pumps but, when placed in a water current, breathing slowed down and eventually stopped at a water velocity between 60 and 80 cm/sec. *Remora* then adjusted ventilation by variation of the cross sectional area of mouth, buccal cavity, and opercular openings.

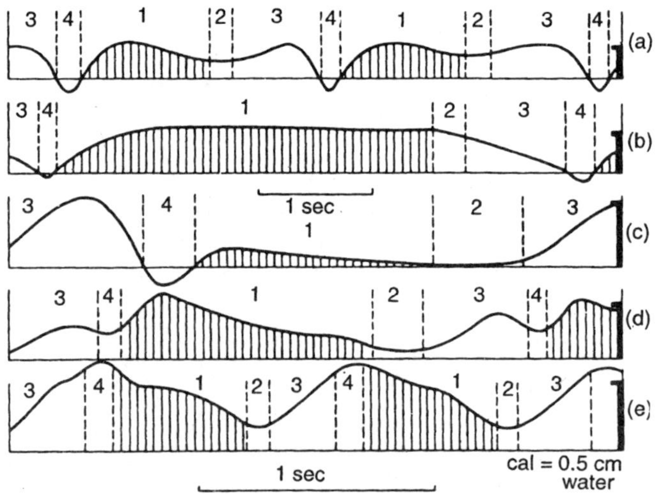

Figure 5.12 : Time course of the differential pressure between buccala nd opercular cavities during the respiratory movements of five marine teleosts: (a) Crenilabrus melops, (b) Callionymus lyra, (c) Conger conger, (d) Pleuronectes platessa, (e) Microstomus kitt.

Fish in the second group of Baglioni's classification are largely bottom-living animals in which the branchiostegal apparatus is of greater importance in breathing. In general there is a relative increase in the size of the opercular cavity and in the predominance of the opercular pump in fish of this group. There is also a tendency to reduce the size of the exhalent openings, to increase the duration of the inspiratory phase, and therefore to eject water over a short period of time and at high velocity. These features can be seen in several fish; *Callionymus is* a good example.

Baglioni's group III animals are those in which the branchiostegal apparatus is best developed and are true benthic forms. Hughes examined two pleuronectid fish and again found a domina .it opercular pump, although it was not obviously, different from tl.at cf group II fish. The main point of interest was the absence of a reversed differential pressure in phase 4, and Hughes attributed this to an active branchiostegal valve mechanism. However, a number of types of fish have been found to possess patterns of this sort, and it may be that timing of the movements is of more significance than active valve mechanisms.

The relative importance o' upper and lower opercular cavities in pleuronectids is a matter of some interest. Hughes cited the earlier work and concluded from his own experiments that water left through both opercular openings. He also demonstrated that the gill area was identical on the two sides. A channel connecting the two opercular cavities was described by Yazdani and Alexanderwho found that, although both gills were ventilated, water left only from the upper operculum. This view was corroborated by Arnold who observed the respiratory currents when the fish were in still and in moving water. He found that water left only from the upper operculum even when the ventilation movements were large, provided the fish was in contact with the substrate. Actively swimming plaice, observed in a flume, breathed through both opercular openings.

The fish in group IV are a heterogenous collection of families between which there is no clear ecological relationship. They are characterised by the loss of the branchiostegal apparatus, but this is not sufficient to impose any uniformity in breathing mechanism. The buccal pump or opercular pump may be dominant in different species.

The Gill Resistance

It was suggested that the concept of the dual pump rests on a gill curtain offering appreciable resistance to water flow. This resistance must be known before differential pressure records can be used to assess flow patterns and rates through the gills. Although analysis would be much easier if gill resistance remained constant, it seemed from the outset that this would be unlikely . m ne geometry of the gills changes constantly during a single breathing cycle since they are bridging a gap of changing dimensions and are attached ventrally to the basihyal which also moves rhythmically. Parts of the gill sieve could thus be alternately exposed and protected or, more simply, there could be variations of the dimensions in the pores through which the water flows.

Interdigitation of secondary lamellae could, as Hughes points out, have an enormous effect on resistance. In addition, it is uncertain that contact can be maintained under all conditions between the tips of gill filaments from adjacent gill arches. A number of workers have suggested that the gill filaments do separate during some stage in the opercular cycle. Saunders reported separation as the operculum was maximally abducted, Hughes during opercular adduction, and Pasztor and Kleerekoper during all phases but principally during adduction. The latter workers were satisfied that the movements were not a result of the force of the ventilating currents but were produced by rhythmic contraction of gill adductor muscles. They also suggested that the filament tips parted to protect the gills against excess water flow, although it is by no means clear that an animal could produce a 'gill damaging current. In tuna and some other oceanic fishes there is considerable fusion of gill filaments and secondary lamellae to give a more solidly constructed hemibranch. Even in these forms, which usually rely on "ram" ventilation associated with fast swimming, water can still bypass the gills if the filament tips part.

Such determination of gill resistance as have been made support the conclusion that it cannot remain constant. Hughes and Shelton reported experiments on tench in which the ventilation volume was changed by adding carbon dioxide to the water. As the volume increased so the ratio (mean pressure) / (minute volume) decreased, and this was taken as strong evidence that the gill resistance was decreasing. It also seemed reasonable to conclude that if resistance changed with overall flow rate it could quite easily vary within a single cycle since the pressure gradient was not constant. Later experiments in which the gills of deeply anesthetised fish were artificially perfused gave results in which the resistance varied in an unpredictable way, sometimes rising, sometimes falling, and sometimes remaining constant with increased flow. It was quite common for the resistance to remain constant over certain parts of the pressureflow range, but invariably discontinuities were found in the relationship were found int eh relationship. The most usual pattern was one in which the gill resistance remained constant down to a certain flow rate, below whihc the resistane increased in these experiments agreed reasonablty well with those fromt eh actively ventilating tench. Increase in resistance with decreasing ventilation could be explained in terms of change in gill geometry, for relationship could result from some new factor such as size of mouth and opercular openings, assuming greater importance at high flow rates.

Experiments on the relationship between ventilation and differential pressure across the gills have also been done on *Callionymu*. Because

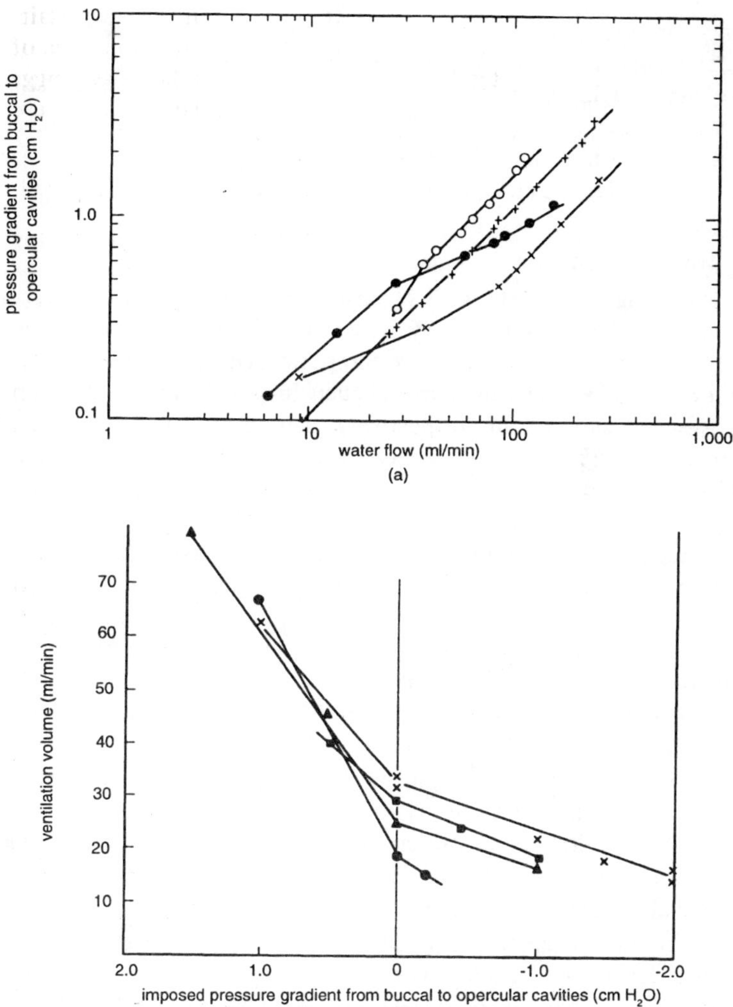

Figure 5.13 : The relationship between water flow over the gills and experimentally applied pressure gradients. (a) In tench (4 animals each of approximately 100 g) which were anesthetized to the point where breathing movements had ceased. The flow range, but invariably discontinuities were found in the relationship.

of the convenient anatomy of this animal it is possible to glue rubber connections around the opercular openings and impose external changes of pressure across the gills of otherwise normally breathing fish. Hughes and Umezawa found a change in slope of the pressure-volume relationship at the point where the imposed pressure gradient was zero. Larger changes in ventilation were produced by increasing the gradient from buccal to opercular cavities than were seen when similar gradients were imposed in the reverse direction. The change in slope is not necessarily to be interpreted in terms of a change in gill resistance because the fish is actively pumping water and may respond to the imposed gradient. A change in activity may be more important than a change in gill resistance in this case.

The Work of Breathing

Breathing muscles do work against three main categories of force, the magnitude of which may vary greatly with the environment in which an animal lives. The flow resistive forces offered by the gill curtain of fish or by the airways in a mammalian lung will depend upon the viscosity of water and air, respectively (these are related in a ratio of approximately 55:1) ; similarly inertial forces will depend on the mass of water or air (densities related approximately 840:1) and of the tissues being accelerated. Only the third category, the elastic forces developed in the tissues of the ventilating system, may be independent of the medium and related more directly to the nature of the ventilating system. These considerations have led many workers to conclude that the cost of breathing in aquatic organisms is very high, although precise figures are not usually available.

In man the total energy needed for breathing can be measured by determining the overall oxygen consumption at different forced ventilation levels. The cost in resting man is about 2% of the overall oxygen consumption, although there is some variation in different workers' results. Pressure-volume measurements have been used to determine the mechanical work of breathing, and comparison of the two types of determination gives efficiencies in respiratory musculature of 19-25%.

There are difficulties in persuading fish to ventilate their gills at different levels by means which will not otherwise affect oxygen consumption. Conditions of carbon dioxide excess or oxygen lack will produce the necessary changes but it is known, for example, that oxygen depletion ultimately limits uptake. Figures given by van Dam, who measured ventilation in eels and trout by enclosing their opercular region in a chamber separated from the mouth, enable one to calculate that the eel uses 10-12% and the trout 20% of its oxygen uptake for the work of breathing. There are too few determinations to give any degree of confidence in these calculations, added to which the fish, although fastened, may have been variably active. However, calculations based on the results of other workers also give figures for metabolism of breathing muscles between 10 and 15% of the resting oxygen uptake. Increased ventilation certainly has a marked effect on oxygen uptake as Beamish demonstrated. He determined standard oxygen consumption in fish by extrapolating the relationship between oxygen uptake and activity back to zero activity. These extrapolations showed that the standard level of oxygen uptake increased in goldfish, carp, and trout, as they breathed more vigorously in lowered oxygen tensions. Surprisingly, carbon dioxide had no effect on standard oxygen consumption. Since ventilation volume was not measured, however, a more quantitative relationship cannot be determined from these results. Schumann and Piiper enclosed tench in the same type of opercular chamber as used by van Dam in order to determine both ventilation volume and oxygen uptake. Spontaneous changes in ventilation were used to determine the cost of breathing together with measurements of oxygen uptake during artificial ventilation of animals paralysed by succinylcholine. The results showed that 18-43% of the total oxygen uptake at resting ventilation volumes and 44-69% at ventilation volumes three times resting were used by the respiratory musculature.

Unfortunately, calculations on the mechanical work of breathing give somewhat different answers. In fact, simultaneous pressure and volume measurements have not been made. The separate determination of volume changes in buccal and opercular cavities is a difficult proposition, and, although volume changes of the whole system could be more easily measured, no data are available. However, making some simplifications, it is possible to use the ventilation volume as a measure of the latter as Alexander has pointed out. If the system is represented by a simplified model then it is clear that on inspiration the work done is

$$W_i = -P_i(V + v) + (P_i - p_i)V$$

and on expiration

$$W_e = (P_e - p_e)V + p_e(V + v)$$

where V and v are the volume changes of buccal and opercular cavities, respectively, P_i and p_i the pressures in these cavities on inspiration, and P_e and p_e the pressures on expiration. If it is assumed that the differential pressure across the gills remains constant throughout the cycle so that

$$P_i - p_i = P_e - p_e = D$$

the total work done per cycle is

$$W = (-P_i + \mathrm{D} + p_e)(V + v)$$

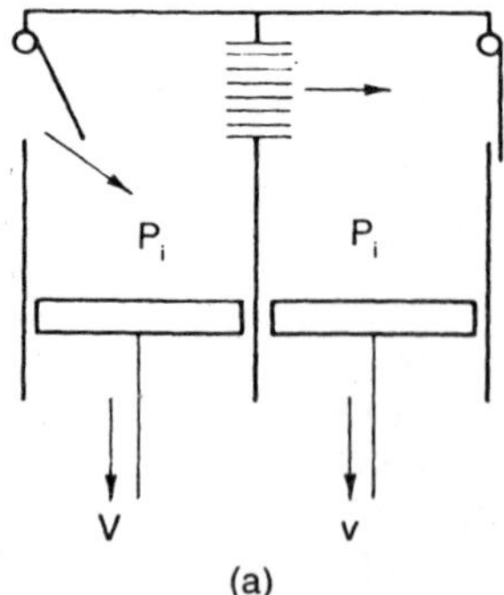

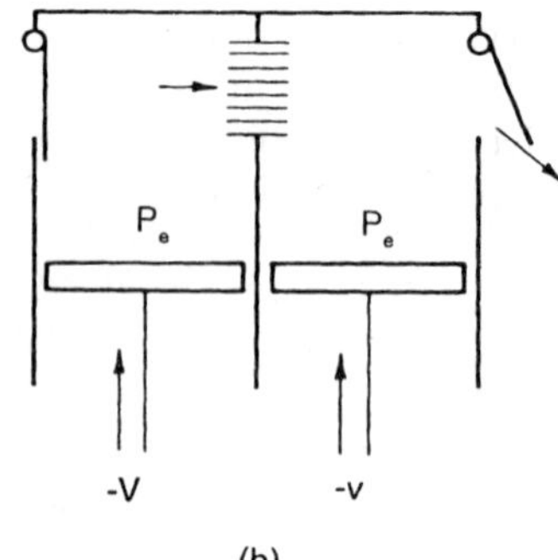

Figure 5.14 : Model of ventilating system showing buccal and opercular pumps operating during (a) inspiration and (b) expiration.

The work done in breathing a unit volume of water would be $-P_i + D + P_e$. If the data available in work cited previously are used, this total would usually be between 1×10^6 and 2×10^6 ergs/liter of water ventilated. Assuming 20% efficiency in the respiratory muscles, 0.025-0.05 ml of oxygen would be needed to pump 1 liter of water. Under normal conditions this would represent only 0.5-1% of the oxygen taken from that amount of water by the fish. Considerable increases in these figures are found if the calculations are repeated for active fish or for animals in a poorly aerated environment.

The fact that the pressures in all parts of the system vary during both inspiration and expiration will make the real system less efficient than the model. In addition, there is often considerable reflux through the mouth during expiration and through the opercular openings during inspiration. This means that the ventilation volume is not the same as the overall volume changes. It seems very doubtful that these factors could account for the discrepancy between oxygen consumption measurements and the calculations.

The Pump Musculature and Skeleton

Several accounts have been published of the head skeleton and its musculature in teleosts, some of which refer specifically to ventilation. Interactions occur between the skeleton, muscles, and tendons so that movements of one component tend to cause complementary movements in many others. Because of these interactions it is very difficult to determine the precise role of any muscle in the overall pattern. Anatomical studies, extirpation of muscles, and movement recordings all provide valuable information, but the recent studies of Ballintijn and Hughes and Hughes and Ballintijn using electromyography give the most convincing picture of the sequences of muscle contraction. The level of electrical activity in a muscle can also be used as a guide to the relative contribution of that muscle in the total mechanism, and in *Callionymus* Hughes and Ballintijn have demonstrated a clear relationship between stroke volume and electrical activity as shown by the height of the integrated electromyogram.

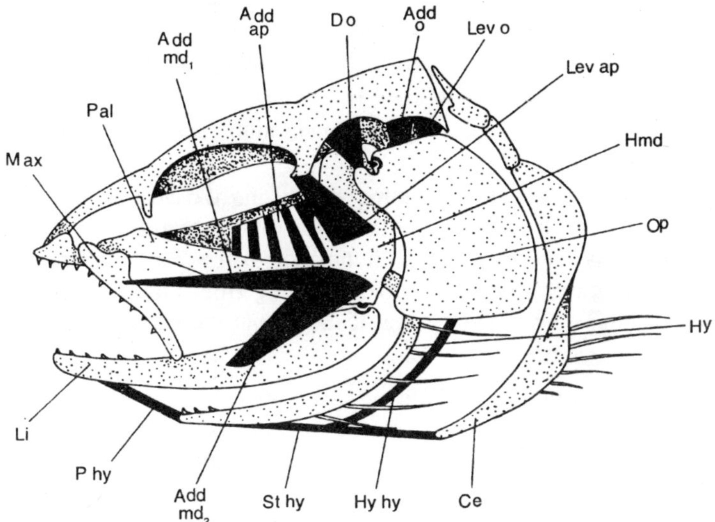

Figure 5.15 : Lateral view of the skull and important respiratory muscles in a typical teleost: Max, maxilla; Pal, palatine; Add md, and md2, adductor mandibulqe (two divisions); Add ap, adductor arcus palatini; Do, dilator operculi; Add o, adductor operculi; Lev o, levator operculi; Lev ap, levator arcus palatini; Hmd, hyomandibula; Op, operculum; Hy, hyoid; Ce, cleithrum; Hy hy, hyohyoideus; St hy, sternohyoideus; P hy, protractor hyoidei; and Lj, lower jaw.

The following brief account is based on a generalised scheme of the teleost respiratory musculature. Details of the musculature obviously vary from species to species as will their mode of action. One of the interesting conclusions of Ballintijn and Hughes is that the same breathing pattern is maintained even when different sets of muscles are involved. Thus the change from shallow to deep breathing with no other differences in the overall pattern is achieved by recruiting new, and quite important, muscles and not merely by increasing activity in the ones already working. In a system as labile as this it would be unreasonable to expect uniformity.

Expiration

During expiration the mouth and opercular flaps are closing, and as the whole system decreases in volume, water leaves through the opercular openings. The muscles involved are as follows:

(1) Adductor mandibulae, closing the mouth
(2) Adductor arcus palatini, adducting the palatal complex and hyomandibula, thus laterally decreasing the volume of the buccal cavity
(3) Adductor operculi, adducting the operculum and so decreasing the volume of the opercular cavity
(4) Protractor hyoideus, protracting the hyoid and thus adducting the ventral end of the hyomandibula
(5) Hyohyoideus, adducting the operculum and folding up the branchiostegal apparatus
(6) Levator operculi, adducting and lifting the dorsal border of the operculum. This has the effect of rotating the operculum on its articulation with the hyomandibula so that the operculum base is retracted. A ligament running from this region to the angular causes a lowering of the jaw at the end of expiration

Inspiration

During inspiration the whole system is increasing in volume and, since the opercular openings are closed by their valves, water is taken in through the mouth. The muscles involved are as follows:

(1) Levator arcus palatini, abducting the palatal complex and hyomandibula by rotating the latter outward on its articulation with the skull
(2) Sternohyoideus, retracting the hyoid arch thus expanding the branchial arches and abducting the hyomandibula
(3) Dilator operculi, abducting the operculum by pivoting it on its articulation with the hyomandibula

Interaction of Elements

Because the hyomandibula is articulated with the neurocranium, the quadrate, the operculum, and the more ventral hyoid elements, it forms a vital coupling element in the interactions between the skeletal and muscular components. The lower end of the hyomandibula moves freely and the mechanical linkages are such that its abduction is accompanied by retraction of the more basal elements. This expands both the buccal and opercular cavities, opens the mouth and, because the quadrates are also moved out laterally, widens the mouth. The branchial arches also expand. In the opposite sense, closure of the mouth will protract the hyoid which in turn adducts the hyomandibula and quadrate and swings in the operculum. A fuller account of this type of coupling is given by Ballintijn and Hughes.

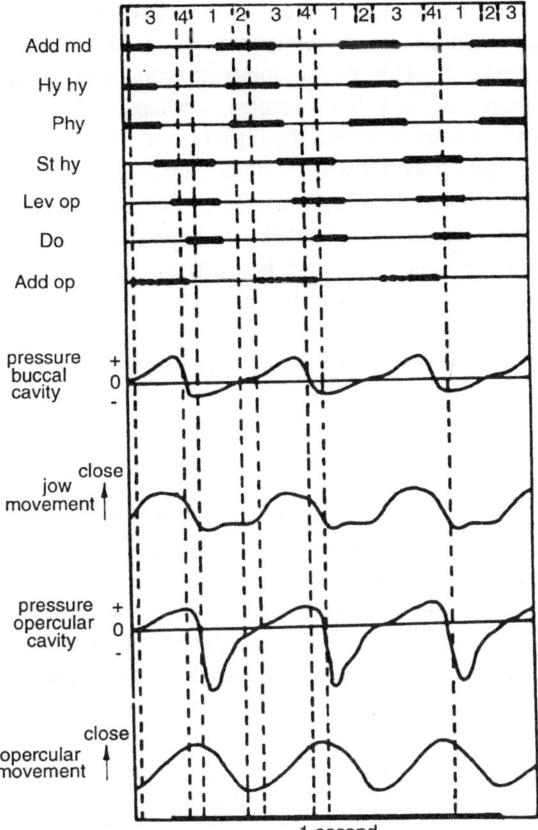

Figure 5.16 : Breathing movements and pressure changes in the buccal and opercular cavities of the trout, together with a diagrammatic representation of the time relations of activity in seven respiratory muscles. The respiratory cycle is divided into the four phases described in the text.

The precise timing of activity in seven of the main respiratory muscles of the trout as determined by the electromyographic studies of Ballintijn and Hughes. It might seem reasonable to suppose that in a smoothly oscillating system there would be temporal dispersion of overall activity throughout the cycle, and overlap between different units, certainly as inspiration took over from expiration and vice versa. This appears to be the case over most of the cycle, but there is a period during the final stages of opercular abduction when little activity is found in any muscles. It may be that elastic elements continue the movements smoothly during this period. It is interesting that all the muscles are active over at least two and sometimes three of the four phases proposed by Hughes and Shelton. This is desirable to produce the continuity of movement as already mentioned, but in fact the overlapping activity does not usually occur during the transitional phases. It is clear from this that there are very few muscles, perhaps none, that can be conveniently and specifically ascribed to buccal pressure pump or opercular suction pump as defined by manometric studies.

In general the most important muscles in the trout are the levator arcus palatini, the adductor arcus palatini (which also incorporates the adductor operculi in this animal), and the adductor mandibulae. During shallow breathing these muscles only are active, the sternohyoidens, protractor hyoideus, and hyohyoideus being brought into action as the breathing deepens. The dilator operculi only operates during hyperventilation.

Elasmobranchs

The Pattern of Water Flow

The nature of the relationship between water flow and the gill filaments in elasmobranch fish is less well defined than it is in teleosts. There are no direct observations of the gill configuration during breathing although it has been assumed that filaments from adjacent gill arches are, to some extent, in contact. The fact that a measurable pressure gradient exists from orobranchial to parabranchial cavities suggests that this is the case. Since the visceral arches themselves are participating in the pumping process to a much greater extent than in teleosts, there can be little doubt that changes in gill position and hence gill resistance occur during the breathing cycle. Flow direction over filaments and lamellae, particularly in relation to blood flow through them, is also poorly established. Robin and Murdaugh, using indirect methods of calculating oxygen uptake and the amount of oxygen removed from the ventilation stream, came to the conclusion that countercurrent gas exchange did not occur in dogfish gills. However, other investigations have shown that in many cases the arterial oxygen tension was higher than the mean expired oxygen tension. This can be explained by some sort of countercurrent system, although the path of water across the secondary lamellae is ultimately obstructed by the median septum. The possibility that water must flow along the length of the filaments in elasmobranchs has led Piiper and Schumann to propose a multicapillary model which is also effective in giving a negative value for the difference between expired and arterial oxygen tensions.

Anatomical differences between teleost and selachian are marked, since in the latter the respiratory current passes in through the spiracle as well as the mouth and out through a variable number of gill slits not covered by an operculum. The flow pattern from mouth and spiracle is curious. Water entering the spiracle of a dogfish will leave through the three anterior gill slits of the same side, whereas water entering through the mouth leaves through the last three slits. There is clearly not much turbulence in the orobranchial cavity. The pattern is not so obvious in the skate.

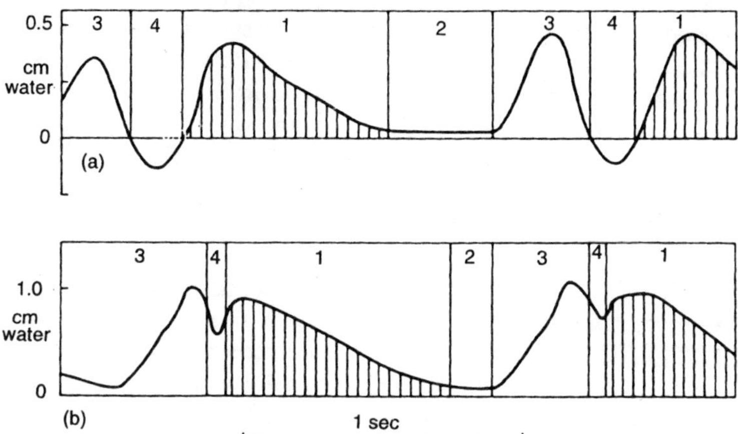

Figure 5.17 : Differential pressures between orobranchial and third parabranchial cavities of Scyliorhinus (a) and Raia (b). A positive differential pressure indicates that the orobranchial pressure is greater than the parabranchial pressure.

These problems and differences notwithstanding, the general picture of gill ventilation in selachians as it emerges from studies of movement and pressure does not appear to be radically different from that seen in the teleosts. A pressure gradient from orobranchial to parabranchial cavities exists for most of the respiratory cycle in dogfish and for the whole cycle in skates. Again four phases of activity are recognizable. A parabranchial suction pump predominates in phase one and an orobranchial pressure pump in phase three, and there are transitional phases in between. These distinctions are most obvious on the differential pressure curve. There are no direct measurements of gill resistance in selachians, and the problem of determining relative volume changes in different parts of the system is made even more difficult than in teleosts because of the large number of parabranchial chambers. Since the latter are small in volume compared with the orobranchial chamber, it might seem likely that the changes in volume would also be small. The implication of this would be that the suction pump mechanism would be relatively unimportant in selachians. However, the pressure measurements do not bear this out if it can be assumed that the gill resistance does not change too drastically during the breathing cycle. Even if the resistance does change it seems likely that it will fall during the suction pump phase since this is the time when the gill chamber is expanding in volume. Under these conditions the flow of water during the suction phase would be considerable. The differential pressure curves obtained by Hughes suggest that the parabranchial suction pump is at least as important as the orobranchial pressure pump in dogfish and of greater importance in the skate. Teichmann, who measured the pressures achieved before and after functional disturbance of different parts of the system, claimed that the pressure pump dominated in *Scyliorhinus,* there was a balance in *Mustelus,* and the suction pump was of greater significance in *Torpedo.* In this work the time course of the ventilation pressures was not recorded. The whole relationship between pressure and flow is an extremely difficult problem which is still virtually unresolved in these animals.

The dominance of the parabranchial suction pump in skate and rays was correlated with the bottom-living habit by Hughes. The absence of gradient reversal in the differential pressure curves of these animals was also associated with their ecology and was thought to be the result of active closure of the external openings to the gill slits. This could ensure that no sand entered the gills by reflux through the external gill openings. Water intake took place through the spiracle when the animals were partially buried in sand, but both mouth and spiracles were used if the animals were swimming or resting with the snout raised. The more pelagic selachians also showed respiratory adaptations. Some sharks stopped their breathing movements when swimming actively and ventilated their gills simply by opening their mouths in the water stream. It may be that, as in teleosts of similar habits, ventilation is controlled by the extent to which the mouth is opened, but so far this type of regulation has not been confirmed by observation.

The Pump Musculature and Skeleton

The head muscles of selachians are, with one or two exceptions, poorly differentiated but those that are defined, where the arrows on each muscle give some indication of the effect produced by contraction. With the exception of the hypobranchials (coracohyoideus, coracobranchialis), the muscles shown are adductor or adductorlike in function. Their contraction reduces the volume of orobranchial and parabranchial cavities. In addition to the muscles appearing in the diagram, there is a sheet of superficial constrictor muscle covering the

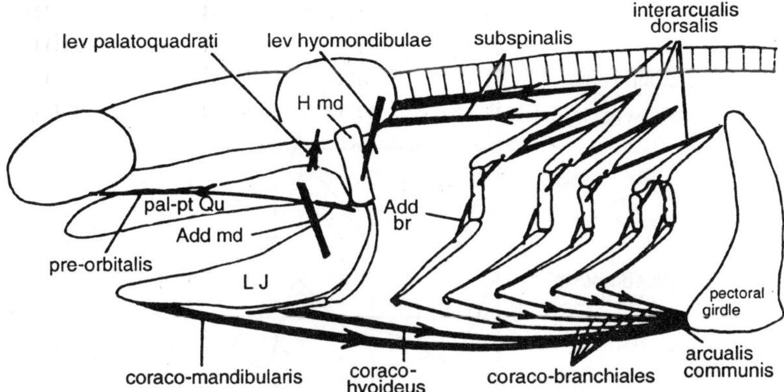

Figure 5.18 : Lateral view of the skull, visceral skeleton, and main respiratory muscles of the dogfish. The superficial constrictor muscles are not shown. Add br, adductor branchialis; Add md, adductor mandibulae; Pal-pt Qu, palato-pterygoid; LJ, lower jaw.

whole head region and being particularly well developed in the branchial region. The constrictor system is formed of several overlapping sheets, each sheet associated primarily with a branchial arch.

The timing of activity in the respiratory muscles, based on the electromyographic results obtained by Hughes and Ballintijn. It is clear from this figure that most of the activity occurred as first the orobranchial and then the parabranchial cavities decreased in volume. Following the period of major activity, the orobranchial cavity began to expand slowly and passively. When the dogfish was breathing quietly the whole of this part of the cycle was a result of elastic recovery. If the dogfish was made to hyperventilate, only a short period of purely elastic recovery occurred, followed by a more rapid expansion caused by contraction of the hypobranchial musculature. Even when these muscles were active there was usually a pause after their contraction before the next cycle began. The only muscle to be excited during the whole of this period was the adductor mandibulae in which a certain amount of tonic activity probably served to check the opening of the mouth. Hughes and Ballintijn conclude that contraction of the constrictor muscles, reducing the volume of both orobranchial and parabranchial cavities, is the prime mover in normal ventilation and that recovery is largely the result of elastic recoil of the visceral skeleton.

Cyclostomes

An account of the skeleton and musculature of the branchial region in the lamprey, *Lampetra,* has been given by Roberts. The gills are found in pouchlike chambers supported laterally by the adjacent branchial

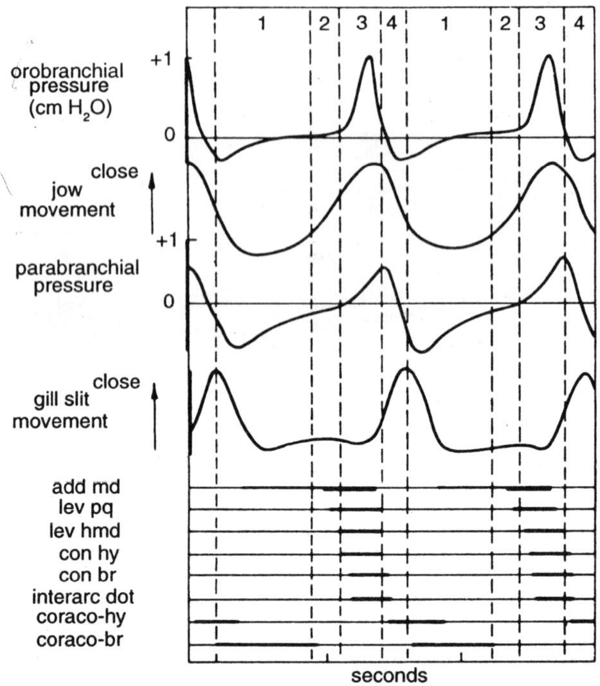

Figure 5.19 : Dogfish. Pressures in the orobranchial and parabranchial cavities in relation to movements of lower jaw and gill region. The main phases of activity in eight respiratory muscles are also shown. Add rod, adductor mandibulae; Lev pq, levator palatoquadrati; Lev hmd, levator hyomandibulae; Con by, constrictor hyoideus; Con br, constrictor branchiales; Interarc dot, interarcualis dorsalis; Coracohy, coracohyoideus; Coraco-br, coracobranchiales.

arch skeleton. The pouches are also supported top and bottom by extensions from the branchial arches forming two lateral commissures. The whole complex forms an unjointed network, the enclosed volume of which can be changed by bending reentrant curves in the cartilage of which it is made up. The important muscles are the constrictors running from the dorsal to ventral part of the branchial skeleton. In addition, a pair of diagonal muscles span the dorsal and ventral re-entrant curves of each branchial arch; there are some smaller muscles in the interbranchial septa and the gill sacs. The constrictor and diagonal muscles were shown, by observation and experiments in which electromyograms were recorded, to function together in expiration. No muscles were active during the inspiration of water, which in the adult lamprey occurs through the external gill openings. Just as in the dogfish it appears that expansion of the gill chambers is caused by elastic recoil, although Roberts noted that this was not a very forceful movement in the cartilaginous branchial skeleton of *Lampetra.*

Gill ventilation in myxinoids is very different from that described in the lamprey. Water is taken in through the single nostril, propelled down the pharynx by the rolling and unrolling action of a velum working in a velar chamber, and eventually leaves through paired branchial apertures formed by the union of efferent ducts from the gills of either side. A pharyngocutaneous duct which connects the pharynx directly to the outside, thus short circuiting the gills, also empties into the common opening. Normally this duct is closed off by a sphincter, but it occasionally opens to clear the pharynx of large particulate matter.

The pumping action of the velum has been established by studies on preserved material, by examination of live specimens, and by cineradiography. The velar skeleton and musculature is well described by Strahan, and a full review is given by Johansen and Strahan. The velum pulsates rhythmically. At this time the velar chamber would begin to decrease in volume and the velum would roll up. This activity,

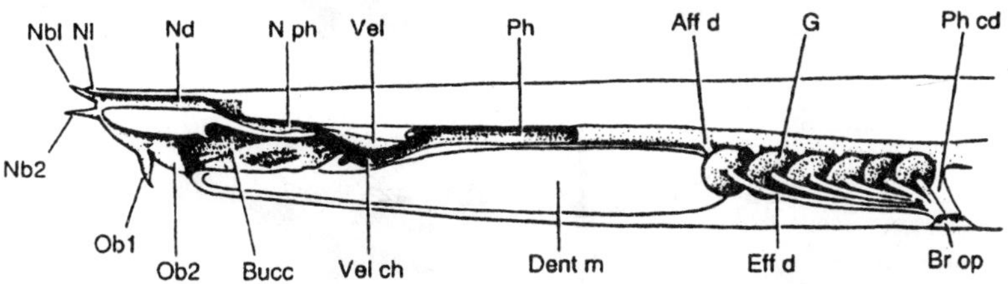

Figure 5.20 : Respiratory tract of blyxine glutinosa. *Diagrammatic half-section of anterior part of body: Aff d, afferent gill duct; Br ap, common branchial and pharyngo-cutaneous aperture of left side; Bucc, buccal cavity; Dent m, dental muscles; Eff d, efferent gill duct; C, gill; Nbl and Nb2, first and second nasal barbels; Nd, nasal duct; N1, nasal lip; N ph, nasopharyngeal duct; Ohl and Ob2, first and second oral barbels; Ph, pharynx; Ph cd, pharyngo-cutaneous duct; Vel, velum; Vel ell, velar chamber.*

propels water backward as indicated by the arrow and draws water in anteriorally through the nostril. It has been demonstrated by radiography that constrictor muscles of the pharynx and branchial regions also take part in pumping water. The gill pouches themselves contract rhythmically, and synchronised activity can be seen in the sphincters of afferent and efferent gill ducts operating in such a way as to ensure unidirectional flow.

CENTRAL FACTORS IN THE REGULATION OF BREATHING PATTERNS

Electrical discharges recorded, as described above, in the respiratory musculature of the head may be considered to be closely representative of activity passing down the motor nerves from its origins within motor neurons of brain stem nuclei. The fact that motor activity is not synchronous in all these final pathways but occurs in different phases of the breathing cycle in a very regular and specific way underlines the complexity of the respiratory oscillator. This mechanism must be capable of maintaining an appropriate basic rhythm and in addition be able to initiate activity of different duration and pattern to individual muscles at different times in a breathing cycle. The processes which lead up to the complex and highly integrated discharges in the respiratory musculature can be considered in two convenient but not totally independent categories. In the first place it is necessary to understand the interaction of large numbers of neural elements, most of them situated centrally, whose function is to initiate the basic respiratory process and produce the final coordinated motor pattern. The aggregation

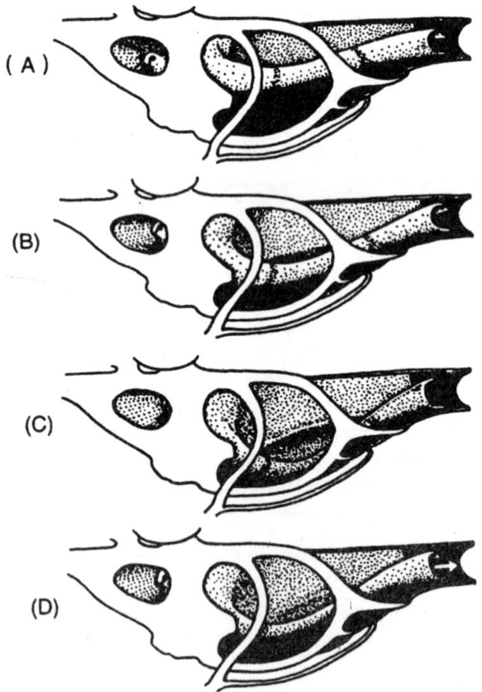

Figure 5.21 : Four successive stages in the velar cycle. The velar chamber is shown from the left side with the left pharyngeal wall removed: (A) resting; (B) velar scroll beginning to unroll; (C) velar scroll unrolled to full extent; and (D) velar scroll beginning to move dorsally, arrow showing direction of water flow.

of neurons within the medulla oblongata involved in this function is usually referred to as the respiratory center, although implications in such a term of a fixed group of neurons with precise and restricted anatomical location are unfortunate. In addition, there are many processes involved in respiratory homeostasis. That is to say, the basic process must be modified in an appropriate way to meet the demands of changes in internal state such as the transition from exercise to rest or of changes in external environment of great range and variety. The second category is thus one of feedback processes which ensure that the motor pattern produced by the center is appropriate to the conditions. The two categories will be considered in turn.

Experimental Techniques for Investigating Central Respiratory Mechanisms

Investigations of neural coordinating mechanisms rely in general upon the three classical methods of destruction, stimulation, and detection of activity. Their use in experiments on the central coordination of respiration in a variety of animals has led to the present position in which a great deal of information (and controversy) exists about the large-scale arrangement and interaction of neuron groups, but very little is known about the role of individual neurons; for example, expiratory and inspiratory groups which are functionally, if not morphologically, distinct form the conceptual basis of many hypotheses about the medullary respiratory center in mammals. In some cases they are considered adequate to coordinate rhythmic breathing; in others they are thought to interact with further neuron groups such as apneustic and pneumotaxic centers before normal activity is produced. The differences and problems result from studying a system of dispersed yet interacting neurons with methods which to some extent are all inadequate.

The extent of damage caused by transection or electrical coagulation is seldom easy to determine. In addition, it is not clear whether the changes in breathing always result from the destruction of regions or neurons exclusively associated with this particular function. Thus Wang and Ngai, following a number of earlier workers, found that pontine transection provided evidence for both a pneumotaxic center and apneustic center in the pontine region of cat brains. However, Hoff and Breckenridge, using similar techniques on the dog, were led to conclude that apneusis was the result of a generalised activity in the facilitator system of the reticular formation. They thought that pontine transection produced the necessary facilitator-suppressor imbalance to give inspiratory cramps. Nor are stimulation experiments free from these limitations.

Stimulus spread, using either chemical or electrical methods, can be controlled with care but not, so far, to the point where drugs can be administered into the immediate environment of a neuron or a single unit can be stimulated electrically. Furthermore, the precise status of the region stimulated, whether exclusively respiratory or more general in function, whether neuron or interconnecting nerve tract, is often in doubt. In the medulla of mammals, for example, electrical stimulation has always led to a wider and, to some extent, a different area being implicated in respiratory control than determinations made by recording electrodes. The recording technique, on the other hand, can only be used to identify neurons as respiratory when they are rhythmically active and synchronised precisely with the breathing movements. Cells, discharging in ways different from this may be involved in respiratory coordination but cannot be recognised. Moreover, the technique has provided little information about neuron interaction because single neurons or, with less selective electrodes, single groups of neurons only have been studied. Some workers have tried to combine the stimulation and recording techniques. For example, Gill and Kuno studied the effect of medulla stimulation on activity in phrenic motor neurons, Keder-Stepanova and Ponomarev stimulated the medial region of the medulla and examined the effect this had on respiratory activity in more laterally situated neurons, and Hori recorded the changes produced in respiratory center cells by mesencephalic stimulation. Multielectrode preparations recording from several respiratory neurons, or stimulating some and recording from others, should provide more information on neuron interaction. The practical difficulties are obviously considerable and interpretation is not easy as studies on the integrative properties of decapod heart ganglia have shown. Even though the anatomy and physiology of the interconnections in this potentially simple, nine-neuron group are well established, a complete analysis of the ganglion's rhythmic activity is not yet possible.

The Site and Extension of the Respiratory Center

The teleost medulla is capable of coordinating rhythmic breathing movements independently of the rest of the brain and spinal cord. In this respect the respiratory centers of fish and mammals are similar, since it is usually agreed that in the latter basic control of respiratory rhythmicity is a property of the medulla. However, in fish, no extramedullary centers have been identified, although there is, of course, no differentiated pontine region. Transections immediately in front of the motor nuclei of the Vth and VIIth cranial nerves did not substantially change the breathing movements of tench. Similarly, transections through the posterior parts of the Xth motor nucleus, just behind the facial lobe, suggested that the more caudal regions of the central nervous system were inessential. The fish center thus appears to extend over a wide area of the medulla below the cerebellum and facial lobes. No recognizable subunits, either anatomical or functional, were located in the area.

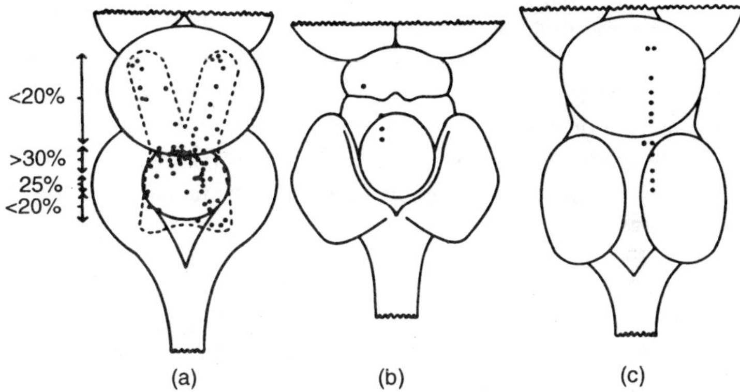

Figure 5.22 : Respiratory regions of the teleost brain: (a) dorsal view of the tench medulla showing the regions where successful electrode insertions were made. Broken line delimits the proposed respiratory area. The percentages of total electrode insertions which were successful at various levels within this area are indicated, and give some idea of the relative density of respiratory neurons. The regions from which activity has been recorded in (b) the carp medulla and (c) the goldfish medulla are given for comparison.

These general conclusions are supported by a number of different investigations in which electrical activity was recorded in the brain of teleosts. All the activity was recorded from the area outlined above, although the search was often extended over much wider areas of the brain. The results of the recording experiments are also in quite good agreement as to the more detailed location of active site. The motor neurons within the nuclei of the Vth, Vllth, IXth, and Xth cranial nerves were frequently found to be active. This is not surprising since these are the neurons whose axons innervate the musculature of the head and gill arches. It is perhaps worth making the point that motor neurons are not usually considered to be part of the respiratory center in mammals, to some extent because they are anatomically distinct from the medullary center. The work of Eccles *et al.*, Sumi, and Sears has shown that integrative processes go on in the spinal cord, down to the level of the motor neurons, so that there is only slight justification for making a distinction on functional grounds. The central respiratory drive potentials recorded intracellularly in spinal motor neurons are almost certainly derived directly or via interneurons from activity within the medullary center. To the extent that the latter is independently rhythmic it is therefore possible to treat it separately.

In the teleosts examined, rhythmic discharges of a respiratory nature also occurred in the smaller cells of the motor nuclei and in the reticular cells around the nuclei. There was evidence that the active cells were not uniformally distributed throughout the respiratory area. In general, the region which was most likely to result in a successful insertion of the searching electrode was immediately behind the cerebellum slightly to either side of the midline. If the search was carried on anterior or posterior to this region, the chances of making a successful electrode insertion were decreased. This could account for some of the discrepancies in different investigators' results. The full

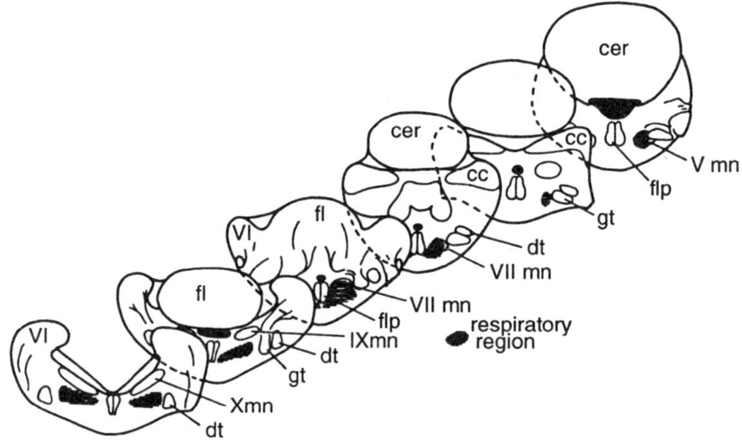

Figure 5.23 : Distribution of active respiratory sites within the tench medulla. Series of cross sections, taken from just behind the facial lobe and running forward to the midcerebellar region, with the respiratory regions indicated. cc, crista cerebellis; cer, cerebellum; dt, descending root of Vth cranial nerve; fl, facial lobe; lip, fasciculus longitudinalis posterior; gt, secondary gustatory tract; mn, motor nuclei of cranial nerves; vl, vagal lobe.

extent of the area defined would depend very greatly on the number of electrode insertions made and, if the distribution varies between animals, on the number of fish studied. There was, in fact, reason to suggest that the respiratory area was not constant from fish to fish or in the same fish at different times. For example, it was often the case that activity once located could not be relocated by positioning the electrode in the same region at a later time. Von Baumgarten and Salmoiraghi suggested that the extent of the area depended upon the depth of respiration.

Certainly there is good evidence that the behaviour of neurons can alter with this type of variation. Electrodes chronically implanted in the brains of tench or African perch detected changes both in activity, and in the number of units active, when breathing was stimulated by lack of oxygen or high levels of carbon dioxide. It seems very likely that the number of neurons, if not the general area they occupy in the medulla, reflects changes in ventilation, both tending to increase or decrease together.

There is very little information available on the central coordination of respiration in elasmobranchs. The early work of Hyde and Springer, based on ablation and transection experiments, has not been extended. Hyde believed that independent segmental units associated with the VIIth, IXth, and Xth nuclei existed in the skate medulla but this was questioned by Springer. Both workers showed that the medulla was capable independently of coordinating rhythmic breathing, although Springer claimed that some supramedullary influences were important. Satchell's more recent transection experiments demonstrated that the mid- and forebrains could be removed without causing marked change in the breathing pattern of the dogfish. After transection the medulla was more susceptible to reflex inhibition, but in the absence of reflex effects the medulla maintained rhythmic activity.

The Functional Organisation of the Respiratory Center

It has been assumed in the foregoing account that the population of respiratory neurons in the fish medulla is a large one. The fact that a number of widely separate, active sites with several neurons at each site can be found in a single fish suggests that this is so. Since the population is a changing one, however, it is very difficult to reach an accurate estimate of its size. Indeed, on this evidence it could be quite small if one is prepared to accept rapid shifts from one locus to another. Fortunately, it is possible to show that there is a little more stability than this and some units can be held for hours, or even days, without showing substantial change. Furthermore, it is possible to destroy respiratory neurons around the tip of a recording electrode by electrocoagulation and for rhythmic breathing to go on normally. Localised coagulation of this type was never effective in bringing breathing to a halt, other than momentarily, even though as many as five such sites were destroyed in a single fish. This argues in the favor of a large rather than a small population.

Intrinsic Rhythmic Activity in the Respiratory Centre

The early work of Adrian and Buytendijk is often cited as demonstrating the autonomous nature of the fish center. They recorded slow potential changes from the medulla of a completely isolated goldfish brain. The similarity in frequency between these potentials and breathing movements of an intact fish was slender evidence for the former's respiratory function. More recently, Schade and Weiler also detected slow potentials on the surface of the goldfish medulla. In the intact animals they occurred in rhythm with the breathing movements and persisted after the fish had been paralysed by Intocostrin injection. Although the precise nature of these slowly changing surface phenomena is still in doubt, their connection with some aspect of respiratory activity seems now to be well established. Using needle electrodes, Hukuhara and Okada demonstrated autonomous activity more directly within the cells of catfish and carp respiratory centers. The rhythmic bursts of action potentials, which identified the cells as respiratory in the intact fish, continued after medulla isolation by nerve and brain transection.

As further evidence of the center's independence of mechanoreceptor feedback, a number of workers have described the effects of paralyzing doses of neuromuscular blocking agents such as curare, succinylcholine, and gallamine. In such preparations not only can rhythmic proprioceptive feedback be eliminated but also the ventilation can be controlled entirely by the experimenter. There seems to be little doubt that the respiratory neurons of mammals continue to fire rhythmically after paralysis of the musculature, although some workers have shown a change in the pattern of activity. In fish there seems to be more variation. Von Baumgarten and Salmoiraghi sometimes recorded an unaltered discharge and sometimes a slight increase in the frequency of rhythmic bursts from respiratory cells of goldfish after succinylcholine paralysis. Satchell found a similar increase in the frequency of bursts in efferent fibers from the dogfish center after administration of curare. Precisely opposite results were described by Serbenyuk working on carp. Paralysis of the respiratory muscles produced either by curare injection or division of the motor nerves at first caused no change in respiratory neuron activity. After a few minutes the normal rhythmic pattern disappeared and was replaced by activity of a different type at a much lower frequency. Serbenyuk suggested that sensory activity from a variety of sources could affect cells of the center by raising both their general excitability and their frequency of oscillation. Removal of any one of his "reflexogenic zones," such as muscle proprioceptors, had a depressant effect on the center more or less equivalent to the removal of any other of the zones, such as gill or swim bladder receptors.

These results on immobilised fish are not easy to interpret for several reasons. They depend on a pre- and postparalysis comparison of activity from a single recording site within the medulla. It has therefore been difficult to decide whether the changes are representative of the center as a whole. Ballintijn has recently described three categories of response in respiratory neurons after gallamine induced paralysis. He found that some cells continued to fire rhythmically, some stopped firing completely, and a third category stopped but could be started again by appropriate, but not necessarily rhythmic, proprioceptor stimulation. Clearly the relationships that different cells have to the central oscillator and to the proprioceptive feedback are quite variable. Moreover, immobilisation of an animal affects many systems in addition to the sensory feedback from muscles. For example, the pattern of

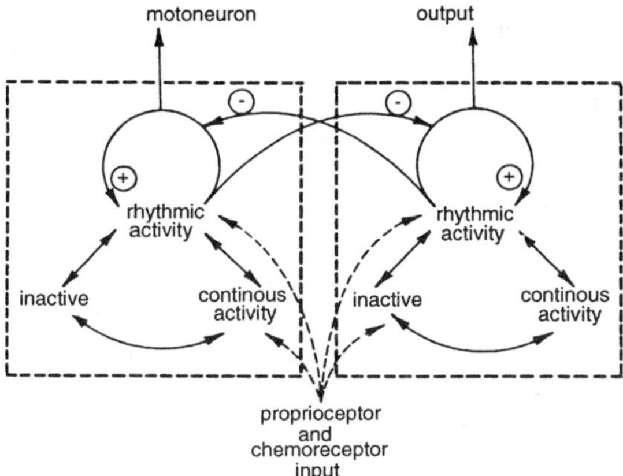

Figure 5.24 : Diagram showing a hypothetical functional arrangement of the respiratory center. Two populations of rhythmically active neurons are shown, each population being self reexciting and having an inhibitory action on the opposing population.

water flow through the gills is obviously changed, as is the level of oxygen and carbon dioxide in both gills and blood. Shelton has done some preliminary work to examine the effect of such changes in tench paralysed by curare or succinylcholine. No respiratory units were found in the medulla that were not susceptible to alteration in the flow of water through the gills from a cannula placed in the mouth. Some discharged rhythmically when the flow was stopped, others only when it was restarted. Usually neurons were silent when the rhythmic activity was not produced, although occasionally a continuous discharge was seen. A smooth transition from rhythmic to nonrhythmic state could usually be produced by appropriate adjustment of flow. The influence of different gas tensions in the water has not been fully examined. Although there is little doubt that rhythmic discharges can be detected in the paralysed fish, the evidence that the functional integrity of the respiratory center is preserved is not convincing. In fact, the reverse seems to be the case in that different parts behave in quite different ways. It would be interesting to know how this variation affects motor activity to the respiratory musculature.

The Production of Rhythmic Activity

A satisfactory theory cannot yet be produced to account for the generation of the normal breathing rhythm. A single pacemaker neuron or neuron group cannot be responsible since breathing cannot be stopped by destruction in any particular region. There must therefore be interaction between a number of dispersed respiratory neurons. Since two fairly distinct neuron populations have been described in mammals and in fish, it has been suggested that the interaction takes the form of alternating and reciprocally inhibiting discharges between the two groups. A parallel can be drawn with the two triodes in a multivibrator circuit: when one becomes maximally conducting the other is cut off. Each group is thought to be self reexciting so that activity increases in individual neurons and spreads throughout the group. In order to explain how the balance alternates it must be supposed that activity is in some way self-limiting; in the multivibrator analogy there must be some phenomenon which corresponds to the extreme positions of the triodes and to the discharge of the coupling condensers setting off the opposing cycle. In some way the active group must reach saturation and its activity thereafter decline. It is possible to suggest a number of ways in which this could happen. Salmoiraghi and von Baumgarten described a small but progressive accommodation of the postsynaptic membrane in both inspiratory and expiratory neurons of the cat. As a burst of activity proceeded, slightly greater depolarisation of a neuron membrane was required to initiate an action potential. If all the cells of an inspiratory or expiratory group behaved in this way there would be a substantial fall in excitability throughout an activity cycle and eventually it would be brought to an end. These observations, which were made on very few cells because of the difficulty of impaling respiratory neurons with intracellular electrodes, do not exclude other ways of terminating an inspiratory or expiratory discharge.

These suggestions are extremely speculative and raise a number of difficulties. In the mammal there is evidence conflicting with that given above to show that rhythmicity arises, not by an equivalent interaction of two opposing neuron groups, but by periodic inhibition of a primary, tonic activity occurring in the inspiratory neurons. Centers other than the expiratory and inspiratory groups of the medulla are thought to be involved in these processes. In fish there seem to be no neuron groups other than the two medullary ones, but one of these groups (that which fires as the mouth and opercula are adducted) is numerically very much the dominant one. It may be that two opposing groups are unnecessary for oscillation; certainly there is no a *priori* reason why a single neuron group should not show alternating periods of activity and quiescence. At some level of integration bursts of activity, which do alternate in a general way, must be produced since the discharges from motor neurons are known to be of this nature. None of the suggestions made so far accounts satisfactorily for the extensive variation in behaviour occurring in different neurons in both the normal and the immobilised animal. This variety is particularly difficult to explain in a scheme which assumes two fairly uniform and opposing populations. Some of the components in the oscillator must be different to the extent that they can become silent or continuously active under some conditions without causing complete disintegration of the rhythmic system.

THE ROLE OF MECHANORECEPTORS IN RESPIRATORY REGULATION

Although the evidence reviewed above suggests that the respiratory center can function in a rhythmic manner in the absence of feedback from mechanoreceptors, there is no doubt that they have a significant part to play in the normal fish. In elasmobranchs it has been known for some time that the breathing rate is in some way related to the volume of water which is allowed to flow over the gills from cannulae in the mouth or spiracles confirmed these results and showed that the response was produced as a result of mechanoreceptor stimulation by the movements of ventilation. It was totally independent of the oxygen content of the inspired water.

Sharks decrease their breathing rate by prolonging the period of oro- and parabranchial expansion and extending the pause after this inspiratory period; the opposite occurs when the breathing rate goes up. Satchell showed that stimulation of the branchial branches of the IXth and Xth cranial nerves, or the prespiracular branch of the VIIth, stopped or slowed breathing by increasing the duration of the pause. Inflation of a balloon inside the pharynx produced a similar result. It was suggested that these experiments demonstrated the existence of inhibitory pathways to the respiratory center in the branchial nerves. It also seemed clear that the pathways originated in pharyngeal mechanoreceptors, and support for this came from recordings of sensory discharges in the branchial nerves during contraction of oro- and parabranchial cavities. Satchell and Way examined activity from the pharyngeal mechanoreceptors in more detail. They were found on the branchial processes lining the inner opening of the gill slits and were slowly adapting sense organs with both static and dynamic characteristics. The effective stimulus was thought to be displacement of the pharyngeal wall rather than water flow itself. The branchial processes were thus deflected by being pressed against the gill bar in front. Satchell and Way suggested that the receptors were involved in a breath by breath regulation of breathing rate, the duration of the inspiratory period and the following pause being determined by the level of activity during the preceding expiratory period. Thus in an almost completely curarised animal the breathing rate became largely independent of imposed ventilation volume, suggesting that the mechanoreceptors were not excited by abnormally small breathing movements. In the normal animal, however, the decrease in rate as ventilation was progressively restricted resulted from greater excitation of the mechanoreceptors because the amplitude of breathing increased. The original observations of the relationship between water flow and breathing rate must now be explained in terms of a change in amplitude of breathing. The receptors and pathways involved in this response are not known. The inhibitory reflex is unusual in that the receptors are excited during orobranchial contraction and decrease breathing rate by delaying the onset of the next expiration phase after a longer inspiration and pause.

Section of the branchial branches of the IXth and Xth cranial nerves in the shark caused an increase in breathing rate; inspiration was quicker and the pause disappeared. This correlates well with the proposed inhibitory role of the mechanoreceptors. In teleost fish the effects of IXth and Xth nerve section are not so well established. Shelton showed that the immediate change caused by sectioning the nerves in tench was an increase in amplitude of the breathing movements. The nature of the receptors and the function of the afferent pathways involved in this response were not defined. Powers and Clark had earlier done similar experiments and had found that breathing failed after section of nerves IX and X. The gill blood vessels were also sectioned in their experiments. Effects other than the direct ones of nerve section may therefore have been complicating the results, although the respiratory center seems to be quite resistant to circulatory failure. Serbenyuk *et al.* and Serbenyuk found that rhythmic activity persisted in the center after section of the branchial nerves or novocaine anesthesia of the gills in carp, although there was a considerable change in breathing pattern. There can be little doubt that important afferents, probably from mechanoreceptors as well as other sense organs, are carried in the branchial nerves of teleosts although their role cannot at present be precisely defined.

THE CHEMICAL REGULATION OF RESPIRATION

The metabolic rate in fish, as measured by oxygen consumption, can be influenced by a large number of internal or external factors. Temperature and size, pollution of various types, season, and activity have been examined in some detail. Although they all change the animal's demand for oxygen there is no evidence, except perhaps in the case of changes in activity, that they are directly involved in the modifications of ventilation and perfusion which result in the demand being met. It seems likely that factors such as these operate primarily, although not exclusively, at the level of the cell. Other changes, most notably of oxygen concentration, carbon dioxide concentration, and pH level in the ventilation stream and/or the blood, also influence oxygen consumption but in a more complex way. All three factors may be involved to varying degrees in mechanisms important in the regulation of gas exchange. The degree of internal change in the oxygen, carbon dioxide, or pH levels as a result of environmental fluctuations in these parameters would therefore depend on the characteristics of the total regulatory system. Variations in metabolic rate in this case would be a function of the metabolic cost of regulation, the overall efficiency of regulation, and ultimately the direct effect of any changes at the tissue level, although these would be small if the regulation were efficient. It is probable that regulatory systems of this general type, sensitive to respiratory gas concentrations within the body, and perhaps in the environment as well, would be directly involved in the responses which cope with the respiratory requirements of the cells under a wide variety of different conditions, including those of temperature, size, season, and activity mentioned above.

Oxygen

Effect of Oxygen Concentration on Oxygen Consumption and Gill Ventilation

Respiratory Dependence

Progressive depletion of oxygen in the water breathed by a fish at first produces no change (respiratory independence) but eventually leads to a greater and greater restriction of the animal's oxygen consumption

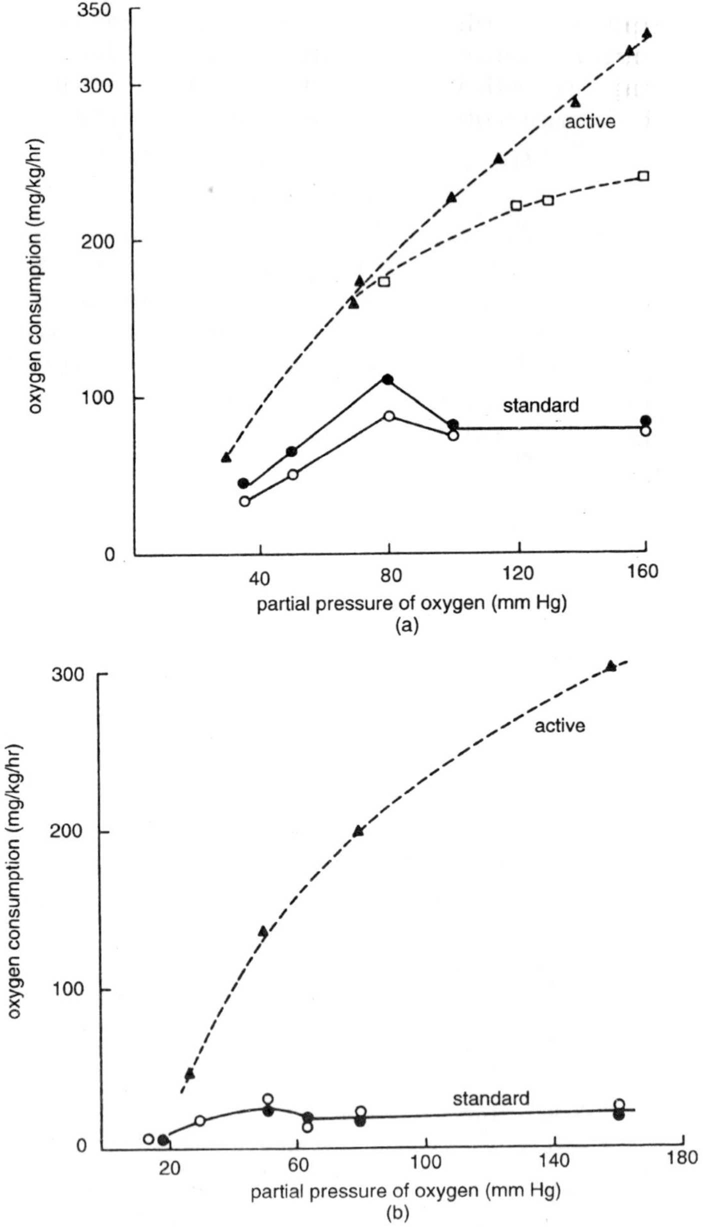

Figure 5.25 : The effect of changes in partial pressure of environmental oxygen on standard and active oxygen consumption of fish.

(respiratory dependence). Activity is usually included in the concept of respiratory dependence thus increasing its ecological significance by giving some insight into the equilibrium between environment and active fish. Determinations can be made over the whole range of activities shown by the animal, but for comparative purposes the maximally active and the standard (resting) level are most useful pointed out that a decline in oxygen consumption of a maximally active fish would indicate a similar decline in activity if the animal were to remain in oxygen balance. This would be a progressive effect from the "incipient limiting level" at which reduced oxygen concentration just produced a change to the "level of no excess activity" where oxygen consumption equaled the standard rate. In the case of a resting animal, independence should be maintained at a steady level over a wider range of oxygen concentrations, and at lower levels the consumption might be expected to go up as the work of ventilation increased. Fry suggested that the cost of ventilation could continue to rise until the standard oxygen consumption equaled the reduced active rate.

It has been widely accepted that limitations in the gas exchange system are largely responsible for the appearance of respiratory dependence. There are, however, a number of other factors which will complicate a possible relationship of this type. Rates of oxygen consumption, both at the active and standard levels, are different for different species of fish. Suggestions have been made that the species having the highest rates are also those which show dependence in the highest concentrations of oxygen so that the onset of dependence is related to the ability of the fish to be active. Determinations of active oxygen consumption made for any particular species by different authors show considerable variation, and it is difficult to substantiate this suggestion, reasonable though it seems. The standardisation of activity levels in the experimental animals constitutes a major difficulty in these determinations. Another difficulty is that fish such as the dragonet, *Callionymus,* and the toadfish, *Opsaunus,* which are inactive, bottom-living forms, show dependence at oxygen values only just below air saturation.

Anaerobiosis and acclimation to different oxygen tensions may also affect respiratory dependence curves. Beamish demonstrated that at very low levels of oxygen a fish's standard rate of oxygen consumption increased as predicted, but never to the active level. Furthermore, the increase was usually less in animals acclimated to the low tensions concerned than in those acclimated to air saturation values. Beamish concluded that anaerobic respiration occurred in reduced oxygen concentrations. He was able to show that fish which were active to the same extent consumed less oxygen at lowered tensions than they did in air-saturated water. Fish acclimated to low oxygen tensions not only had lower standard rates of metabolism at those tensions but also showed reduced consumption for any level of activity when compared with their fellows acclimated at air saturation values. The acclimation appears to have some influence on the extent to which a fish can indulge in anaerobic respiration. The accumulation of lactic acid in the muscles and bloodstream of trout and sunfish during exercise and hypoxia and its subsequent return to normal during a prolonged period of elevated oxygen consumption are well established. Blazka produced evidence for anaerobiosis during hypoxia in the crucian carp with no oxygen debt or accumulation of lactate.

Ventilation Volume and Utilisation

Although the factors outlined above complicate the relationship between oxygen concentration, activity, and respiratory exchange, it is very likely that the gills and ventilating mechanism do tend to become limiting during respiratory dependence. The failure may result from deficiences in the measurement and control systems or from physical limitations on the part of the respiratory pump and exchanging surfaces. The augmented volume of water must be sufficient not only to cope with the reduced oxygen content but also to compensate for the increased work involved in ventilation and the reduced utilisation of oxygen in the ventilation stream. Indeed, it has been argued that the increased metabolic cost of breathing and the fall in utilisation may together determine the effective upper limit for gill ventilation rather than the capacity of the pump itself. This suggestion was based on an unsubstantiated estimate of ventilation work and an extrapolation of van Dam's data from a single trout recovering from exercise in air-saturated water. The fall in utilisation in this case could very well lead to a maximum oxygen intake being reached at ventilation volumes 8-10 times the resting level, but the conclusion rests precariously on results which van Dam himself regarded as a "first orientation."

At this point it may be useful to digress briefly and discuss the measurement of ventilation and utilisation, their usefulness in assessing respiratory function, and their relationships to one another and to other parameters which have been used in respiratory studies on aquatic animals. Ventilation has been measured by three methods, all of which have inherent faults. The first method depends on separating inspired from expired water, usually by using a rubber membrane stretched

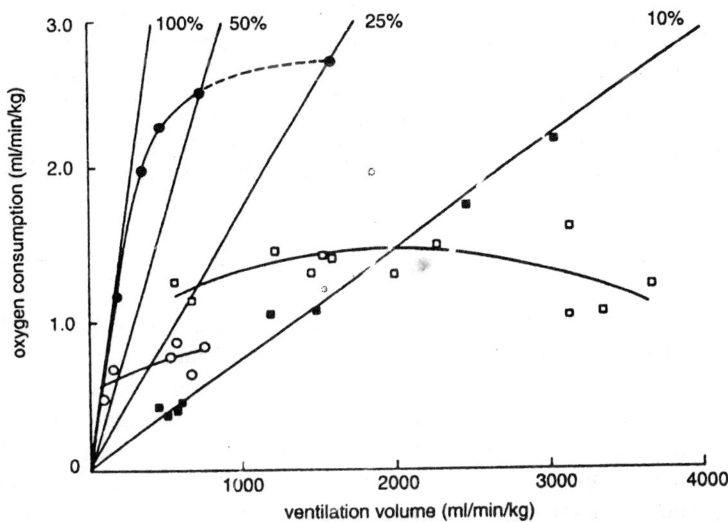

Figure 5.26 : Relationship between oxygen consumption and ventilation volume in fish under different conditions. (●) Trout recovering from exercise in aerated water at 12°C. The lines indicate relationships between ventilation and oxygen consumption which would exist in fully aerated water at this temperature for values of utilization of 100, 50, and 25%. van Dam's results are extrapolated to the 25% utilization line as explained in the text. (O) Eel during hypoxia at 18°C. (■) Trout before, during, and after a 10-min period of exercise in aerated water at 5°C. The 10% utilization line is drawn for fully aerated water at this temperature. (□) Trout during reduction of environmental oxygen to a partial pressure of 30 mm Hg at 15°C.

across part of the head or glued over the external gill opening. Using suitable siphons the expired water can be collected for measurement. The method has the disadvantage of restricting the fish and placing a load on the muscles of the branchial pump. A considerable effect of this type of system on the oxygen consumption of *Callionymus* has been demonstrated by Hughes and Knights. Ventilation can be calculated from measurements of the oxygen consumption of a fish if the oxygen content of the inspired and expired water is also known. The method is suitable for work on freely moving fish from which samples can be taken through buccal and opercular catheters. Inaccuracies will occur if the samples are not representative, and relatively small changes in oxygen concentration in considerable volumes of water are not easily measured. A dye dilution technique has been described by Millen *et al.*. A bolus of dye is injected from a syringe into the buccal or orobranchial cavity, and continuous sampling from opercular or parabranchial chambers into a densitometer gives a dye dilution curve. From the downslope of this curve the ventilation volume can be calculated. For this technique to be accurate, the injection and sampling sites must be suitably located and the mixing of the dye, once injected should be complete and preferably instantaneous. Millen *et al.* achieved good mixing by implanting a stirrer in the orobranchial cavity.

Utilisation (U) is the ratio between oxygen extracted from the ventilation stream and oxygen contained in the inspired water, often expressed as a percentage:

$$U = \frac{P_{I_{O_2}} - P_{E_{O_2}}}{P_{I_{O_2}}} \times 100$$

where P is the partial pressure of gas, in I the inspired water, and E the expired water. Its greatest use is probably in examining ventilation and ventilation control because at known oxygen concentrations it relates oxygen consumption and ventilation in a direct way without involving the determination of any parameters in the circulatory system. The relationship between ventilation volume and utilisation is not a simple one. Thus the exercising trout maintained a steady utilisation of 10% and showed a progressively increasing oxygen consumption as ventilation increased in air-saturated water. The hypoxic trout also increased its ventilation over a similar range, but it had a higher overall utilisation falling to about 20% and eventually may have shown a declining uptake of oxygen as the environmental oxygen fell. Although utilisation usually falls as ventilation increases, the extent of the fall depends on the factors responsible for the ventilation change and the way they affect the oxygen requirements of the animal and the oxygen content of the medium. Saunders measured utilisation in fish which were exercised or exposed to low levels of oxygen in the presence or absence of carbon dioxide all of which affect oxygen requirement and/or oxygen content differently. The utilisation and ventilation changes were also different as his results.

It is clear that utilisation will be affected by any change which alters the characteristics of the total exchange system. Thus a falling oxygen concentration, in the absence of compensatory changes, has an adverse effect on utilisation because the oxygen diffusion gradient goes down at the same time as increased water velocity reduces diffusion time and probably alters gill geometry. In fact, it is extremely likely that some compensation does occur as ventilation increases, thus making the relationship between ventilation and utilisation even more complicated. A measure of the gas exchange capacity of the respiratory surface is given by the transfer factor which is the constant of proportionality relating total gas exchange and the mean partial pressure gradient between water and blood. Randall *et al.* have shown that the transfer factor in trout increases gradually during hypoxia, and to a lesser extent during exercise, so that the gas exchange capacity of the gills must also be increased.

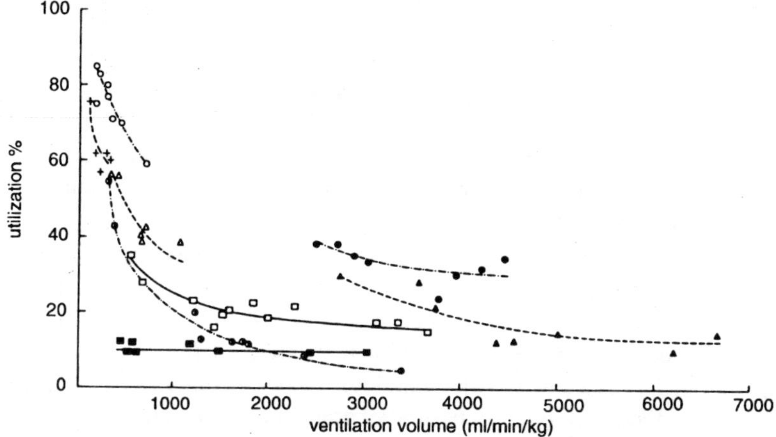

Figure 5.27 : Relationship between utilisation and ventilation volume in fish under different environmental conditions.

Although utilisation is determined by a combination of ventilation volume and the gas exchange properties of the gills, it is so all-embracing as to say little about the details of the exchange. For example, it takes no account of the role of the bloodstream in gas exchange. In this respect effectiveness (E), which is the ratio between actual and maximum possible gas transfer in a countercurrent exchange system, can give a more useful measure. Effectiveness can be calculated for oxygen with reference to the transfer either from water or into blood, and similar figures can be obtained for carbon dioxide moving in the opposite direction. The relationship between utilisation and the effectiveness of oxygen removal from the ventilation stream would be

$$E_{W_{O_2}} = \frac{\dot{V}_{O_2}(\text{actual})}{\dot{V}_{O_2}(\text{maximum})}$$

$$= \frac{\dot{V}_G}{\dot{V}_G} \cdot \frac{\beta_{O_2}}{\beta_{O_2}} \cdot \frac{\left(P_{I_{O_2}} - P_{E_{O_2}}\right)}{\left(P_{I_{O_2}} - P_{\bar{V}_{O_2}}\right)} \times 100 = U \frac{P_{I_{O_2}}}{\left(P_{I_{O_2}} - P_{\bar{V}_{O_2}}\right)}$$

where P is the partial pressure of gas in I (inspired water), E (expired water), and $\bar{V}$ (mixed venous blood; V_G is the volume flow of water over the gills in unit time, $\dot{V}_{O_2}$ the oxygen uptake in unit time, and

(3O_2 the solubility coefficient of oxygen in water. In this case transfer which is 100% effective can be achieved by a system having no water shunt, a very high transfer factor (low diffusion resistance), and any countercurrent arrangement which brings venous blood and outgoing water into contact. If the effectiveness of oxygen uptake by the blood is also to be 100%, the requirements are more stringent still, since a perfect countercurrent arrangement then becomes necessary. In addition, the flow of water and blood must be appropriately balanced so that outgoing blood is either in equilibrium with incoming water or fully saturated. For complete utilisation the same conditions will obtain as for effective withdrawal of oxygen from water, but in addition the venous blood must be devoid of oxygen.

The data at present available do not allow firm conclusions to be reached about the effective upper limit of ventilation which would restrict the availability of oxygen to the body. If the needs of the whole body, except for the respiratory musculature, are considered then falling utilisation, increased work of breathing, and limitations in the stroke volume and rate of operation of the branchial pump all operate to determine that upper limit. Little is known of the work factor which in any case will differ in moving and nonmoving fish. Even if a simpler view is taken and the changes in the work of breathing are ignored the situation is still far from simple. Since ventilation-utilisation relationships are variable, the balance between factors will change with conditions. In none of the experiments in which hyperventilation was the result of low environmental oxygen did utilisation decrease more rapidly than ventilation increased, as would have been the case if utilisation alone had been limiting. In order to maintain oxygen consumption at a steady level it is necessary for the ventilation volume to go up at the same rate as the product of oxygen concentration and utilisation declines. This just failed to be the case in the trout examined by Holeton and Randall which showed signs of dependence below oxygen tensions of 80 mm Hg. It is significant that ventilation was not maximal but increased almost by a factor of two as the oxygen concentration was decreased further. The inference from this is that the control systems may also participate in the failure to maintain oxygen availability before the pump has reached its maximum capacity.

Coordination of the Responses to Changes in Oxygen Concentration

The evidence cited above shows that in the majority of teleosts examined depletion of oxygen in the medium results in an increased ventilation volume over a wide range. Other reactions, for example, avoidance behaviour, have also been described in a number of teleosts; but it seems likely that the respiratory response is the primary one. The situation is not so clearly resolved in the case of elasmobranchs. According to Ogden a decrease of environmental oxygen produced no change in the ventilation volume of the dogfish, *Mustelus,* although the animal struggled violently under these conditions. Satchell showed that perfusion of dogfish, *Squalus acanthias,* gills with completely deoxygenated water caused only a very slight increase in amplitude, together with a decrease in the rate of breathing. Because of the extreme nature of the stimulus this work did not rule out the possibility of more appropriate ventilation responses to moderate anoxia. However, the more recent work of Hughes and Umezawa showed only a small increase in breathing rate in *Scyliorhintis canicula* during a gradual reduction of oxygen tension in the inspired water. Similarly, Baumgarten and Piiper found only small changes of ventilation volume in *Scyliorhinus stellaris* under these conditions. Because simple experimental procedures have so far been ineffective in causing substantial and reproducible modifications of ventilation in elasmobranchs very little is known of its range and control. There is obviously not a fixed ventilation volume; Baumgarten-Schumann and Piiper reported marked changes in resting dogfish and commented that the animals were not nearly at their maximum ventilation levels. They attributed the ventilation changes they observed to repayment of an oxygen debt after excitement and activity. Whether some product of metabolism constitutes the effective stimulus to ventilation change in elasmobranchs cannot be decided on present evidence. It is possible that the level of activity in the animal is in some way directly coupled to breathing, but this cannot be the whole explanation since an inactive fish shows variations in ventilation. The whole field of respiratory regulation in elasmobranch fish is poorly understood and clearly requires closer study, particularly on freely breathing preparations.

In those fish in which a fall in oxygen concentration does result in increased ventilation, the environmental changes must influence the respiratory center in such a way as to cause changes in the motor output and increase the rate and/or the amplitude of breathing. The relationship between rate and amplitude has been shown to vary considerably from species to species. Very little is known of the underlying neurophysiology directly responsible for these alterations in breathing. Modifications in the oscillatory activity of cells making up the respiratory center are obviously produced by incoming sensory information related to environmental change. The receptors for this information could be in one or more

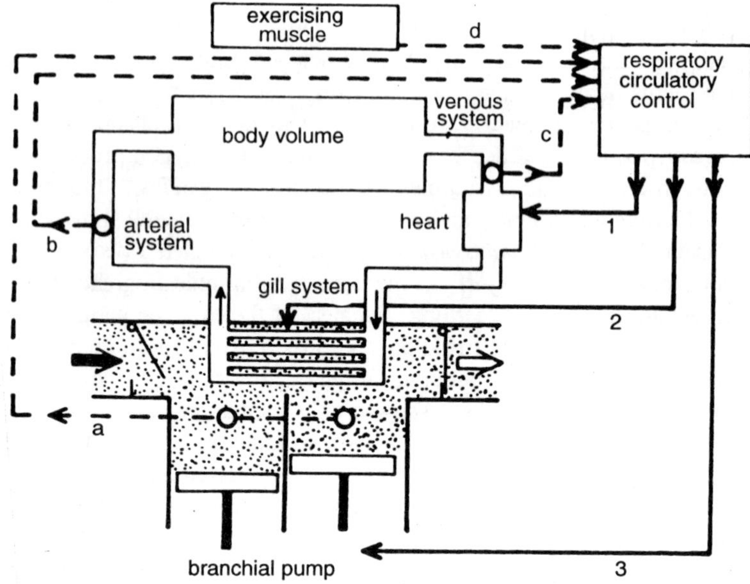

Figure 5.28 : Diagram of the peripheral relationships of control systems regulating ventilation, perfusion, and gas exchange in fish. Three possible receptor sites for gas tension changes are shown, located in (a) ventilation stream, (b) arterial, and (c) venous blood as suggested in the text. Receptors detecting exercise levels (d) may also be important. Information provided by these sensory channels may lead to the modification of (1) cardiac output, (2) functional gill area, and (3) ventilation volume.

of three general regions:

(1) On the outside of the animal, on the walls of the pharynx and the surface of the gills. They would detect oxygen concentrations in the water of the environment or ventilation stream.

(2) On arterial blood vessels containing blood after oxygenation in the gills or in tissues closely associated with arterial blood. They would detect oxygen concentration in the bloodstream.

(3) On blood vessels carrying venous blood or in tissues closely associated with venous blood. They would detect oxygen concentration in the blood after it has passed through the tissues of the body. This region would include the ventral aorta and afferent branchial arteries, taking in sites which are homologous with those where carotid and aortic glomi are found in the mammal.

It is not easy to decide either experimentally or theoretically which of these possible sites is in use in the fish. Satchell, working on dogfish, and Randall and Smith, working on rainbow trout, claimed that oxygen receptors were to be found in the buccal cavity or on the gills. The evidence in both cases was that ventilation with oxygenpoor water caused a reflex slowing of the heart which was still seen after the ventral aorta had been clamped. The receptors detecting the low oxygen concentration must have been directly available to the ventilation stream since no blood could flow to more remote sites. Hughes and Umezawa also proposed that the receptors for bradycardia induced by oxygen lack were located on the gills. In addition, Hughes and Ballintijn thought that the increased breathing amplitude during hypoxia resulted from activation of receptors which detected oxygen concentrations in the inspired water. De Kock has described "end buds" in the buccal cavity and on the gills of trout and has suggested that they have a chemoreceptor function. However, control based solely on detection of environmental oxygen would entail a precise relationship between different concentrations of oxygen and the corresponding ventilation volumes with no feedback adjustment. Even supposing that this relationship could be modified appropriately at different levels of exercise, it would be unlikely that such a system could give stable control of oxygen supply under all conditions, especially with varying gill geometry and perfusion.

It is probable that such stability can only be conferred by a control system with receptors in the bloodstream, and the situation in mammals argues strongly for some such system. If the receptors were located in the vessels of the gills, control would be susceptible to local variations in geometry and perfusion rates. Integration of activity from receptors throughout the gills would overcome this difficulty. More satisfactorily, the receptors could be situated downstream from the gills so as to measure oxygen concentration in mixed arterial blood. The pseudobranch has frequently been considered as a region where receptors might occur. Indeed, Laurent demonstrated a sensitivity to changes in oxygen and carbon dioxide concentrations by recording from the nerve of an isolated and perfused tench pseudobranch. This does not mean that all the receptors need be located in a single site. Shelton showed that bilateral section of the IXth cranial nerves did not abolish the ventilatory response to oxygen lack although it was considerably modified. This experiment was repeated with section of the Xth cranial nerves and IXth and Xth nerves together, and in all cases the fish was able to respond to low oxygen levels. There is thus clear evidence that oxygen receptors exists outside the areas innervated by the IXth and Xth cranial nerves.

A suggestion that the receptor system could be located in the venous blood system was made by Taylor *et al.*, who based their work on an analysis of the equilibria between environment and blood under a variety of conditions. These workers described a digital computer simulation of the gas exchange system in trout, using the results of Randall and his collaborators. Equations were derived to describe oxygen transport and storage in all parts of the system, together with three feedback equations which controlled ventilation volume, heart output, and gill area. All the equations could be simultaneously solved for blood and environmental gas tensions, ventilation volume, and heart output, by using different input driving values for metabolic rate or inspired oxygen tension. Realistic simulations could not be achieved if arterial or venous carbon dioxide tensions were used as reference factors in the feedback equations. Equally, the use of arterial oxygen tensions produced responses in the model which did not correspond with those of the fish. However, if venous oxygen tension was used as a reference to control ventilation, blood flow, and gill area, the model responded in a way which resembled the resting and exercising trout and had a qualitative similarity to the hypoxic animal. The model has some shortcomings, for example, in relating cardiac output directly to ventilation, but in spite of these it is reasonably consistent with experimental evidence. It is not yet clear that the venous location for the receptor system is justified. The dissociation curve would normally be steepest at venous oxygen tensions, and there may be disadvantages in locating the error detector where the error is smallest.

There are therefore reasons to believe that oxygen receptors may be found in the ventilation stream and in arterial and venous blood. Much of the evidence needs more detailed experimental examination but, if confirmed, points to a control system of greater complexity than hitherto supposed.

Carbon Dioxide

Effects of Carbon Dioxide Concentrations on Oxygen Consumption, Gill Ventilatyion, Gill Ventilation, and Utilisation

The precise role played by changes in carbon dioxide concentration, either in the environment or within the body of a fish, in regulating breathing is not yet understood. Various effects of increased environmental carbon dioxide have been described. At very high levels the gas is toxic and completely inhibits oxygen consumption. Below these levels fish survive for varying periods of time, although the ability of an animal to extract oxygen from the environment is reduced to an extent which depends on the amount of carbon dioxide present or on hydrogen ion

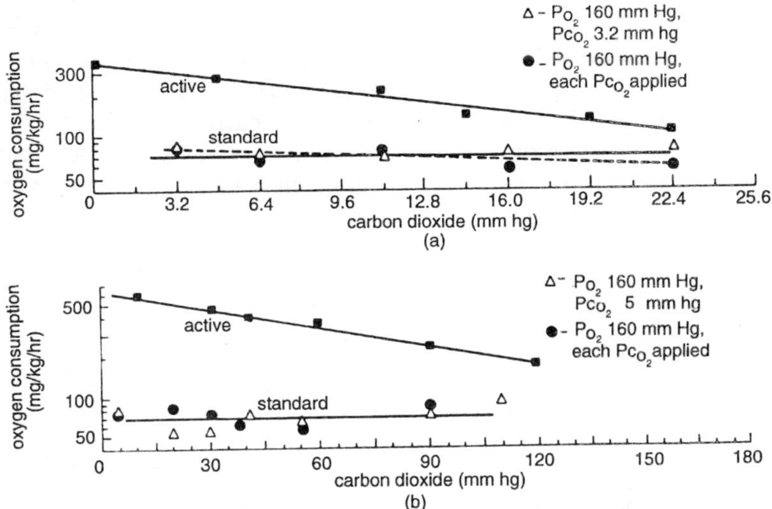

Figure 5.29 : Standard and active rates of oxygen consumption in (a) brook trout at 10°C and (b) carp at 25°C in relation to elevated carbon dioxide tensions in the environment.

concentration. Basu found that the rate of oxygen consumption by active suckers, bullhead, and carp fell as carbon dioxide levels increased, the logarithm of the consumption rate being linearly related to carbon dioxide tensions up to 50 mm Hg. The same linear relationship was found for trout, using carbon dioxide tensions up to 25 mm Hg. The results on carp were confirmed by Beamish, who also examined the effects of similar ranges of carbon dioxide on the standard rates of oxygen consumption. He found that carbon dioxide usually produced no significant change in the standard metabolic rate.

It is not easy to reconcile these results with those of other workers who examined the direct effect of carbon dioxide on gill ventilation. The early experiments of Olthof and Meyer showed that carbon dioxide stimulated an increase in breathing frequency in minnow, perch, and stargazer. In the tench and trout the amplitude of breathing movements went up, causing significant changes in ventilation volume in both cases. Saunders also described increased ventilation in the sucker, bullhead, and carp but only with high tensions (25-56 mm Hg) of carbon dioxide. At more moderate tensions (15-35 mm Hg) there was an initial increase in ventilation with a later return to normal. The utilisation of oxygen was found to be quite substantially reduced in the presence of carbon dioxide. The rate at which utilisation fell with increased ventilation was also increased in the presence of carbon dioxide, compared with a solely hypoxic stimulus. The effect of these ventilation and utilisation changes on the work of breathing might reasonably be expected to increase the standard rate of oxygen consumption. That they did not do so in Beamish's experiments could be explained by his suggestion that carbon dioxide does not in fact stimulate the truly resting fish to hyperventilate. Certainly many fish become excitable and active if carbon dioxide is introduced into the environment and measurements on resting animals become difficult.

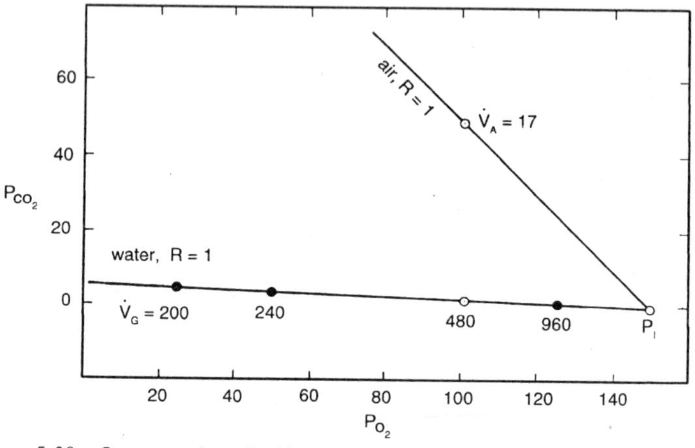

Figure 5.30 : Oxygen-carbon dioxide tension diagram showing the simultaneous oxygen and carbon dioxide tensions which occur when air is breathed (air, R = 1 line) and when water is breathed at a temperature of 20°C (water, R = 1 line).

The significance of these responses can best be examined in the light of some of the physical properties of carbon dioxide. In an important contribution to the theory of aquatic gas exchange, Rahn demonstrated that because of the differences in solubilities of oxygen and carbon dioxide in water the change in carbon dioxide tension between inspired and expired water must necessarily be small. Dejours *et al.* went on to show that carbon dioxide differences were even smaller than Rahn had predicted if fish breathed in well-aerated, high carbonate water, because of the buffering action of the carbonatebicarbonate system. For example, the carbon dioxide tension in water expired by a goldfish at 25°C was increased by less than 1 mm Hg. The relationships vary in different types of water, but in all cases the tensions are low and are not substantially increased by gas exchange at the gills. Carbon dioxide tensions in the tissues and bloodstream of fish would also be expected to be low, and this has been repeatedly confirmed by experiment.

Low tensions in the blood are also attributable to the high permeation coefficient of carbon dioxide. The diffusion gradients between blood and water are very much smaller for carbon dioxide than they are for oxygen. Indeed, Kylstra *et al.* have derived diffusion equations for a gill model from which they suggest that diffusion dead space for carbon dioxide may be negligible; that is to say that persisting gradients of tension within exchange units would be very small, assuming realistic values for diffusion time and distance. This would not be the case for oxygen, however, and Kylstra *et al.* argue that a major difference of this type would relate the rates of oxygen uptake and carbon dioxide elimination volume in quantitatively different ways. The regulation of ventilation and oxygen uptake by reference to arterial tensions of carbon dioxide would therefore be extremely unsatisfactory.

It can be argued that carbon dioxide concentrations as high as 50 mg/liter do occur in some polluted waters and that increased ventilation would assist in reducing the gradient from blood to environment under such conditions. But the gradient is very small in any case so that hyperventilation would not affect the blood concentration significantly. This being so, Randall and Shelton suggested that the response to environmental carbon dioxide was inappropriate. The ventilation volume increased beyond the point that was necessary to satisfy the animal's oxygen demands unless the oxygen concentration in inspired water was simultaneously reduced to very low levels. This superfluous increase in ventilation would account for the large fall in utilisation that was usually attributed to the effect of carbon dioxide on affinity of the blood. It seems more likely that the carbon dioxide response is coupled primarily to increased levels in the tissues or in the venous blood system, changes which would normally signify increased metabolism. The tension changes would be relatively small, and the evidence on responses to low level effects is conflicting. Moreover, a control system of this type could not be universally effective and would, for example, be unsatisfactory during hypoxia. There may be some role for it, however, in conjunction with an oxygen control system such as is known to exist.

Coordination of the Responses to Changes in Carbon in Carbon Dioxide Concentration

Very little is known about the sensory mechanisms involved in the response to carbon dioxide. The relative importance of carbon dioxide tension or pH change as the effective stimulus has not been established. Receptors could be in the three general areas suggested earlier for oxygen receptors, but the arguments above suggest that environmental and arterial blood receptors would have very limited use in respiratory control. Carbon dioxide sensitivity has been shown in the pseudobranch. However, Shelton demonstrated that substantial ventilatory changes to carbon dioxide persisted after total denervation of the gill area by section of the IXth and Xth cranial nerves. Injection of small quantities of saline containing carbon dioxide into some regions of the teleost medulla caused changes in the breathing movements, but the precise nature and location of the stimulated area was not decided.

Respiratory Homeostasis in Resting and Active Fish

The validity of experiments on the effects of changes in environmental gas concentrations on some species of fish is occasionally questioned if that species is only found in well-aerated environments. One justification for such work is that the changes may elicit responses from the same overall system as would be activated by minor internal fluctuations in the resting animal, or major alterations in demand as the animal became active. The results at present available would suggest that homeostasis is achieved in the resting animal by monitoring oxygen tensions in the blood with carbon dioxide not being involved in any significant role. The precise location of the oxygen receptors is not settled.

Differences in the level of activity are the main cause of short-term variation in oxygen consumption; the relationship has been well documented by Brett. Adjustments in ventilation which meet increased oxygen demand during exercise are also accompanied by increasing transfer factor as Randall *et al.* have shown in the trout. They may or may not involve a decrease in utilisation. By virtue of the increased transfer factor, active trout maintain a steady utilisation, whereas less active fish such as carp and suckers show a marked fall in utilisation as ventilation goes up.

The regulation of ventilation during exercise presents some problems to the physiologist if not to the fish. The changes in both arterial and venous oxygen tensions are very small, whereas ventilation changes are marked. This impressive stability of tension might be expected in a well-designed feedback control system. However, in hypoxic conditions there are, necessarily, larger fluctuations in blood oxygen tensions; but not until these tensions become very low are they accompanied by ventilation changes as large as in the active fish. The same feedback system could not operate in both cases since the error in oxygen tension is differently related to ventilation volume. During hypoxia, the tension changes are smallest in venous blood, which is why the computer simulation of Taylor *et al.* for both exercising and hypoxic trout works best with the error detector in the venous circuit. Stevens and Randall suggest that ventilation in swimming fish could be regulated by the changes in muscular activity in some way. It is well known that there is a direct effect of exercise on respiratory activity in mammals. Another possibility might be that the increased levels of carbon dioxide in the active tissues and venous blood could act in conjunction with the oxygen receptor system, for example, to reset the level of feedback to work with a smaller error than during hypoxia. In the absence of further experimental evidence, oxygen, carbon dioxide, and muscular activity must be assumed to have some part to play in the regulation of ventilation during exercise.

THE RELATIONSHIP BETWEEN VENTILATION AND PERFUSION

In order for gas exchange to occur, water and blood must be brought into intimate contact. The provision for this at the level of the gill lamella is discussed elsewhere (see chapter by Randall, "Gas Exchange in Fish," this volume). In addition, there must be an adequate flow of both blood and water to transport the respiratory gases to and from the exchanging surface. The problems involved in the provision and control of blood flow to the gills, especially in relation to the ventilation volume and gas exchange have been discussed by Hughes and Shelton, Hughes, Rahn, Randall *et al.*, and Piiper and BaumgartenSchumann. Hughes and Shelton introduced the concept of capacity-rate ratio in an attempt to quantify some aspects of the relationship. This ratio can be determined as the quotient of the changes in oxygen tension in water and blood as they pass through the gill, viz.:

$$\text{Capacity-rate ratio for oxygen} = \frac{\dot{V}_G}{\dot{Q}} \cdot \frac{\beta_{O_2}}{\gamma_{O_2}} = \frac{P_{a_{O_2}} - P_{\bar{V}_{O_2}}}{P_{I_{O_2}} - P_{E_{O_2}}}$$

where P is the partial pressure of gas in a (arterial blood), $\bar{V}$ (mixed venous blood), I (inspired water), and E (expired water); $\dot{V}_G$ is the volume flow of water over the gills in unit time, $\dot{Q}$ the volume flow of blood through the gills in unit time, β_{O_2} the solubility coefficient of oxygen in water, and γ_{O_2} the slope of the oxygen dissociation curve for blood between the arterial and venous points. Hughes and Shelton suggested that capacity-rate ratios of approximately unity were likely because oxygen transfer (carbon dioxide does not constitute a problem because of its high solubility in both blood and water) could then be achieved with minimum total flow of water and blood. This would be desirable if large amounts of energy were expended in pumping the two media. Since the solubility of oxygen in blood is very much higher than it is in water, a capacity-rate ratio of one would argue for high ventilation : perfusion $(\dot{V}_G / \dot{Q})$ ratios in fish. Rahn also reached this conclusion.

The precise value for both capacity-rate and ventilation: perfusion ratios in different fish will depend on many factors such as gill area, affinity of blood and oxygen, and activity characteristics of the animal. A considerable range of capacity-rate ratios have been found in practice, with values from 5.0 in resting and exercising trout to 0.4 in dogfish and carp. The ratio is high in trout largely because of the relatively enormous ventilation volume $(\dot{V}_G / \dot{Q} = 70)$ and this is accompanied by a correspondingly low oxygen utilisation of 10%. Conversely, the low values in carp and dogfish are the result of much lower ventilation volumes ($\dot{V}_G / \dot{Q} =$ 12 in carp, 9 in dogfish) with high utilisations of 60% and 50-70% respectively.

In the resting elasmobranch, evidence of a centrally coordinated relationship between breathing and heartbeat has existed for a long time. The early work was confirmed by Satchell when he showed that the heart rate was always some simple fraction, e.g., one-half, onethird, or one-fourth of the breathing rate. He suggested that the coordination was achieved by a reflex mechanism, proprioceptors in the pharynx being connected to inhibitory fibers running in the cardiac branch of the vagus. The reason for such a precise relationship, as opposed to a more general one between bulk flows as outlined above, was thought to be one of providing synchronisation of high flow rates in the two pulsatile systems. The outcome would be a moment-to-moment balance of ventilation and perfusion and an increase in efficiency of gas exchange. This proposal was reexamined by Hughes and Shelton who combined Hughes' results on ventilation in *Scyliorhinus* with those of Satchell on the circulatory system. The temporal relationships of blood pressure and water pressure differentials across the gills. If it is assumed that flow is proportional to differential pressure in both systems, it can be seen that maximum flow of blood coincides roughly with the important events of a single respiratory cycle. Because the heart rate is always less than the breathing rate it is difficult to draw any firm conclusions about the total value of synchronisation to the fish.

Teleosts in well-aerated environments do not show the same exact coordination between heartbeat and breathing movements, although there is often a greater tendency for the heart to beat during mouth closing than in other phases of breathing in the tench. Although the relationship is not so clear cut, there is still a great deal of evidence for control in teleosts. Shelton and Randall, working on eels and tench, and Peyraud

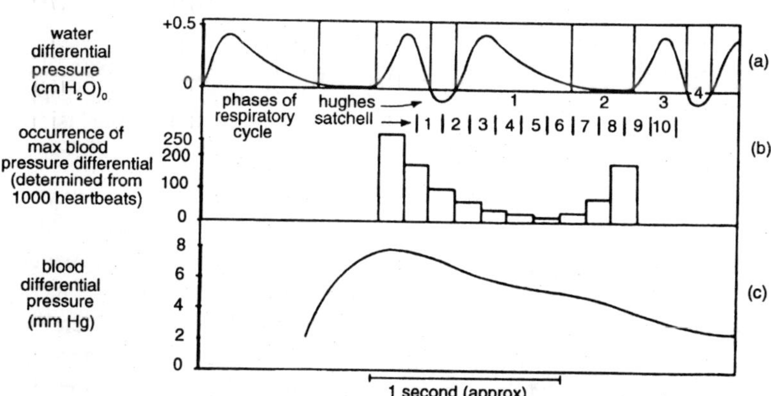

Figure 5.31 : Time relationships between the respiratory cycle and heartbeat in a dogfish. (a) Differential water pressure between orobranchial and third parabranchial cavities. From Hughes. (b) The occurrence, in ten phases of a respiratory cycle, of the maximum blood pressure differential across the gills. From Satchell. (c) Differential blood pressure across the gill circulation produced by a single heartbeat. The curve is plotted so that the maximum differential coincides with the period of maximum occurrence as shown in the histogram.

and Serfaty, working on carp, showed that periodic bursts of breathing movements were accompanied by an increase in heart rate, whereas the rate fell in periods of apnoea between bursts. A very clear synchrony developed in trout if the oxygen content of the environment was reduced, and, as in the dogfish, it was shown that the receptors were in the branchial region and the efferent fibers were cardio-inhibitory ones carried in the vagus. As in the case of elasmobranchs, it was suggested that phasic cardiac inhibition allowed the heart to beat during the period of maximum water flow over the gills.

Different environmental conditions and variations in activity place different demands on heart output just as they do on the gill ventilating mechanism, although the responses of the two systems need not necessarily be similar in nature. Bradycardia occurs as a response to hypoxia in both elasmobranchs and teleosts and is attributable to an increase in vagal tone in both cases. Randall and Shelton suggested that the bradycardia represented a fall in cardiac output, but Holeton and Randall have recently shown that this is not so. In the trout, hypoxia produced no substantial change in cardiac output although there was a shift from high rate-low stroke volume to low rate-high stroke volume. While the gills can maintain complete saturation of the hemoglobin during hypoxia, no change in cardiac output is necessary to maintain a uniform oxygen consumption. The significance of the bradycardia is not clear. If the arterial oxygen tensions fall below the level for complete saturation, then the heart output must go up, or there must be greater extraction in the tissues giving lowered venous oxygen tensions. The latter seems to be the case in trout.

Exercise causes substantial increase in oxygen consumption and will therefore require greater perfusion rates to transport the oxygen from gills to tissues. Increased cardiac output and ventilation volume should proceed together in the absence of any very great tension changes in the different parts of the system. This was in fact the case in the trout, heart rate and stroke volume both going up to produce the increased cardiac output. The control of this response is as difficult to understand as that of the ventilation change already discussed. Ventilation and cardiac output are not directly coupled together as the hypoxic animal

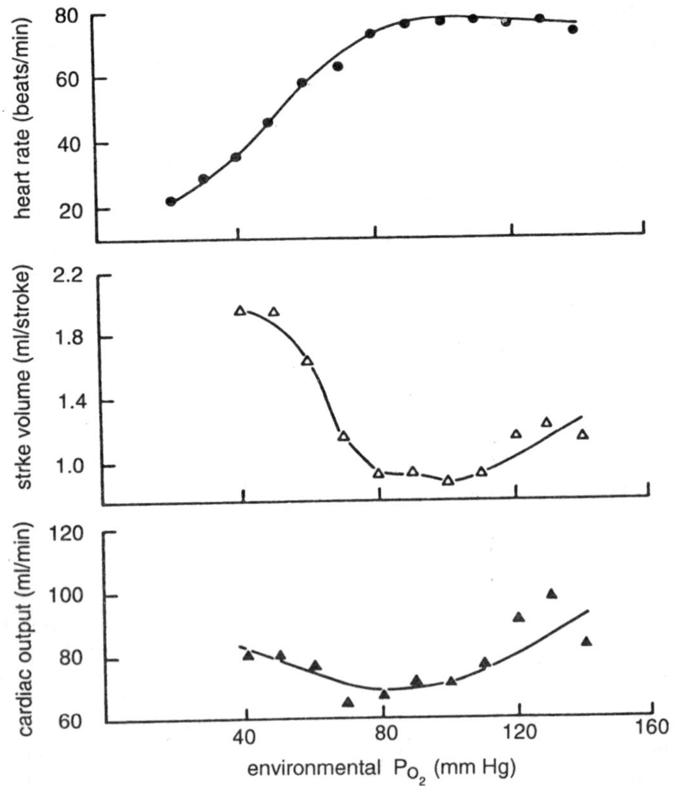

Figure 5.32 : The effect of hypoxia on cardiac output, stroke volume, and heart rate in the rainbow trout. Although the hypoxia results in a bradycardia, the cardiac output is maintained by a compensating increase in stroke volume.

shows. There must therefore be some specific signal in the exercising animal, and it seems most likely that it is a direct monitoring of the exercise itself which results in the increased heart output. Exercise will also increase venous return to the heart, but this can hardly be responsible for the total effect. Oxygen tensions cannot constitute the effective sensory signals because they change most in the hypoxic animal where the cardiac output changes least. The only evidence which implicates oxygen receptors is that which shows the deoxygenation of the gill region produces a bradycardia, not an increase in rate. The efferent side of the cardiac response during exercise also presents problems. It has been widely accepted that there are no sympathetic fibers innervating the fish heart. The whole nervous supply was thought to be cholinergic and inhibitory, although recent electron microscope studies by Yamauchi and Burnstock suggest that nerves liberating catecholamines do innervate the fish heart. Stevens and Randall found that injections of atropine had no effect either on the heart rate of resting trout or on the responses during exercise. They concluded that increased output as a result of exercise was aneurally mediated. Injection of adrenalin produced cardiovascular changes which were similar to those observed during increased activity. Some fish may regulate heart rate by changes in vagal tone, but it now seems likely that a great deal of cardiovascular adjustment in swimming, and perhaps in resting, fish is accomplished by changes in the level of circulating catecholamines. The source of these catecholamines and the mechanism regulating their release have not been identified. In this, as in all fields of respiratory and circulatory physiology, there have been tremendous advances in recent years only to reveal further problems for experimental analysis.

AIR BREATHING IN FISHES

Occurrence and Bionomics of Air-Breathing Fishes

Natural selection of air-breathing habits in fishes has occurred many times and in diverse ways during vertebrate evolution.

The majority of extant fishes showing structural adaptations for direct use of atmospheric oxygen are tropical freshwater or estuarine forms. Most air-breathing fishes are teleosts but typically the dipnoan, chondrostean, and holostean fishes are dominated by air-breathing forms. Reports of air-breathing habits among elasmobranch fishes have not been adequately confirmed.

A smaller number of air-breathing fishes live in temperate regions, and in exceptional cases such as the bowfin, *Amia calva,* they occupy waters seasonally frozen over.

A shortage of dissolved oxygen constitutes the primary environmental condition which has stimulated the development of air-breathing in fishes, although such adaptations may also have been correlated with behaviour or activity patterns. A few forms, inhabiting oxygen-deficient tropical swamps, have acquired the faculty of sustained air breathing when entirely removed from water during droughts. Such periods are passsed in an estivating condition. Air breathing has also evolved among fishes living in torrential mountain streams in Asia. Here, the air-breathing habit is an adaptation to drought only since these streams carry well-oxygenated water during the wet season. These fishes either estivate or migrate over land to neighboring streams. Many tropical swamps and stagnant tributaries to larger rivers have very acid and CO_2 rich water. Adverse conditions for aquatic breathing with gills also result when water gets excessively turbid with suspended material which may cover the gill surfaces.

The principal causes of oxygen depletion in tropical freshwaters are related to the abundance of organic matter which provides a substrate for rapid oxygen consumption by microorganisms. The prevailing high temperatures of the tropics accelerate oxygen depletion by increasing the O_2 requirements of the microfauna as well as the other inhabitants of the swamps. Diurnal temperature variations are small, and thus little vertical water movement occurs to promote reentry of atmospheric O_2 and the swamps become stagnant. Oxygen release from photosynthetic activity is very slight because of the low intensity of light below the dense foliage of the prolific vegetation in regions of tropical swamps and rain forests. The same foliage cover prevents wind disturbance from effectively stirring and thus aerating the water.

Naturally tropical waters show a vertical gradient in O_2 concentration owing to a more well-oxygenated surface layer. However, as close as 1 cm to the surface, O_2 concentrations can be as low as 2% of saturation value. Several inhabitants of tropical swamps show behavioural adaptations for utilisation of the more O_2-rich surface layer by breathing at or near the surface. Often O_2 > availability is so limiting that all species not adapted for direct air breathing must utilise the surface water as a source for oxygen.

Among air-breathing fishes, those occupying brackish and marine habitats are rare because of the well-stirred turbulent conditions of these waters. However, several members of marine gobiid and blennid families show structural adaptations for air breathing. Estuarine waters may show large spatial, diurnal, and seasonal fluctuations in oxygen availability. More exceptionally, fishes like *Periopthalinus* have developed air-breathing habits in support of behavioural acts such as feeding and escape reactions.

NATURE OF THE STRUCTURAL ADAPTATIONS FOR AIR BREATHING

Structural adaptations for air breathing in fishes are extremely diverse. The extent of modifications from structures typical of an aquatic breather often reflects the relative role of air breathing in overall gas exchange. Some species rely predominantly on water breathing with gills and only supplement gas exchange with air breathing when adverse respiratory qualities in the water make aquatic breathing insufficient or too costly for extraction of the required oxygen. Others are obligate air breathers and succumb if denied access to air for a short time.

All air-breathing fishes have a hollow compartment as a receptacle or holding space for air. The air chamber may have evolved as a modification of an already existing structure or it may be a neomorphic structure. It is richly supplied with blood vessels, and fine capillaries or lacunar spaces invest the membranes bordering the air space. The interface between blood and air is specialised to reduce the diffusion distance and other diffusion barriers between air and blood. Frequently, various forms of structural changes such as foldings, papillations, or arborisations of the gas exchange membranes have resulted in a considerable surface expansion that enhances the diffusion exchange. Such structural specialisation is more prominent in species utilizing air breathing as a major means of gas exchange. A concurrent trend is for the primary aquatic gas exchange organs, the gills, to degenerate progressively to few, coarse, and sparsely perfused gill filaments as air breathing assumes a more important role. In some species, such as the electric eel, *Electrophorus electricus,* the gills have become vestigial. Air-breathing fishes also show structural specialisations for rhythmic ventilation of the air-breathing organs, while the muscle and skeletal parts involved with purely aquatic breathing have become reduced or modified for the new purpose of breathing air.

Species of air-breathing fishes arranged according to the type of their structural adaptation. The list is not exhaustive, and the interested reader is referred to more specialised morphological texts.

Structural Derivatives of the Mouth and Pharynx as Air-Breathing Organs

By far the most common and yet the most diversified structural adjustments for air breathing are those associated with the branchial, opercular, and pharyngeal cavities.

Piscine gills are designed to function in water where lack of net gravitational forces prevents collapse of the fine filaments and lamellae. However, the collapse of these in air makes the gills of most fishes useless as organs of aerial gas exchange. A very few fishes, such as *Symbranchus* and *Hypopomus,* are able to use their gills for air breathing. They fill the buccal cavity with air regularly at the surface, and the used air is voided from the mouth or the opercular openings. The gills of these fishes are structurally modified to prevent their collapse in air. Most notably they show a high ratio between the breadth and height of the secondary lamellae. The ratio is more than twice that in purely aquatic fishes or fishes possessing other types of structural adaptations for air breathing. The epithelium of the buccal and pharyngeal cavities of these two fishes is richly vascularised and plays a supporting role in aerial gas exchange.

The common eel, *Anguilla anguilla*, and species of *Periopthalmus* also employ gill breathing successfully during air exposure.

A further development of aerial mouth breathing is seen in fishes with foldings and papillations of the buccal epithelia projecting into the buccal cavity and forming extensive vascular surfaces for aerial gas exchange. The electric eel, *Electrophorus electricus,* offers a striking example. In this fish the papillated projections are distributed over both the floor and the roof of the mouth, while smaller prominences are present on the branchial arches and the lateral branchial walls. The system of papillae forms a labyrinth of air passages. Studies on formalin fixed material revealed a surface area of the mouth respiratory organ about 15% of the total body surface. This is considerably less than for either gills in purely aquatic fishes or lungs of higher vertebrates. Yet the mouth organ in *Electrophorus* functions well enough to have permitted the gills of adult fishes to become almost vestigial.

The next step in a progressive structural adjustment to air breathing includes the formation of an air chamber extending from the buccal or pharyngeal cavities. In the Ophiocephalidae with several air-breathing species in tropical marshes of Asia and Africa, the air chamber lies dorsal to the branchial area and becomes partly enclosed within the cavity of the skull.

In *Amphipnous cuchia,* another air-breathing fish living in India, the paired air chambers referred to as pharyngeal lungs are more free and extend several centimeters behind the head. The internal walls of the air chamber are very vascular and may show surface expanding structures

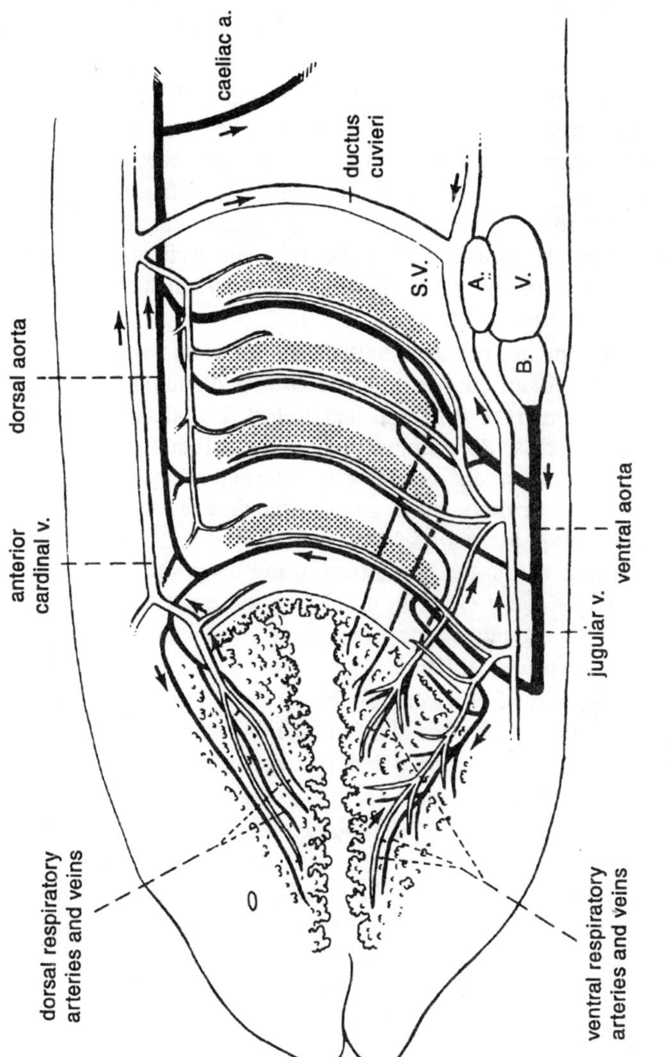

Figure 5.33 : Schematic drawing of the mouth respiratory organ and its vascular connections with the heart and central circulation in the electric eel. The arrows indicate the direction of the blood flow.

in the form of rosettelike arborisations protruding into the lumen. In others, internal trabeculation may give the air chamber a resemblance to an amphibian lung.

Another structural variation is seen in fishes where the opercular cavities have become richly vascularised and serve as receptacles for air. When air is lodged in such chambers they stand out in balloonlike fashion. The opercular walls have become thin and elastic and their bone structures have regressed. The opercular openings are either fused to one small channel ventrally or have otherwise become reduced in size. These modifications and others involving the intrinsic muscles of the opercular walls permit ventilatory movements and renewal of air inside the opercular chambers. Examples of such specialisation are seen in *Hypopomus brevirostris, Monopterus sp.*, and several species of gobiid fishes (e.g., *Pseudapocryptes lanceolatus*).

More elaborate use of the opercular cavities in air breathing results when specialised diverticula develop from their dorsal side commonly between the hyoid and the first branchial arches. Two distinctly different types of such opercular air chambers or lungs are found. In some cases they remain free and extend posteriorly. Internally they possess ridges and trabeculations or other protrusions expanding the highly vascular internal walls. Often these air sacs extend half the length of the animal and lie embedded within the myotomes of the body wall, a fact which undoubtedly bears significance in the filling and emptying of the sacs. Indian fishes of the genus *Saccobranchus* are well-known examples of this type of structural adaptation.

The other variety of opercular air chamber results when the opercular diverticulum extends inside the skull to take the form of a labyrinthlike organ (*Anabantideae*) or coralike dendritic formation as seen in species of Clarias.

Members of the Anabantidae (*Anabas, Macropodus, Trichogaster,* and others) typically show the air labyrinths starting as outgrowths from the epibranchial section of the first branchial arch. Laminated projections contribute to the surface expansion inside the chambers. The detailed structure of the labyrinthine organ is extremely complicated. The diffusion distance from air to blood in *Trichogaster* has been measured as the smallest of all recorded in respiratory organs, showing a minimal value of 600 Å, a medium range of 1300-1700 Å, and a maximum value of 1.2 μ. Based on electron microscopy, Schulz reported that capillaries of some areas of the labyrinthine organ lack both basal membrane and endothelium, thus leaving only a thin epithelium intervening between blood and air.

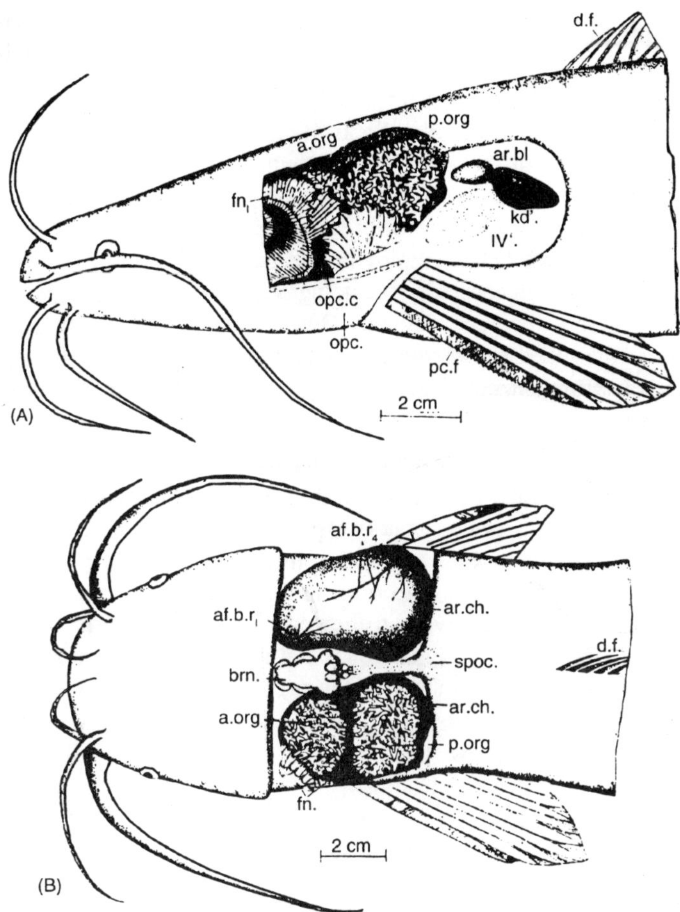

Figure 5.34 : (A) Lateral and (B) dorsal views showing the gross structure and relative positio i of the air chamber in Clarias lazera. *In A and to the left in B the wall of the air chamber has been removed to expose the surface expanding arborescent organs. From Moussa. af.b.r., 1 and 4, ramifications of the first and fourth afferent branchial vessels in the wall of the air-chamber; a.org., anterior arborescent organ; ar.bl., air or swim bladder; ar.ch., air-chamber; brn., brain; d.f., dorsal fin; fn-, serial order of fans; kd'., part of kidney; lv'., part of liver; opc., operculum; opc.c., opercular cavity; pc.f., pectoral fin; p.org., posterior aborescent organ; spoc., supraoccipital.*

In *Clarias* the two remarkable vascular rosettelike trees which fill the air chambers on each side receive structural support from gill arches 2 and 4 and their vascular supply from the first and fourth afferent branchial arteries. The gross appearance of the labyrinth in *Clarias lazera.*

Structural Adaptations of the Gastrointestinal Tract for Air Breathing

In some fishes portions of the gastrointestinal tract, notably the stomach or part of the small intestine, have been modified to serve in gas exchange between the blood and rhythmically swallowed air. The used air is voided at the anus or the mouth. Members of the families Loricaridae, e.g., *Plecostomus plecostomus*, and Callichthyidae, e.g., *Callichthys* and *Hoplosternum* offer typical examples of gastrointestinal air breathers. Most fishes employing the gastrointestinal tract for gas exchange use it as an accessory breathing organ when aquatic conditions become severely oxygen deficient. Others depend on it during short migrations over land *(Plecostomus),* and a few are obligate air breathers that need to supplement the aquatic gas exchange with air breathing at all times.

The Air Bladder as a Respiratory Organ

It is now generally conceded that the air bladder in fishes originally functioned as an accessory respiratory organ. In most extant actinopterygian fishes the air or swim bladder serves primarily in buoyancy control or sound production or sound detection. In dipnoan fishes the air bladder has retained and further evolved the respiratory function it allegedly possessed in the crossopterygian fishes that also gave rise to amphibians and land vertebrates. Members of the primitive orders

Chondrostei *(Polypterus)* and Holostei *(Amia* and *Lepisosteus)* have also retained a respiratory function of the air bladder but fall short of the refinement in structural adjustment for air breathing seen in the dipnoans. Among modern teleosts some have secondarily developed the air bladder into a respiratory organ such as seen in the South American fishes *Erythrinus uniteniatus* and *Arapaima gigas*.

Only in the lungfishes, and of these *Protopterus* and *Lepidosiren* in particular, has the air bladder evolved structurally to resemble a lung as it exists in lower vertebrates. Internal septa, ridges, and pillars divide the air space into smaller compartments opening into a median cavity. Further subdivisions of these compartments by progressively smaller reticulated septa terminates in alveoli-like pockets richly covered with blood vessels. *Neoceratodus* has one lung, while *Protopterus* and *Lepidosiren* have two lungs fused in their anterior region where the pneumatic duct and the pneumatic sphincter connect the lungs with the esophagus. The short airway duct, which in no way resembles the air distributing system of the trachea and bronchii in terrestrial vertebrates, is richly supplied with smooth muscle as is the lung parenchyma itself.

The smooth musculature is important for the mechanics of breathing and may exert additional influence in distributing the air inside the lung. Electron microscopy has revealed that the fine structure of the lung in *Protopterus* does not differ basically from that in higher vertebrates. Klika and Lelek, reporting on lung structure in *Protopterus,* state that the respiratory epithelium adjoins the basement membrane which in turn is separated from the endothelial basement membrane by a thin layer of collagen fibrils. The approximate blood to air diffusion distance in *Protopterus* is about 0.50 μ. The *Protopterus* lung is much more similar to the amphibian lung than that of *Polypterus* which has a smooth inner surface much less vascular than in the lungfishes. Figure 3A shows the gross structure of the *Protopterus* lung.

Structural specialisations in the perfusion pattern through the gas exchange organs and the vascular linkage between them and the general circulation is of utmost importance for efficient gas exchange. In airbreathing fishes presumably the highest degree of efficiency will prevail when the air-breathing organ is perfused with systemic venous blood and when the respiratory efferent blood is dispatched directly for systemic arterial distribution. Most air-breathing fishes also employ aquatic breathing and efficient perfusion of the gills and/or skin is important since these areas are the primary sites for CO_2 elimination. The vascular system should hence ensure that both the air- and waterbreathing respiratory organs are perfused before the blood is turned over to the nutritive circulation.

The various types of circulatory arrangements seen in air-breathing fishes. All types fall far short of matching the ideal perfusion pattern existing in purely aquatic fishes (type A) or in the highest vertebrates where the respiratory and systemic vascular beds are coupled in direct series. All air-breathing fishes show a parallel linkage between the two types of vascular beds, and only the lungfishes are structurally adapted to minimise shunting between these

The heart and vascular beds in relation to the aquatic (gills) and aerial respiratory organs in air-breathing fishes. The amount of white and black represents only the approximate level of oxygenated and deoxygenated blood carried in the vessels. (A) General piscine arrangement with the branchial (gill) and systemic (tissues) vascular beds in direct series. (B) Airbreathing organ derived from pharyngeal and/or opercular mucosa. Afferent vessels to air-breathing organ arranged in parallel with branchial circulation and derived from afferent branchial vessels. Efferent vessels from air-breathing organ connected to systemic

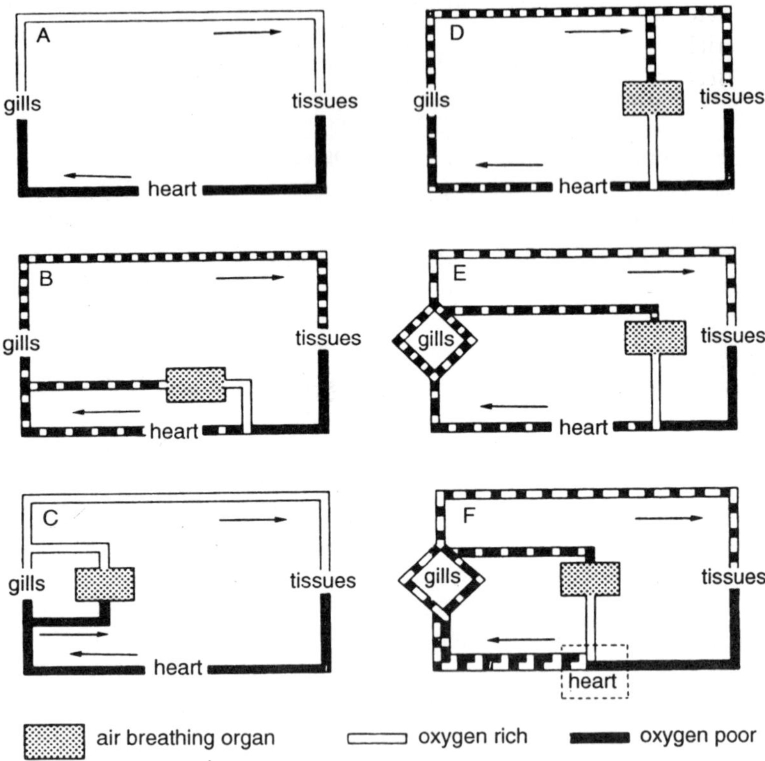

Figure 5.35 : Schematic representation of the heart and vascular beds in relation to the aquatic (gills) and aerial respiratory organs in air-breathing fishes. The amount of white and black represents only the approximate level of oxygenated and the deoxygenated blood carried in the vessels.

veins. Arrangement typical of *Motopterus, Ophiocephalus, Electrophorus, Amphipnous, Periopthalmus,* and *Anabas.* (C) Gills, buccal mucosa, or chambers extending from the opercular cavity serving as air-breathing organs. Afferent vessels to air-breathing organ arranged in parallel with the branchial vascular bed. Efferent circulation largely connected with the efferent branchial circulation. Systemic arterial blood highly oxygenated. *Clarias, Saccobranchus,* and in part, *Symbranchus* and *Hypopomus* typical of type C. (D) Air-breathing organ associated with the gastrointestinal tract. Afferent circulation derived from the dorsal aorta (systemic arteries). Efferent circulation connected to systemic veins. Arterial blood mixed. Arrangement typical in *Hoplosternum, Plecostomus,* and *Ancistrus.* (E) Air bladder serving as air-breathing organ. Specialised afferent vessel derived from the most posterior epibranchial arteries. Efferent circulation connected to systemic veins. Arterial blood mixed. Arrangement typical of *Polyterus, Amia,* and *Lepisosteus. (F)* Air bladder structurally advanced to resemble amphibian lung. Specialised afferent and efferent circulation in parallel with systemic arterial circulation. Partial septation of the vascular beds. In the electric eel (type B), the entire gain in oxygenation resulting from perfusing the mouth is shunted back to the systemic veins before blood is redistributed to the systemic and respiratory arteries. This deficiency prevails to various degrees in all types except the very few fishes able to use their gills alternately for aerial and aquatic breathing [e.g., *Symbranchus* or the few fishes employing air breathing with specialised air chambers extending from the opercular cavities such as *Clarias* (type C)].

Only in fishes employing the air bladder for aerial gas exchange have specialised vessels developed as connections between the primary branchial circulation and the gas exchange organ (types E and F). These vessels take origin from the sixth aortic arch and are homologue with the pulmonary arteries of higher vertebrates. In addition to the dipnoans, *Polypterus* and *Amia* possess these vessels. The dipnoans are unique among fishes by also having a specialised efferent circulation from the air bladder connected directly with the left side of the heart. *Gymnarchus niloticus,* a teleost using the air bladder for accessory respiration, has an arterial supply to the bladder from the dorsal aorta but a venous drainage connecting directly with the atrium. No septation of the heart is evident. However, the heart in dipnoans is in part structurally adapted for accommodation of two bloodstreams by showing partial atrial and ventricular septa, structures which clearly forecast a further development of the heart and circulation toward conditions in terrestrial vertebrates.

The mechanics of air breathing in lungfishes has been carefully studied. The mechanics of inhalation and exhalation of air is derived from a series of basically aquatic breathing cycles typical of fishes modified to serve a specific step in the air-breathing cycle. McMahon emphasises the role of the buccal force pump as the main ventilatory mechanism in purely aquatic as well as transitional bimodal breathers among fishes and amphibians.

PHYSIOLOGICAL, ADAPTATIONS IN AIR-BREATHING FISHES

Respiratory Properties of Blood

Studies of respiratory properties of blood from air-breathing fishes have disclosed adaptive features which correlate with environmental factors and the relative importance of air and water breathing.

A large variability is apparent in hemoglobin concentration and O_2 capacity, with no adaptive trend evident in air-breathing fishes as a group. However, a comparison of tropical fishes studied by Willmer disclosed a definite tendency for O_2 capacities to be higher in the air-breathing forms. This difference probably reflects the nature of the fishes' habitats, particularly the degree of O_2 deficiency rather than the habit of air breathing per se. The fast flowing rivers and creeks where the purely aquatic breathers *Myelus setiger* and *Hoplias malabaricus* live are well aerated, whereas the habitat of the air-breathing fishes *Hoplosternum* and *Electrophorus* can be severely O_2 deficient.

In evaluating this type of adaptive adjustment, one must recognise the relative importance of air and aquatic breathing in the total gas exchange of these fishes. If water breathing is the dominate mode and air breathing is only accessory or supplemental, adaptive adjustment of the blood to meet the conditions in the water may be expected. An early documentation by Krogh and Leitch showed that aquatic species tended to show higher O_2 capacities when living in O_2 deficient water. The obligate air breathers, among fishes, have an ample reservoir of atmospheric oxygen available. However, these fishes may be compared with the diving animals among higher vertebrates which, by means of high oxygen capacities, increase their O_2 stores and consequently extend the time between surfacings. Sluggish or active habits also may influence hemoglobin levels in fishes. Still another type of selection pressure for development of high O_2 capacities in air-breathing fishes has been suggested and exemplified in the electric eel, *Electrophorus*, and applies to most other air-breathing fishes. This relates to the shunting of respiratory efferent blood into systemic veins which results in systemic arterial perfusion with mixed blood. Thus arterial blood never gets fully saturated, a condition which obviously reduces the efficiency of the gas transport system. The situation in lungfishes differs considerably in that recirculation of the well-oxygenated blood from the lung is largely prevented by a separate pulmonary vascular circuit and a partial division of the heart and its outflow channels.

The general shape of O_2-Hb dissociation curves conform to those of other fishes. Oxygen affinity generally decreases in vertebrates with increasing dependence on aerial gas exchange. Another generalisation relates an increased O_2 affinity to the ability of a species to survive in an O_2 poor medium. The applicability of these generalisations to air-breathing fishes depends on the relative importance of air breathing in their bimodal gas exchange. Among species emphasizing water breathing, there is a tendency for higher O_2 affinity to correlate with the most O_2-

deficient habitats. No clear-cut adaptive trends are apparent from a comparison of the obligate air breathers and those practicing occasional supplemental air breathing. In fishes like *Electrophorus* in which the O_2-rich blood is admixed with O_2-deficient blood before perfusion of the arteries, a high O_2 affinity would improve the efficiency loss caused by the mixed condition of arterial blood.

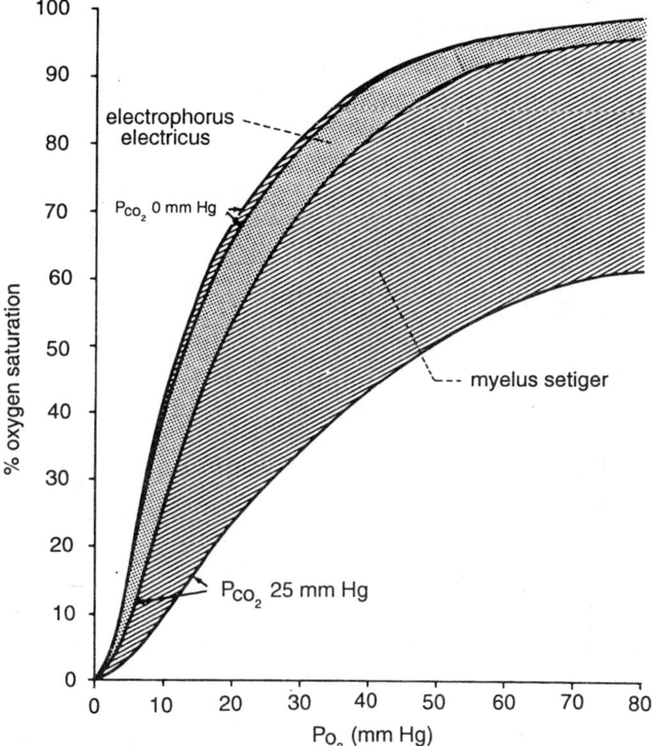

Figure 5.36 : Oxygen dissociation curves showing a conspicuous Bohr effect in Myelus living in well-aerated CO_2-free water. A slight Bohr effect is present in Electrophorus living in stagnant oxygen-deficient, CO_2-rich water.

Several authors have claimed evidence of a relationship between the CO_2 sensitivity of the O_2-Hb dissociation and the habitat and air-breathing habits of fishes. Willmer offered convincing data to show that tropical fishes from larger rivers and fast flowing creeks with well aerated water of low CO_2 content have blood which is adversely affected by CO_2 and pH changes. Conversely, blood from fishes living in stagnant swamps and creeks with O_2-deficient, CO_2-rich water were distinctly less sensitive and often even insensitive to CO_2 changes.

For instance, if *Myelus setiger,* normally living in well-aerated CO_2-free water, was transferred to a typical swamp habitat, its blood could never be more than 40% saturated with oxygen because of the large Root and Bohr shifts. It has been argued that the diminishing sensitivity to CO_2 in fishes normally living in a CO_2-rich habitat is an important adaptation needed for adoption of air breathing since this change inevitably will entail an elevated concentration of CO_2 at the gas exchange surfaces because of the bidirectional airway of air breathers.

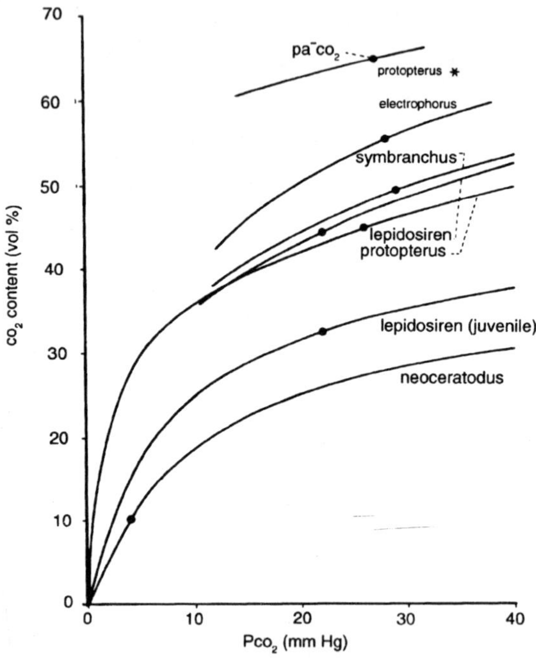

Figure 5.37 : Comparison CO_2 dissociation curves from representative species of air-breathing fishes. The black dots indicate the normal levels of arterial Pco_2.

The CO_2 combining power of blood shows a clear increase with increased importance of air breathing in overall gas exchange. A comparison of CO_2 dissociation curves with normal values of arterial P_{CO_2} indicated for representative air-breathing fishes.

The increase in CO_2 combining power correlates with an increase in the average level of arterial P_{CO_2}, thus meeting the obvious need for a higher capacity in CO_2 transport associated with the air-breathing habit. A composite plot of overall buffering capacity in air-breathing fishes, illustrates the same adaptive increase in buffering capacity with the increased importance of air breathing.

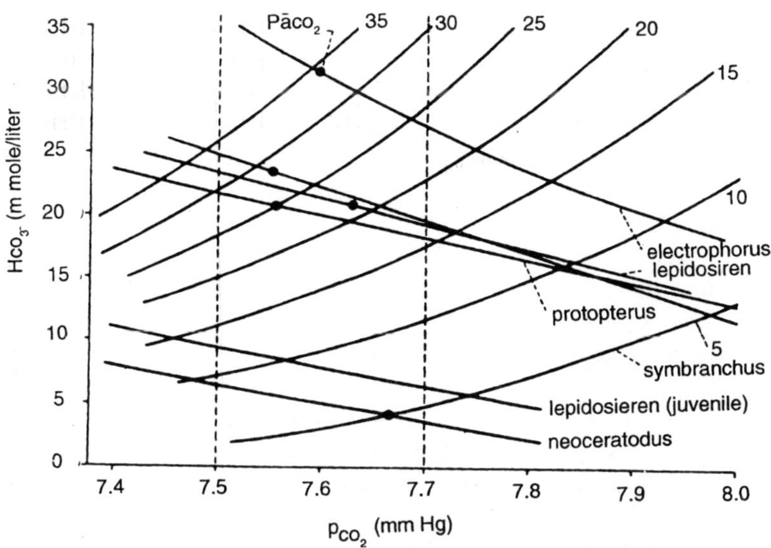

Figure 5.38 : A comparison of standard bicarbonate and buffering capacities in representative air-breathing fishes.

Few studies have assessed the influence of temperature on the respiratory properties of blood in air-breathing fishes. A comparison of blood from *Neoceratodus* and *Protopterus* shows that the former, which experiences large annual temperature shifts, is notably insensitive to temperature changes, while *Protopterus,* living in a temperature-stable habitat, has far more temperature-sensitive blood.

Gas Exchange in Air-Breathing Fishes

All air-breathing fishes possess some capacity for aquatic breathing. Thus gas exchange is bimodal and depends upon exchange diffusion with the water via gills or skin and with air rhythmically taken into, and expelled, from, the air-breathing organ. The fish may employ both means of gas exchange simultaneously or may resort to one method at a time depending upon external circumstances. A very few fishes undertake voluntary migrations over land in moist or wet grass. Aquatic respiration must then be altogether halted except for some possible exchange via the skin. Estivating air-breathing fishes represent a special case. They must rely almost exclusively on aerial gas exchange although retention of a moist microenvironment in their cocoons or burrows in the substratum is essential for survival. This requisite may be as important for fluid balance and osmotic control as for continued cutaneous gas exchange. The literature is deficient of any attempt to study gas exchange in estivating fishes.

Air-breathing fishes experience two vastly different environments. A low solubility of O_2 in water provides an oxygen availability, in aerated water, about one-thirtieth of that in air in a tropical environment. Hence, aquatic breathers must ventilate much larger volumes than air breathers at similar O_2 uptakes. The greater work of breathing in water is further aggravated by the huge density difference between the two media. The high solubility of CO_2 in water along with the high ventilation of aquatic breathers brings about internal CO_2 tensions which may be one-thirtieth of those usually found in terrestrial vertebrates. Air breathing is advantageous in terms of O_2 availability and high diffusion rates, but it is countered by the hazards of desiccation and by the different requirements for mechanical support under the influence of net gravitational forces in air. Exclusive air breathers also have to meet the need for increased efficiency of internal buffering systems because of elevated internal CO_2 tensions. Air-breathing fishes largely avoid these problems by remaining essentially aquatic, only exploiting the atmosphere as a source of oxygen.

All air-breathing organs in fishes show a low gas exchange ratio indicating that their function is mainly O_2 absorption. Gills and skin conversely play their most important role in CO_2 elimination and show accordingly gas exchange ratios exceeding unity.

Table III offers a compilation of the relative role of aquatic and aerial gas exchange in the few fishes that have been adequately studied.

Among obligate air breathers such as *Lepidosiren* and *Protopterus,* aquatic breathing is of little significance in O_2 absorption and contributes less than 10% of the total oxygen uptake in adult fish. Aquatic breathing remains very important in CO_2 elimination and is, in *Protopterus*, about 2.5 times more effective than the lungs in removing CO_2. In many fishes the skin plays an important role in gas exchange, particularly in CO_2 elimination. Cunningham reported a gas exchange ratio in excess of 10 for the skin of *Lepidosiren* in water. During air exposure the skin is also important in gas exchange and marked cutaneous vasodilation has been reported for the common eel, *Anguilla* and *Lepidosiren*.

In *Cobitis fossilis,* an intestinal air breather, and *Gillichtys mirabilis,* a marine gobiid fish practicing accessory air-breathing with the buccopharyngeal cavity, O_2 absorption is the dominant feature of air breathing.

The large variation of total O_2 uptake, may be related to feeding. Oxygen uptake in *Protopterus* has been reported to decline rapidly between 24 and 48 hr after the last food intake, and a constant metabolic rate was maintained only as long as the intake of food exceeded the prevailing oxidation rate.

A labile oxygen uptake was also apparent during air exposure of juvenile *Lepidosiren* and adult *Amia.* In *Lepidosiren* the O_2 uptake fell to 20% of the value in water in less than 6 hr. Similarly, Berg and Steen, working on the common eel, reported on O, uptake in air half of the value measured in water. A different pattern is evident in *Protopterus* during air exposure. A normal level of O_2 uptake was maintained when rebreathing air until the ambient O_2 tension had fallen to 80 mm Hg and CO_2 tension had risen to 35-40 mm Hg.

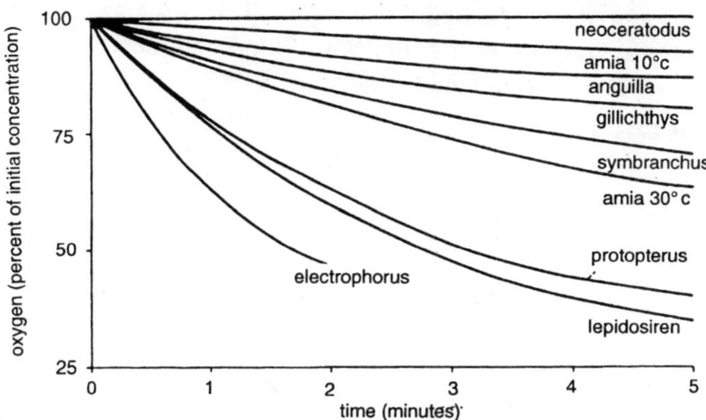

Figure 5.39 : Rate of O_2 depletion from air-breathing organs in various fishes. The slopes represent average values.

The rate of change of gas composition inside airbreathing organs of a number of fishes based on samples obtained when the animals were resting in well-aerated water. The rate of O_2 absorption in the obligate air breathers is conspicuously higher than in the fishes employing accessory or supplemental air breathing. Note, also, the marked effect of temperature on the rate of O_2 depletion from the air bladder in *Amia.*

A comparison of the slopes for O_2 depletion and CO_2 accumulation reveals the former to be much steeper, suggesting a low gas exchange ratio. The slopes for CO_2 accumulation also show a more abrupt change to a gentler slope late in the interval between surfacings for air, a fact suggesting that the low gas exchange ratio becomes even lower, later in the interval between air breaths. In the case of *Protopterus* and *Electrophorus,* this change is particularly apparent. Thus aquatic CO_2 elimination varies and may increase with time, possibly owing to a vasodilatory effect of increasing CO_2, in gills and skin vessels.

The slopes given for O_2 depletion average values, and variations likely reflect changes in the blood perfusion through the exchange organ. The average slopes seem related with the breathing habits of the fishes. Thus, in *Electrophorus,* intervals between air breaths rarely exceeded 2 min, while in *Lepidosiren* and *Protopterus* the intervals varied but were usually between 3 and 6 min, whereas, *Symbranchus* and *Amia* both showed infrequent surfacings often exceeding 25-30 min between air breaths.

Few studies have assessed the total volume and tidal volume of air-breathing organs in fishes. Juvenile specimens of *Protopterus* (weight 57-78 g) are reported to show average lung volumes of 1.7 ml (or 25.3 ml/kg) measured by radiological techniques. Tidal volumes measured by spirometry averaged 1.45 ml (or 19.8 ml/ kg) and, hence, represent a fraction of total lung volume as large as 80%.

Internal Gas Transport in Air-Breathing Fishes

Gas transport between the respiratory organs and the metabolizing tissues depends on the metabolic rate, the pattern and rate of blood circulation, and on the respiratory properties of blood. Very few airbreathing fishes have been adequately studied by frequent sampling of blood and gas to allow an evaluation of internal gas transport. Table IV summarises available information on gas composition in blood and aerial exchange organs.

The Australian lungfish, *Neoceratodus, is* special among the lungfishes by being predominantly a water breather. During rest, in aerated water, the gills are responsible for the entire O_2 absorption which suffices to maintain the arterial blood 95% saturated with O_2. Comparisons of gas tensions in samples from the pulmonary artery, pulmonary vein, and the lung disclose a near equilibrium indicating no role of the lung in gas exchange. These samples were obtained between the sporadic breaths for air which often fell more than one hour apart. The two other species of lungfishes emphasise air breathing. In *Lepidosiren paradoxa,* blood from the pulmonary artery was only 40-50% saturated with O_2 while that in the pulmonary vein was almost completely saturated. The difference in gas tensions between blood in the dorsal aorta and the pulmonary artery demonstrates a partial selective passage of blood through the heart. This selective passage tends to minimise recirculation of blood from the pulmonary vein back to the lungs.

In *Protopterus aethiopicus* passage of blood through the lung increases the O_2 saturation more than 30% to exceed 90% of full saturation.

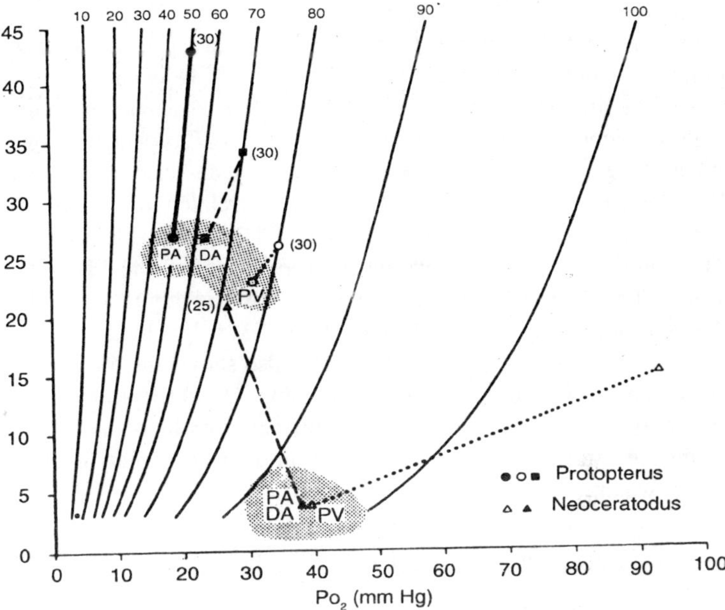

Figure 5.40 : Nomogram comparing blood gas values in Neoceratodus and Protopterus during undisturbed breathing in well-aerated water (values on stippled background) and following air exposure. (Duration of air exposure in minutes indicated by numbers in parentheses.) PA, pulmonary artery; DA, dorsal aorta; and PV, pulmonary vein.

There is a concurrent drop in CO_2 tensions but aquatic respiration still dominates the elimination of CO_2.

The marked differences in the importance and efficiency of air breathing in *Neoceratodus* and *Protopterus* can best be discussed when the blood gas data are described in a nomogram. While resting in aerated water, *Neoceratodus* practices water breathing only, whereas *Protopterus* practices frequent air breathing which results in a sizable gain in O_2 saturation of blood passing the lungs. A significant drop in blood CO_2 tension also accompanies perfusion of the lungs. The much higher CO_2 tensions in *Protopterus* than in *Neoceratodus is* a correlate of the increased importance of air breathing. The behavioural response to air exposure of these two lungfishes differs markedly. While both intensify their air breathing by increased frequency and depth, *Neoceratodus* becomes restless and splashes about in frantic efforts to get back into water. *Protopterus,* on the other hand, normally takes to air exposure with ease. Air exposure of *Neoceratodus* promptly engages the lung in O_2 absorption, and the O_2 tension in blood from the pulmonary vein increases from about 40 mm Hg to more than 90 mm Hg. Yet, it is apparent that this change fails to maintain the systemic arterial O_2 saturation which drops precipitously because of inadequate blood flow through the lung. Note that the changes in oxygenation are accompanied by a conspicuous rise in the blood CO_2 tension. *Protopterus* shows an increased oxygenation in both the pulmonary vein and in systemic arteries. The concurrent increase in CO_2 tension is related to a loss of the important escape route via gills and skin and the inevitable retention of CO_2 resulting from the anatomical dead space of the bidirectional airway. The CO_2 retention is minimised by the high tidal volumes relative to total lung volume. The time course of the blood gas changes during air exposure of the two lungfishes also shows the inadequacy of the *Neoceratodus* lung to cope with both O_2 absorption and CO_2 elimination, while the *Protopterus* lung can handle both tasks even though the internal CO_2 levels stabilise at a much higher level in air than in water.

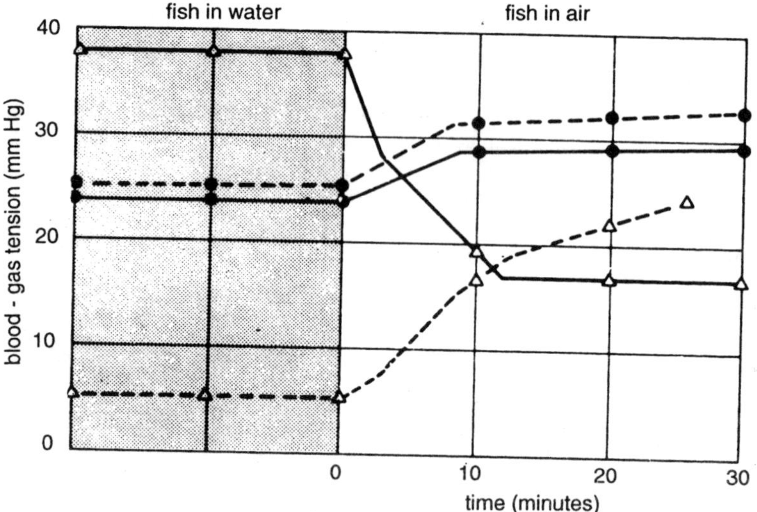

Figure 5.41 : A comparison of the time course of changes in O_2 and CO_2 tension of arterial blood during air exposure of Neoceratodus *and* Protopterus. *Broken lines indicate CO_2 tensions. Solid lines indicate O_2 tensions. Adapted from Lenfant and Johansen.*

Gas exchange and gas transport in *Electrophorus,* which practices air breathing with the uniquely modified oral mucosa. The O_2 tension inside the mouth drops from about 140 mm Hg right after a breath to 60 mm Hg at the end of an average 2-min breath interval. Concurrently, the CO_2 tension increases from about 7 to 30 mm Hg inside the mouth,

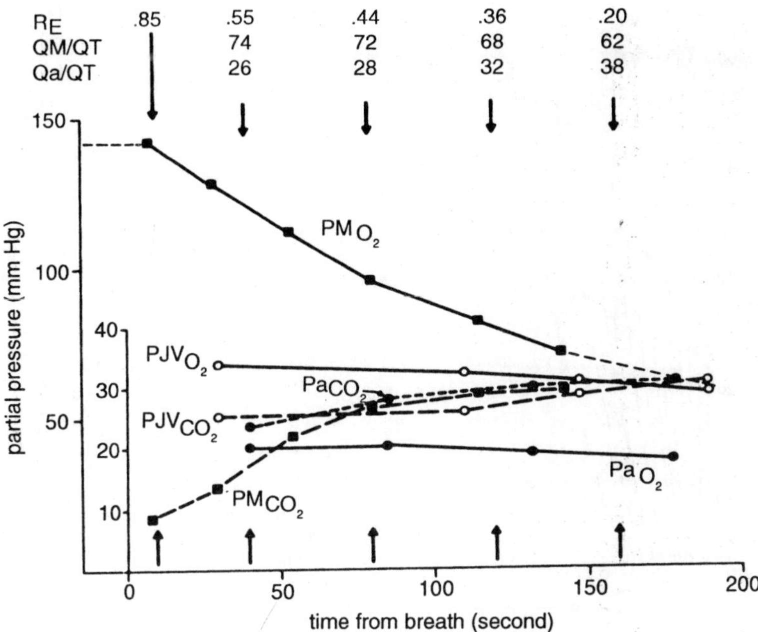

Figure 5.42 : Changes in O_2 and CO_2 tensions in the mouth gas (M), systemic arterial blood (a), and blood from the jugular vein (JV) in Electrophorus *between consecutive air breaths. Listed on top are computed values of the gas exchange ratio (R_E), the fraction of cardiac output going to the mouth QM/QT and to the systemic arteries Qa/QT. The vertical arrows indicate the times at which these calculations were made from the corresponding values of gas partial pressures.*

with most of the increase taking place early in the breath interval. The gas exchange ratio drops from about 0.85 just after a breath to 0.20 just before the next breath. Hence, CO_2 is eliminated to a large extent by aquatic gas exchange through the skin or vestigial gills. The low O_2 tension corresponding to about 65% O_2 saturation in arterial blood results from the shunting of blood from the mouth organ back to the systemic veins before arterial distribution. Blood from the jugular vein draining the mouth organ and the anterior systemic veins showed higher O_2 tensions corresponding to about 90% O_2 saturation. This difference in saturation is caused by the additional admixture of systemic venous blood from the posterior end before the blood is distributed from the heart to the arteries.

The higher level of CO_2 tension in the jugular vein than in the arterial blood suggests that systemic venous blood returning via the posterior cardinal veins, which receive the major cutaneous drainage,

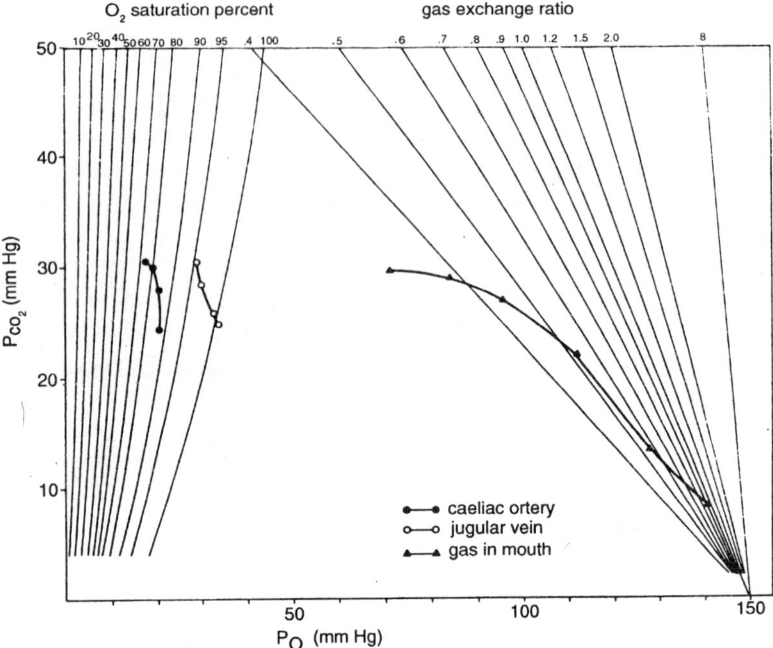

Figure 5.43 : Nomogram summarizing changes in gas tensions in blood and inside the mouth during an interval between air breaths in Electrophorus.

must have the lowest CO_2 tension; otherwise CO_2 must have been largely eliminated during passage through the vestigial gills.

The nomogram summarises results obtained in *Electrophorus* and substantiates the efficiency loss in gas transport caused by the vascular shunt from the mouth organ.

Control of Breathing in Air-Breathing Fishes

Breathing in vertebrates is, in general, geared to satisfy changing metabolic requirements by maintaining stable internal conditions for gas exchange and gas transport. The factors affecting the control of breathing are markedly different for water breathers and air breathers. The aquatic animal is surrounded by an environment of changing gas composition in which O_2 and CO_2 tensions may change in the same or in opposite directions. Also, in aquatic animals the external and internal environments are in intimate diffusion contact at the gills, and external changes will quickly change the internal environment. Air-breathing terrestrial animals, on the other hand, normally experience an external environment stable in its respiratory qualities; in addition, the presence of an internal gas compartment in the lungs will protect the internal environment against external changes. Control of breathing in the air breather can hence be based on negative feedback from receptors sensing internal parameters important to gas transport such as O_2 and CO_2 tension and pH of arterial blood. The aquatic animal has a more difficult problem in regulation because an internal factor like the arterial O_2 tension may change as a result of external, as well as, internal changes. If, for instance, internal hypoxia occurs when a fish enters severely O_2deficient water, an increased breathing effort may further aggravate the internal hypoxia. It would be advantageous for the fish to move to water of higher O_2 content. Control of breathing should hence allow the fish to distinguish between changes occurring internally or externally. This complex scheme calls for chemosensitive mechanisms that can screen external as well as internal changes.

It has long been contended that fishes show preference reactions to external gradients in gas composition and that this is linked with an ability to detect the gas tensions in the water by external sensors. Many authors view the orientation of fish to the higher O_2 availability in a gradient as a nondirective escape reaction caused by incipient suffocation. On the other hand, there is little doubt that many fishes living in severely oxygen-deficient swamp water show a clear chemotaxis for the more oxygen-rich surface layers of the swamp. In fact, the existence of many fishes depends on frequent return to the surface layers to satisfy respiratory needs.

The apparent variance among fishes in their response to O. gradients does not extend to avoidance reactions to P_{CO_2} and pH gradients which reportedly are common and much more acute. Such reactions are presumably triggered by external gustatory receptors. Some authors consider the breathing responses to CO_2 change in fishes to be connected to chemoreceptors located in the gill region. Meanwhile, no experimental work has been reported that conclusively identifies the location of and mechanism of chemosensitive reflex control of breathing in fishes.

A consideration of these problems takes on special significance in air-breathing fishes. From their practice of a dual mode of breathing one will expect that air and aquatic breathing are both individually controlled and mutually coordinated.

Normal Breathing Behaviour

The amplitude and frequency of both air and aquatic breathing in air-breathing fishes are very labile. In fishes employing aquatic as well as air breathing the frequency of branchial pumping generally exceeds the frequency of air breathing. Commonly, the rate of branchial pumping is low just after an air breath, but it increases progressively before the next air breath. The contribution to gas exchange from air breathing evidently reduces the chemoreceptor drive to branchial breathing. Conversely, this drive will progressively increase with time after an air breath with a resultant higher paced branchial pumping and finally evocation of another air breath. The two types of breathing are apparently under a common or mutual control. In many fishes, the rates of both branchial and air breathing show a notable increase during swimming activity. Young specimens of *Neoceratodus* change from sporadic air breaths during rest in aerated water to a steady higher rate of air breathing during swimming. Similarly, *Polypterus senegalus* and *Amia calva,* both of which use the swim bladder for accessory air breathing, show an augmented rate of branchial and air breathing with the onset of exercise. Thus the resting rate of water breathing provides little reserve for increased metabolic activity. The energy expenditure of water ventilation is high, and a definite advantage attends the use of less costly alternative ways of gas exchange during periods of increased demand. Water breathing during low activity, however, permits the fish to exploit its aquatic environment without the need for frequent ascents to the surface for air.

It is generally held that structural adaptations for air breathing in fishes evolved in association with environmental oxygen lack. Yet fishes such as the Australian lungfish, the holostean fishes *Amia* and *Lepisosteus,* and many teleosts breathe air even though their habitats are not oxygen deficient. Thus, internal oxygen deficiencies resulting from activity may have provided a selection pressure for development of air breathing independent of the oxygen availability in the water. Certainly, the exceptional air-breathing habits of marine pelagic fishes such as the tarpon, *Megalops*, support this contention. Recent work comparing the relative importance of water breathing and air breathing in *Amia* at different temperatures has emphasised the importance of air breathing for support of higher metabolic activity. The situation resembles the prompt stimulation of breathing in exercising mammals without an apparent initiating role of arterial chemoreceptors.

A few fishes, such as the electric eel, have come so far in the dominance of air breathing that the muscles used for water ventilation and the gills themselves have atrophied to a point where mechanical manifestation of branchial breathing has ceased altogether. However, aquatic gas exchange continues mainly as CO_2 elimination through the

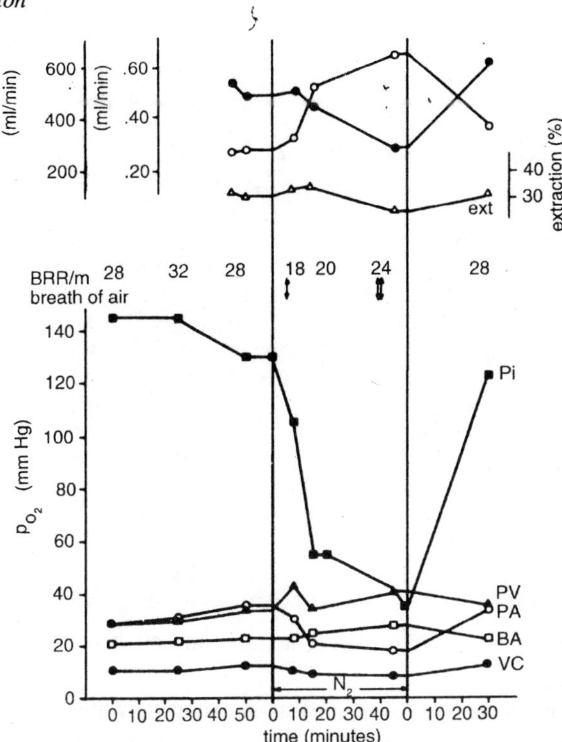

Figure 5.44 : Responses to breathing deoxygenated water in Neoceratodus. Ve, ventilation of water; V_{O2} oxygen uptake from the water; Ext, percent extraction of oxygen from the water; and BRR, branchial respiratory rate. Arrows mark breaths of air. Oxygen tensions followed in: Pi, inspired water; PV, pulmonary vein; Pa, pulmonary artery; BA, an afferent branchial artery; and VC, vena cava.

skin or the atrophied gills. This important part of overall gas exchange is mainly regulated by changes of blood perfusion at the interfaces between water and blood in the skin or gills.

A marked change in the respiratory behaviour of air-breathing fishes likely attend their ontogenetic development. Some species such as the African lungfish are decidedly aquatic breathers during larval and juvenile stages, but air breathing takes on a dominating importance in the adult fish.

Breathing Responses to Changes in External Gas Composition

Changes of Environmental Oxygenation

Exposure to hypoxic water evokes different breathing responses among air-breathing fishes. The response patterns express an apparent

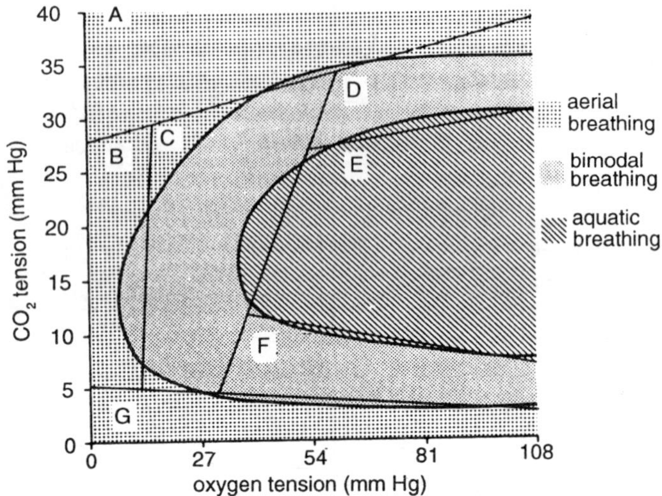

Figure 5.45 : Relationship between the mode of respiration in the yarrow Erythrinus unitaeniatus, and the O_2 and CO_2 tensions in the external water.

relationship to the relative role of aquatic and air breathing in overall gas exchange.

The Australian lungfish is an almost exclusive water breather when at rest in aerated water. It reacts to hypoxic water by promptly increasing its branchial breathing efforts. Water ventilation increased as much as five times when the O_2 tension in the water was reduced to about 80 mm Hg. The result of a typical experiment. Ventilation doubled, while the extraction of O_2 decreased as did the overall O_2 uptake. The increased ventilation was caused by an increase of the volume propelled by each branchial pumping movement and not by an increased frequency of the branchial pumping. The pattern of air breathing changed from no air breaths in the hour preceding the hypoxia to more than three air breaths in the first half-hour of hypoxia. The stimulation of air breathing was much slower in onset than the increase of branchial breathing. The commencement of air breathing, however, initiated important changes in the pattern of blood perfusion. Most notably, the O_2 tensions increased in blood drawn from those branchial arteries, which supply the systemic circulation, while blood in the pulmonary arteries decreased in O_2 tension. The O_2 tension of blood in the pulmonary vein also increased. Thus the lung had assumed a role in O_2 uptake, and important circulatory changes attended this change.

The yarrow, *Erythrinus* sp., is another fish utilizing the swim

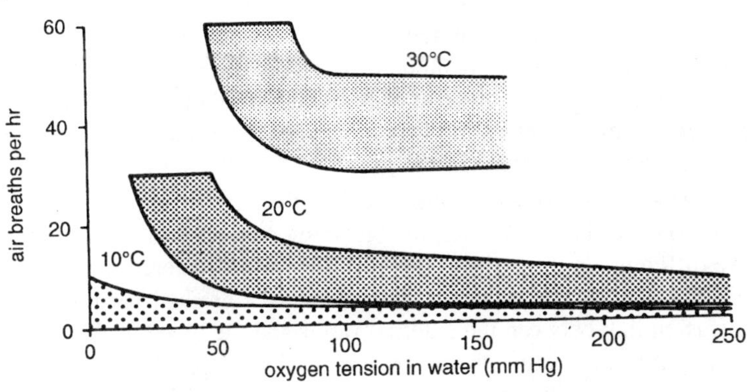

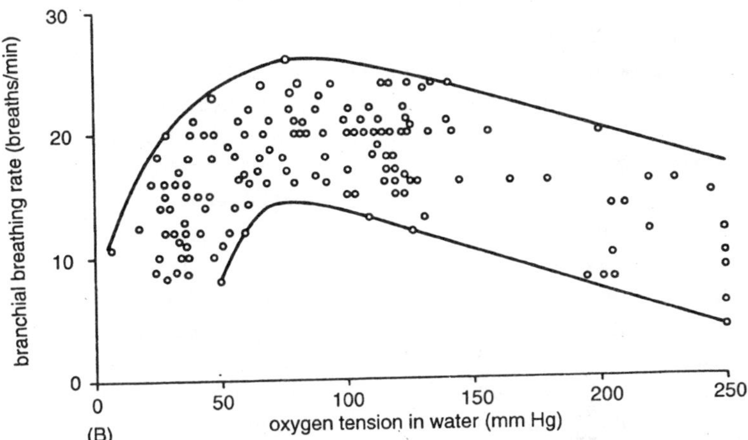

Figure 5.46 : Rate of air breathing in Amia calva (top) at different temperatures and rates of branchial brathing (temp. range, 14°-26° C) both as functions of oxygen tensions in the ambient water.

bladder for accessory air breathing. How the breathing behaviour is related to the external gas composition for *Erythrinus*. Depending on the CO_2 content of the water aquatic respiration suffices to satisfy the respiratory needs at decreasing O_2 tension down to about 40 mm Hg.

The bowfin, *Amia calva,* also utilises the air bladder as an accessory gas exchanger. The relative importance of air breathing in this fish depends in a large measure on the temperature of the water. The fish occupies temperate regions of North America and experiences large annual temperature variations. At water temperatures of 20°-25°C the fish regularly ascends for air even in well-aerated water. Upon deoxygenation of the water the fish initially increases the branchial breathing rate moderately. Further deoxygcnation elicits an increased rate of air breathing. When the latter occurs, both the rate and amplitude of branchial breathing drop below the level prevailing before deoxygenation. This depression of branchial breathing may be an adaptive response which reduces the loss of oxygen from the air bladder to the oxygen-deficient water via the gills.

Among obligate air breathers such as the African and South American lungfishes or the electric eel, the breathing responses to external changes are different. In these fishes the gills are variously degenerated, and the gill surface area is reduced. Thus there is a gradual reduction of the diffusion link between the external water and the blood as the fish shifts emphasis from water to air breathing. Deoxygenation of the water elicits no change in the breathing pattern of these fishes. In a sense they have become relieved of the environmental deoxygenation stress which originally provided the selection pressure favoring the air-breathing habit. Blood gas measurements have substantiated that no change in internal O_2 tension occurs in response to deoxygenation of the water, nor do the fishes exhibit any restlessness suggesting the presence of external chemosensation of O_2 that could guide the fish in avoidance or attraction to the degree of external oxygenation.

Experiments on juvenile specimens of the African lungfish have produced data in conflict with those cited above. Deoxygenated water elicited an increase of both branchial and air breathing in the juveniles. However, these specimens survived on water breathing alone for several days, which contrasts markedly with other observations on adult fish. In addition, the experimental procedure evoked hypoxia simultaneously in water and in the air phase and consequently could stimulate the fish via the lung, as well as the gills. The interesting possibility emerges from these studies that control of breathing changes markedly during the ontogenetic development of the African lungfish.

A logical and simple experiment used to test the importance of air breathing in fish is to deprive them of access to air. Depending on the temperature and gas composition of the external water, such experiments have produced variable results, but they show that the degree of internal oxygenation clearly influences the breathing efforts. The attempts to breathe air increase in all air-breathing fishes, while branchial breathing appears to increase only in species normally dependent on branchial breathing for an important part of their oxygen uptake.

Experiments in which the fish have surfaced into atmospheres of controlled gas composition, while the conditions in the water have remained stable, have been more conclusive. Only in a few experiments have internal blood gas tensions been correlated with such procedures. In the African lungfish, introduction of hypoxic gas into the lung causes a prompt and sharp increase in the rate of air breathing. Simultaneous measurements of arterial O_2 tension show that air breathing is stimulated at much higher values than those recorded when the fish practices undisturbed breathing in normal air (Johansen and Lenfant, 1968). This implies that the chemosensitive areas which trigger the response to inhalation of hypoxic gas are more sensitive to the rate of change than to the actual level of oxygen tension in the hypoxic condition. Experiments on *Electrophorus* support such a contention by showing remarkably prompt responses to changes in composition of inhaled air. How the rate of air breathing was changed after a single breath of air following hypoxic breathing. This implies a location of the chemoreceptors in the buccal mucosa or in the blood path in intimate contact with the gas in the mouth. A location in the systemic arteries is ruled out because of the complete shunt of blood from the gas exchange organ to the systemic veins. The resulting changes in systemic arterial blood are much too slow to represent an important feedback stimulus. Inhalation of oxygen-enriched atmospheres causes a depression of air breathing and is important by suggesting the removal of a tonic P_{O_2} -dependent stimulus. This indicates that normal spontaneous breathing is governed by changes of oxygen tension in the air-breathing organ. The situation corresponds to conditions in higher vertebrates where oxygen inhalation is effective in removing the tonic activity of chemoreceptor cells stimulated by the normally prevailing levels of P_{ot}. Hyperoxic breathing also reduces the rate of air breathing in *Protopterus* and other air-breathing fishes. Species depending largely on aquatic breathing show a depression of branchial breathing in hyperoxic water or following inhalation of hyperoxic air.

Changes of Environmental CO_2 Tensions

Internal CO_2 tensions are very low among purely water-breathing fishes living in well-aerated water. A negative feedback control of breathing based on the level of metabolically produced CO_2 appears to be of little consequence in controlling breathing in such fishes. However, in stagnant tropical swamps, the CO_2 levels can build up high enough to adversely affect the dissociation of oxy-hemoglobin and to disrupt internal acidbase balance. Adoption of air breathing also inevitably leads to a retention and increase of internal CO_2 tensions. For tropical air-breathing fishes the external and internal levels of CO_2 may hence be important factors in the control of breathing.

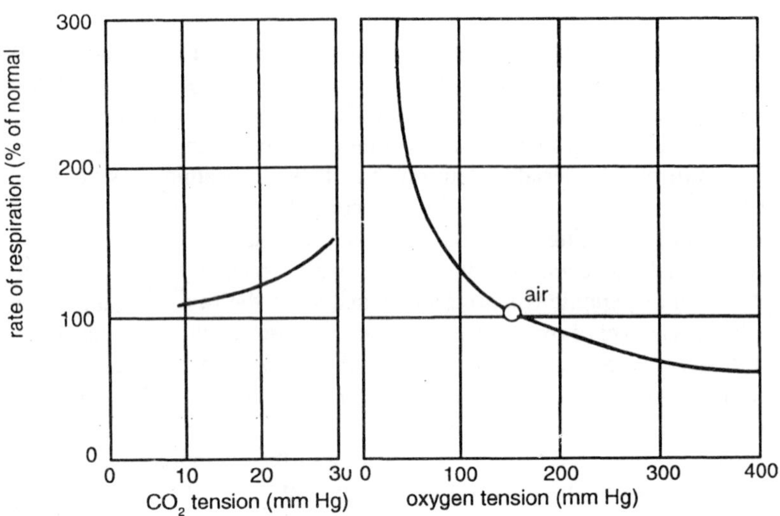

Figure 5.47 : Percent change in the rate of air breathing of Electrophorus in relation to the gas composition of the inhaled air. From Johansen et al..

Exposure to increasing CO_2 tensions in the external water evokes extremely varied breathing responses in different species of air-breathing fishes. This variety and the paucity of controlled experiments complicate the evaluation of the role of CO_2 in respiratory control. It appears that CO_2 at moderately high tensions stimulates both branchial and air breathing. At increasing tensions, however, a depression of branchial breathing occurs while air breathing continues at increased rates. The response pattern also seems to depend on the relative importance of air breathing and water breathing in the normal respiratory behaviour of the fish.

In a few fishes studied adequately, such as the teleost *Symbranchus* and the Australian lungfish *Neoceratodus,* exposure to CO_2 concentration as low as 1-2% caused a prompt depression of branchial breathing while the rate of air breathing was accelerated; CO_2 concentrations below 1% were not tested. In *Amia calva,* which employs accessory air breathing using the air bladder, $CO2_2$ concentrations up to 3% stimulated both branchial and air breathing while higher concentrations depressed branchial breathing. The rapid depression of branchial breathing in *Neoceratoclus* following exposure to 2% CO_2 in the water. Upon removal of the hypercarbic stimulus, the changes were promptly reversed. Atropinisation prevented the CO_2 response. This suggests a cholinergic link in the reflex chain eliciting the response. It is of considerable interest that filling the lung in *Neoceratodus* with a CO_2-rich gas

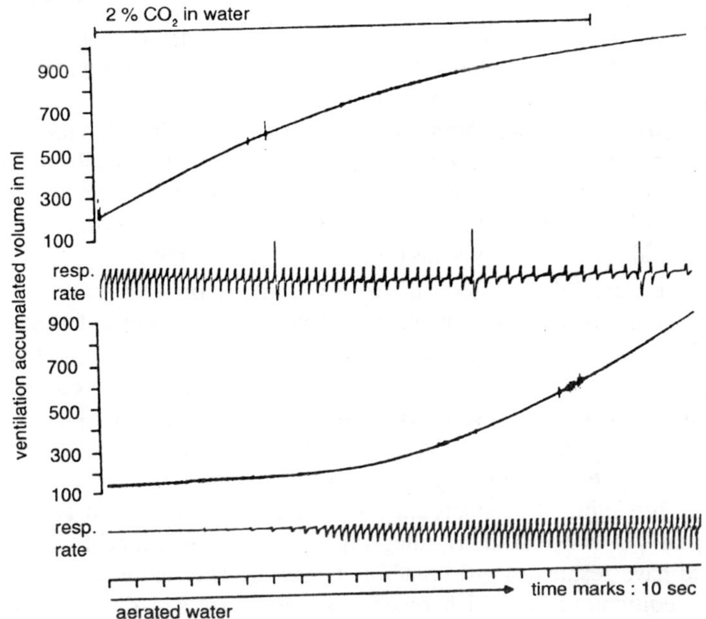

Figure 5.48 : Continuous recording of water ventilation and branchial respiratory rate in Neoceratodus during (top) and after (bottom) an increase in P_{CO2} of the ambient water.

mixture, which greatly elevates the CO_2 level in the pulmonary vein and in blood afferent to the gills, evokes no breathing response whatever. However, blood sampled distal to the gills was very low in CO_2 and attested to a remarkable efficiency of the gills in removing that gas. These data prompted the suggestion that the CO_2-sensitive receptors eliciting the depression of branchial breathing must be external sensors like gustatory or olfactory receptors or they must be located in the blood path efferent to the branchial circulation. Strong corroborative evidence was gained from studies of *Symbranchus,* which is capable of both air and aquatic breathing with its gills. The fish shows inhibition of branchial breathing in CO_2-rich water which persists as CO_2 leaks into the bloodstream. However, when the fish is transferred to CO_2-free water, a high frequency of forceful branchial movements is established. Hence, the fish appears to possess a sensory mechanism that can differentiate between the external and internal CO_2 level. Similarly, in experiments with *Protopterus,* bubbling of 5% CO_2 in the water stopped branchial breathing before there were appreciable changes in CO_2 tension of the arterial blood.

The stimulation of air breathing by CO_2 may be a direct effect, or it may be secondary to the reduced oxygen uptake caused by the depression of branchial breathing. The response of *Protopterus* and *Electrophorus,* in which branchial breathing contributes little or nothing to the oxygen uptake, suggests a direct effect. Also, inhalation of CO_2-enriched gas stimulates air breathing in these fish while internal oxygen levels remain the same or increase. However, in predominantly water-breathing fishes such as *Neoceratodus* or *Amia,* the influence of CO_2 on air breathing may be both direct and indirect.

The variability found in the few studies on CO_2 response in airbreathing fishes was also apparent in the important work of Willmer. The factors controlling breathing in the yarrow, *Erythrinus,* a teleost which uses its air bladder for accessory breathing. Willmer contended that a certain CO_2 concentration in the blood is necessary to drive the branchial breathing. At very low O_2 contents of the water, branchial movements ceased, possibly to prevent loss to the water of O_2 gained by air breathing. The opercula also remain actively closed at higher external CO_2 concentrations, allegedly to prevent internal acidosis which would have resulted if gill breathing persisted. The figure also explains the conditions under which respiratory needs are satisfied by aquatic respiration alone and similarly the factors in the environment which cause the fish to employ bimodal breathing.

Breathing Responses to Mechanical Stimuli

A possible role of external mechanosensitive receptors in control of breathing rhythms is suggested by recent experiments showing a marked acceleration of branchial breathing rate in *Lepidosiren* and *Protopterus* when the water was stirred or agitated. Such a role of mechanoreceptors in fish takes on special significance in the normally O,-deficient stagnant tropical swamps, where stirring of surface water by wind movement or other mechanical agitation should improve the respiratory quality of the water and make branchial breathing more efficient.

Inflation and deflation of the air bladder in *Amia* definitely influence the breathing pattern. A deflation causes the fish to promptly ascend for air, while fishes approaching the surface for air breathing interrupt their ascent and settle to the bottom again if the bladder is artificially inflated. This response occurs whether air, oxygen, or nitrogen is injected. This inflation response shows striking similarity with the inflation reflex described long ago and known as the Hering-Breuer reflex in mammals. In higher vertebrates the inhibition of breathing caused by stretch in the lungs is no longer held to exert any important influence in control of breathing since such changes do not supply information to the respiratory center about the respiratory quality of the blood. In air-breathing fishes a stretch reflex may prove to play an important role since the volume and thus the distention of the air bladder will change during the long intervals between air breaths because of the low gas exchange ratio for the air-breathing organ. Since the air bladder is primarily an O_2 absorber, the change in volume with time provides a measure of the rate of O_2 depletion. If the reflex described in fishes such as *Amia* proves to be homologous with the vagus-mediated inflation reflex in higher vertebrates, the case may exemplify a basic reflex type that exerted a more important regulatory role early in its phylogenetic history.

An important role of the air which a fish takes below the surface is its effect on buoyancy and, indeed, in most teleosts the air bladder is engaged mainly in buoyancy control. In fishes which use the air bladder primarily as a respiratory organ, changes of buoyancy clearly influence breathing behaviour. Ascent for air in such fishes may be provoked simply by tying a weight around them. The airbreathing fishes consequently face the additional task of coordinating the mechanisms for buoyancy control with those engaged in satisfying the respiratory needs.

Breathing Response to Air Exposure

Only a few of the fishes adapted for air breathing make voluntary excursions onto dry land. Estivating fishes are, of course, compelled to subsist on air breathing when entrapped in the substratum during drought.

It seems typical that fishes engaged in voluntary air exposure under natural conditions promptly commence frequent ventilatory movements and show no signs of distress and erratic behaviour such as seen in purely aquatic breathers when air-exposed in the laboratory. Only in the South American teleost *Symbranchus* and in the lungfishes have blood gas values been correlated with breathing behaviour during air exposure. In *Symbranchus,* arterial P_{O_2} and Pea, both increased during air exposure. In *Protopterus,* the onset of the accelerated air breathing was too rapid to have been elicited by internal changes in blood gas composition. The sensory input evoking the response must have been the physical act of air exposure itself. It is of interest that *Protopterus,* while recovering from anesthesia, will attempt to breathe air when air-exposed, even while still unresponsive to tactile or painful stimuli. The blood gas values show an increase in both P_{O_2}. and P_{CO_2} ; the latter obviously results from the interruption of aquatic CO_2 exchange. The sustained rate of air breathing during air exposure could be driven by the increased level of internal P_{CO_2} , although this stimulus could not play a role in the initial elicitation of the response.

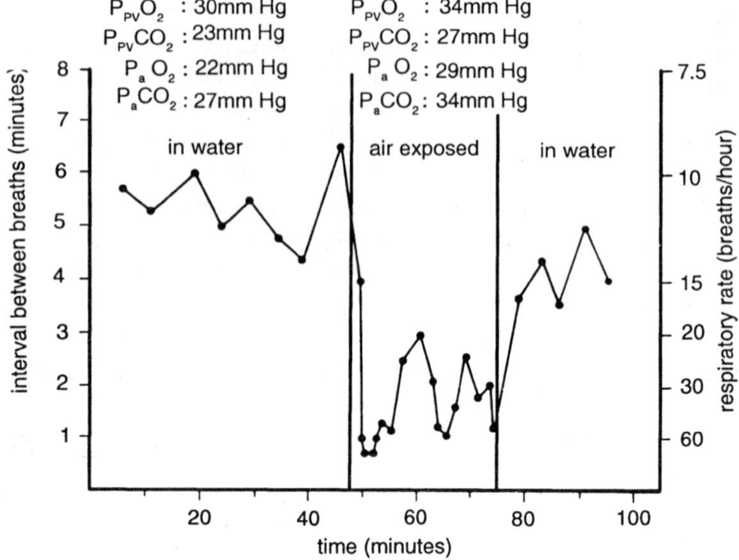

Figure 5.49 : Changes in rate of air breathing and blood gas tensions following air exposure in Protopterus. PV, pulmonary vein; PA, pulmonary artery; and a, systemic artery.

Coupling of Respiratory and Circulatory Events

Cardiovascular and respiratory events are strongly interrelated in fishes. Cardiac output determination by the Fick principle in *Neoceratodus* indicated a marked increase during exposure to hypoxic water. The increase in blood perfusion exceeded the relative increase in the ventilation of water. Consequently, the ventilation perfusion ratio of the gills changed from about 12 before hypoxia to about 4, after 45 min exposure to hypoxic water, then climbed to 17 during recovery. The ratio of pulmonary blood flow to total blood flow rose with increased rate of air breathing. In *Protopterus,* direct blood flow measurements have shown that spontaneous air breathing causes distinct changes in heart rate, total cardiac output, and regional blood flow. Cardiac output increased sharply with each air breath and declined slowly during the subsequent breath interval. Branchial vascular resistance tended to decrease in conjunction with an air breath. The most consistent change was a marked increase of pulmonary blood flow which exceeded that in the vena cava and thus indicated a regional flow shift from the systemic to the pulmonary circuit. Thus, pulmonary flow could vary from less than 20% to more than 70% of the cardiac output.

In *Electrophorus* there is an equally strong interrelationship between circulatory and respiratory events. When the intervals between air breaths were relatively long (exceeding 1 min), cardiac output and heart rate decreased progressively in the late phase of the interval. At the next breath, both heart rate and blood velocity promptly increased and were reestablished at the levels prevailing just after the preceding breath. This cyclic phenomenon occurred as a normal event when the intervals between breaths were long, whereas no time-dependent changes were recorded at short intervals. The sudden changes in blood perfusion elicited by air breaths are at least, in part, controlled by the degree of mechanical distension of the air-breathing organ inside the mouth. Artificial inflation of the mouth via a catheter elicited an increase in heart rate and cardiac output whether nitrogen, air, or oxygen were injected; thus, the chemical composition of the gas in the mouth does not constitute the actual stimulus. A similar interrelation between cardiovascular events and distention of the air-breathing organ has been reported for *Symbranchus*. Based on blood gas values the fractional distribution of cardiac output has been calculated in *Electrophorus*. Just after a breath the blood flow fraction going to the air-breathing organ is at its highest, and it declines steadily later in the breathing interval. Similar experiments during hypoxic and hyperoxic breathing showed that blood will shift toward the respiratory organ when the O_2 tension in the mouth is high. This adjustment has the obvious rationale of promoting the matching process between blood and gas in the air-breathing organ.

Increased perfusion through skin and mucosal linings of the buccal cavity has been frequently reported in air-exposed air-breathing fishes. The marine gobiid fish, *Gillichtys,* displays some remarkable vascular changes when placed in hypoxic water. Under such conditions this fish habitually visits the surface and gulps air which is placed as a bubble between the tongue and the buccal roof. When stimulated to breath air by low O_2 tension in the water, the buccal epithelial capillaries promptly dilate and become engorged with blood, giving the epithelium a corrugated lunglike appearance, bright red from a proliferous circulation. This change in perfusion occurs within a few minutes and is totally absent in related fishes which are purely aquatic breathers. The close coupling between mechanisms controlling breathing and vasomotor responses seems basic to very diverse types of respiratory organs ranging from the gills in fishes to the lungs of both lower and higher vertebrates.

Only in the lungfishes among air-breathing fishes has the vascular circuit to the air-breathing organ gained enough anatomical separation from the systemic circuit to allow separate selective perfusion of the two circuits. During rest in aerated water, the lung in *Neoceratodus* *is* of no consequence to gas exchange, and no apparent tendency for a preferential distribution of blood from the pulmonary vein to the systemic arteries is apparent. Exposure to hypoxic water evokes an active use of the lung and results in a clear preferential passage of blood from the lung to the anterior branchial arteries which carry the major portion of the blood to the dorsal aorta. One analysis showed that more than 83% blood from the pulmonary vein passes this way, whereas recirculation of systemic venous blood to the anterior arches was reduced to about 16%.

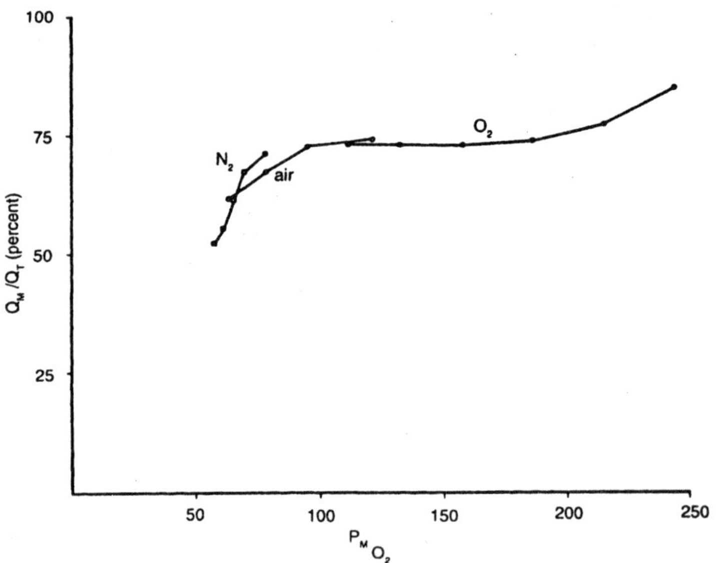

Figure 5.50 : Relationship between the fractional distribution of blood flow to the mouth air-breathing organ and the oxygen tension of the air inside the mouth of Electrophorus. The three curves show the relationship when breathing a hypoxic and a hyperoxic atmosphere and normal air, respectively.

Lepidosiren and *Protopterus* are much more dependent on air breathing and show a consistent gradient of O_2 tension between blood in systemic and pulmonary arteries. This gradient reflects the degree of preferential perfusion which increases with an increased rate of air breathing. In *Protopterus* the proportion of blood conveyed to the systemic arteries from the pulmonary circuit could be as high as 95% shortly after a breath and decline to about 65%-just prior to the next air breath. The factors important in the control of the preferential passage of blood are hence also coupled to the breathing rhythm.

6

Swim Bladder

The swim bladder acts as a hydrostatic organ by replacing part of the fish body, which is heavier than water, with gas. The specific weight of the fish as a whole is thereby brought toward that of the ambient water, and the fish becomes neutrally buoyant.

The function of the swim bladder rests on its ability to maintain a gas space inside a fish and to vary this amount of gas according to changing hydrostatic demands. The problems connected with this function are amplified by the fact that while most natural waters, and consequently the arterial blood of fishes, have a P_{O_2} and P_{N_2} of about 0.2 and 0.8 atm, respectively, the partial pressure of O_2 and N_2 in the swim bladder may be 100 and 20 atm, respectively. This ability of the bladder to concentrate O_2 and N_2 some 500 and 30 times, respectively, is the unique property of the organ. This aspect will therefore be the focus of our interest throughout this chapter. In addition we shall discuss the problems of hormonal and nervous regulation of its activities, as well as the "biological importance" of the organ.

In the swim bladder an intimate relationship between anatomy and function is unusually evident. This holds true for the microstructures but is particularly well displayed in the macroanatomy or, more properly, the architecture of the organ. Since so much of the swim bladder problem hinges on the anatomy, we shall describe this aspect in some detail.

Various aspects of the swim bladder problem have been reviewed at intervals. The articles by Fange, Denton, Marshall, Kuhn *et al.*, Steen, and Alexander are of particular interest.

THE BIOLOGICAL SIGNIFICANCE OF THE SWIM BLADDER

Marshall gives an excellent account of the distribution of those marine fishes that have a swim bladder. The bladder is found most frequently among fishes in the upper 200 meters and is less frequently encountered down to 1000 meters. The organ is not found in fishes living in free water below this depth, but it is found regularly in those living near the bottom down to 2000 meters. The report by Nielsen and Munk of a swim bladder in a fish caught at about 7000 meters depth seems to represent the present depth record. Except for typical bottom fishes most freshwater teleosts possess a swim bladder.

The fact that typical bottom fishes do not possess a swim bladder is reasonable since they presumably rest on the bottom and thus may benefit by a slight overweight. The arguments presented to justify the lack of a bladder in elasmobranchs and in many active pelagic fishes are less convincing. These arguments often imply that since these fishes swim all or most of their lives anyway, the additional energy needed to counteract sinking is insignificant. While the relative energy expenditure for this purpose will be less in active than in passive fishes, it is hardly less expensive in absolute terms. It is more tempting to suggest that certain disadvantages are associated with the presence of a bladder or certain advantages with its absence.

A fish with a certain amount of gas in the bladder is in buoyancy equilibrium at one depth only. Like a Cartesian diver the slightest vertical displacement will bring it out of equilibrium and the force pushing upward or downward will increase with the displacement. Volume changes are checked by an interplay between removal and deposition of gas. However, these processes are slow as compared to the movements of the fish. Few fishes are able to refill their bladder at surface levels in less than a few hours. Thus while a swim bladder saves locomotory energy for the fish, it inflicts upon it an instability with regard to vertical position. This may be particularly inconvenient for active fishes which change depth rapidly, e.g., to chase prey. These difficulties connected with having a bladder will obviously be less marked the deeper the fish lives since the volume change is proportional to the relative change in hydrostatic pressure.

THE ARCHITECTURE OF THE SWIM BLADDER

The anatomy of the swim bladder is, from a functional point of view, very similar in all fishes. It is most instructively displayed in the common eel, *Anguilla vulgaris,* which will therefore serve as an example throughout this section.

General Organisation and Cytology

Embryologically the swim bladder develops as a pouch protruding from the foregut, and in some fishes the connection between the bladder and the digestive tract is retained in adult life. These fishes are called physostomes, and the gas content of their bladder is thus potentially in direct communication with the exterior. This avenue, the pneumatic duct is, however, always guarded by a muscular valve.

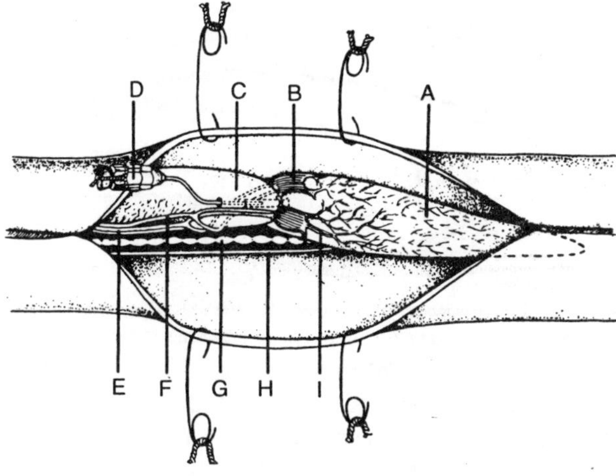

Figure 6.1 : A full-scale drawing of the swim bladder of an eel. A modified cannula is inserted into the secretory bladder. Intestines and gonads removed. A, secretory bladder; B, rete; C, reabsorbent bladder; D, modified cannula; E, prerete artery; F, postrete vein; G, dorsal artery; H, dorsal vein; and I, postrete artery and prerete vein.

In most fishes the proximal part of the pneumatic duct disappears early in their embryological development, whereas the distal part often develops into a well-vascularised, thin-walled resorbent part. The rest of the bladder becomes the site of gas deposition. In accordance with earlier nomenclature this is still termed the secretory area or bladder, and the tissue from which gas emanates is called the gas gland. As we shall see later there is no secretory or glandular function in the strict sense of the terms in this tissue.

Fishes with a closed swim bladder are called physoclists. In the literature one will also find the terms paraphysoclist and euphysoclist. The former refers to fishes where the resorbent and secretory part of the bladder is not sharply separated from one another. In the euphysoclists the two areas are separate.

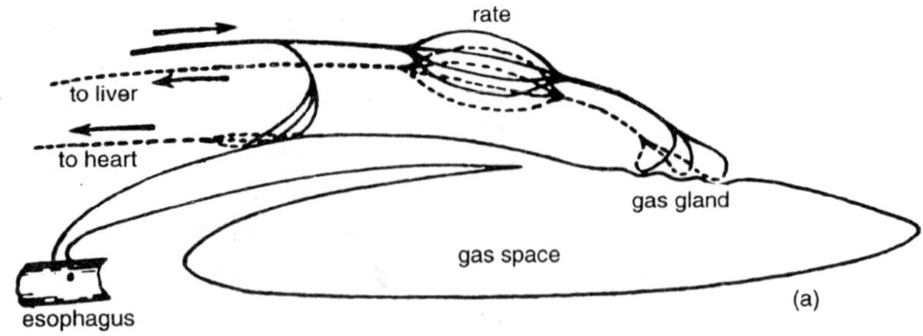

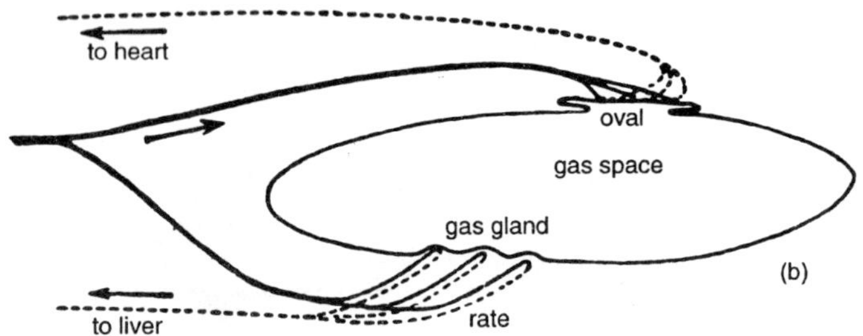

Figure 6.2 : The two main types of swim bladders: (a) a physostome bladder (from the eel, Anguilla vulgaris), and (b) a physoclist bladder.

The swim bladder in the eel displays to some degree the early embryological "physostome stage" of the physoclist bladder. But although the eel is a physostome, its swim bladder has in all essential features the same organisation as that found in physoclists.

The variability in organisation of the different parts of the swim bladder. In some euphysoclists the resorbent part forms a posterior chamber separated from the secretory part by a transverse membrane with a central adjustable opening. This membrane is called the diaphragm. In other euphysoclists the resorbent part is not a bladder but only a wellvascularised dorsal region of the swim bladder mucosa, separated from the rest of the mucosa by a muscular sheath. This is termed the oval. There also exist species provided with swim bladders of undoubtedly physoclist type, but in which neither oval nor diaphragm can be distinguished.

The wall of the secretory bladder consists of an outer silvery layer which is rich in collagen fibers, a middle vascular layer which frequently

The internal surface of the resorbent part is usually covered by flat epithelial cells, while the secretory epithelium is cuboidal. Dorn

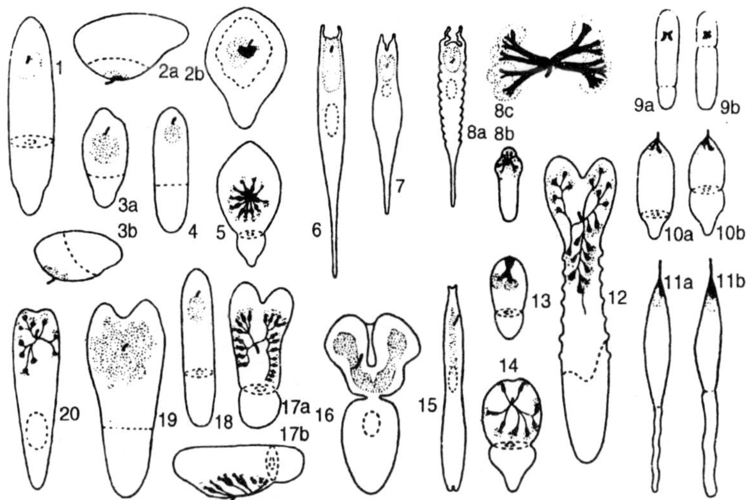

Figure 6.3: The variability in architecture of physoclist swim bladders. Dark areas show the retial structure(s). Dotted areas indicate the area from where gas appears. The dotted line indicates the boundary between areas of gas deposition and gas reabsorption. Species names are found in the original paper. 1. Labrus ossigragus. Diaphragm. 2. Gobius niger. Oval: the secretory bladder is maximally contracted. (a) Left side view; (b) ventral view. 3. Crenilabrus melops: (a) ventral view; (b) left side view. 4. Centrolabrus exoletos. 5. Sebastes viviparus. Diaphragm. 6. Gadus pollachius. Nearly closed oval. 7. Gadus minutus. Closed oval. 8. Gadus callarias: (a) closed oval; (b) embryo. Gas gland and rete mirabile. 9. Gasterosteus aculeatus: (a) normal position of the diaphragm; (b) diaphragm in a more cranial position owing to contraction of the secretory bladder. 10: Spinachia vulgaris: (a) normal position of the diaphragm; (b) the diaphragm translocated in a cranial direction owing to contraction of the secretory bladder. 11. Syngnathus aces: (a) secretory bladder relaxed; (b) secretory bladder contracted. The diaphragm has moved only slightly. The thin-walled caudal part of the bladder is distended. 12. Lota vulgaris. The diaphragm has an oblique position. The secretory bladder is halfway contracted. 13. Onos cimbrius. Diaphragm. 14. Onos mustela. Diaphragm. 15. Caranx trachurus. Nearly closed oval. 16. Raniceps raninus. Nearly closed oval. 17. Trigla gurnardus. Diaphragm: (a) ventral view; (b) left side view. 18. Ctenolabrus rupestris. Diaphragm. 19. Labrus berggylta. 20. Perca fluviatilis. Oval. contains muscle fibers, and an inner epithelial layer.

has done an electron microscopic study of the swim bladder of the eel. The resorbent epithelium is reminiscent of the respiratory epithelium of primitive lungs. The secretory epithelium on the other hand has a microstructure similar to secretory tissues, with an abundance of reticulum, inclusions, and mitochondria. Thus the cytology of the two parts of the bladder clearly indicates two functionally different elements: a metabolically passive one and a metabolically active one.

Circulation

The two main types of circulatory arrangements in the swim bladder. In most fishes the swim bladder receives its blood supply from a branch of the coeliac (pneumogastric) artery. This branch first gives off one or more branches to the capillary network of the resorbent part, then forms one or more retia and next the capillary network of the secretory bladder is formed. Whereas the arterial supply to the two parts of the bladder comes from one single source, the venous drainage is separate. The blood from the secretory part flows back through the gas gland vein to the hepatic portal vein. Thus it passes the liver before entering the sinus venosus of the heart. The blood from the reabsorbent bladder, on the other hand, runs directly to the sinus venosus through a pneumatic duct vein.

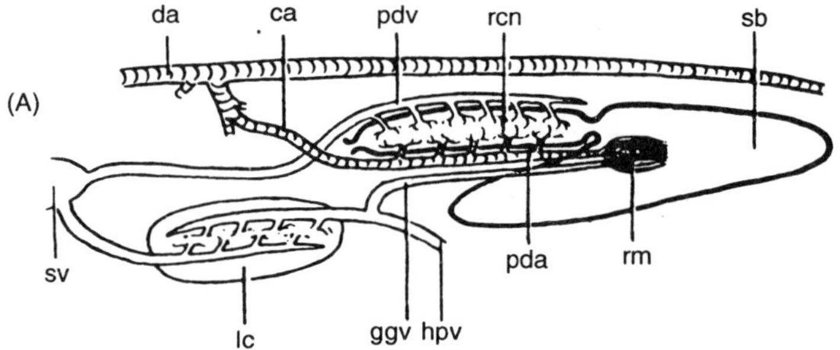

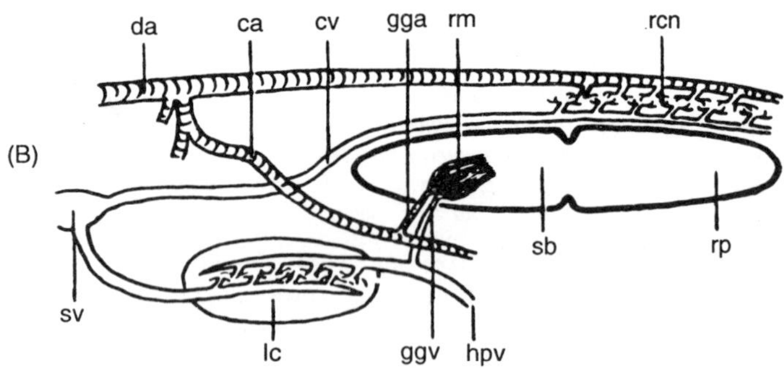

Figure 6.4 : Diagram of vascular supply to the (A) physostome type of swim bladder (the eel) and (B) euphysoclist type. Abbreviations: ca, coeliac artery; cv, cardial vein; da, dorsal aorta; gga, gas gland artery; ggv, gas gland vein; hpv, hepatic portal vein; lc, liver circulation; pda, pneumatic duct artery; pdv, pneumatic duct vein; rcn, resorbent capillary network; rm, rete mirabile; rp, resorbent part; sb, secretory bladder; and sv, sinus venosus.

In some physoclists, the arterial supply emanates from two sources. The secretory part receives blood from the coeliac artery, while the resorbent part is supplied from intercostal arteries.

It is important to note that all the blood to and from the secretory tissue passes through a rete, while that to the reabsorbent area does not.

The Rete Mirabile

Nowhere in biology is the countercurrent principle realised more perfectly than in the rete mirabile of the swim bladder. It is well developed in most physoclists, while in many physostomes it is only weakly developed. The herring is one of the rare species that has a bladder but no rete.

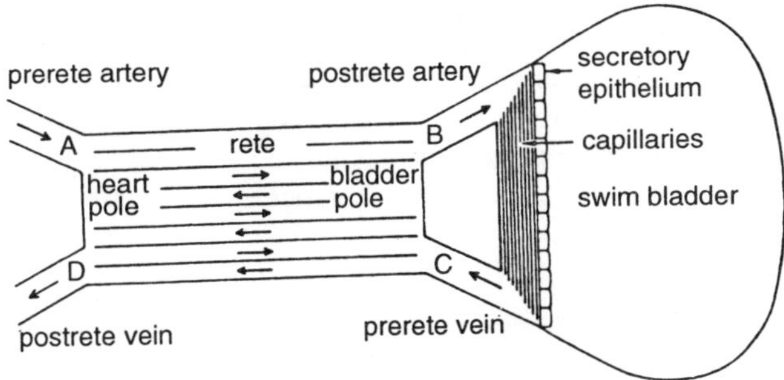

Figure 6.5 : A simplified drawing of the gas gland with rete and swim bladder. Also given are the anatomical designations used in the text.

Usually the rete consists of one or more distinctly discernible bundles of capillaries. The relationship between the bundle and the epithelium of the bladder varies. In all euphysoclist bladders the rete and the secretory epithelium have positions resembling the stem and hat of a mushroom. In these cases the rete is termed unipolar since the retial capillaries continue directly into the secretory epithelium without first running together in larger vessels. The whole structure may properly be called "the gas gland." In the bladder of all stomiatoids and in the eel, *Anguilla,* the retia are bipolar, that is, with its capillaries distinct from the epithelial ones. In these cases the term gas gland is less meaningful since neither the rete nor the epithelium alone has "glandular" capacity. If used, the gland term should include both structures.

In the case of a bipolar rete there are two capillary beds in parallel. This makes the vascular nomenclature somewhat complex. Steen put "pre rete" or "post rete" in front of the artery or vein to distinguish between vessels before and after they had passed the rete.

From a functional point of view the most important aspect of this structure is that the arterial blood to the bladder and the venous blood from the bladder are in intimate diffusion contact with each other. The arrangement is called a hairpin countercurrent system. This term is not strictly an anatomical one, but it expresses the loop nature of the circulation. Krogh counted 88,000 venous and 116,000 arterial capillaries in the two retia of an eel. These had an aggregated length of 352 and 464 meters, respectively. The total capillary surface was 106 meters^2 for the venous and 105 meters^2 for the arterial vessels. All this surface was contained in a volume of 0.064 cm^3 of which two-thirds were occupied by blood. The ratio of the total diffusion area to the rete volume is therefore some 1700 cm^2/cm^3. For the longnosed eel, *Synaphobranchus pinnatus,* the ratio is 1200. By comparison the ratio between the alveolar diffusion area and the volume of the human lung is about 100 cm2 /cm^3. The arterial and venous capillaries are arranged in a checkerboard or hexagonal pattern to provide maximum contact area. The distance between the two bloodstreams in the rete of some species may average 1.5 μ or approximately the same as the distance between air and blood in the human lung alveoli.

The capillary cells of the rete of the eel, *Anguilla,* contain only few and small mitochondria. This observation indicates, when compared to the abundance of large mitochondria in tissues known to have a high aerobic metabolism, that the functioning of the rete does not involve a high expenditure of energy. The electron microscopic appearance of rete capillaries resembles, on the other hand, that of diffusion tissues like lungs. Its abundance of round vesicles and the sparse reticulum gives it a spongy appearance. The microstructure of the rete capillaries thus fits in with the view that their main function is to support passive diffusion between arterial and venous blood. This is also supported by the observation that a high capacity to deposit gases is correlated with the presence of a rete system which gives the two bloodstreams a large, but thin, common exchange area. This aspect will be discussed in quantitative terms together with the functional aspects of the rete.

THE PERFORMANCE OF THE SWIM BLADDER

The Behaviour of the Swim Bladder during Vertical Displacements

The behaviour of the swim bladder during vertical displacements.

As alluded to earlier, a fish with a swim bladder is in buoyancy equilibrium at one depth only. This seemingly inconvenient situation led earlier workers to suppose that the muscles around it must control its volume. Borelli believed that the swim bladder was also an organ of locomotion, the muscles being actively constricted when the fish wished to descend and relaxed when the fish wished to rise. Dispute later arose between the supporters of Borelli's views and those who thought that the muscles around the swim bladder were used only to hold its volume constant. These very rational differences of opinion as to just how the muscles control the volume of the swim bladder vanished when Moreau showed that they exert no control whatsoever. Surprisingly enough, fishes are in fact in a state of unstable equilibrium, as we are when standing upright.

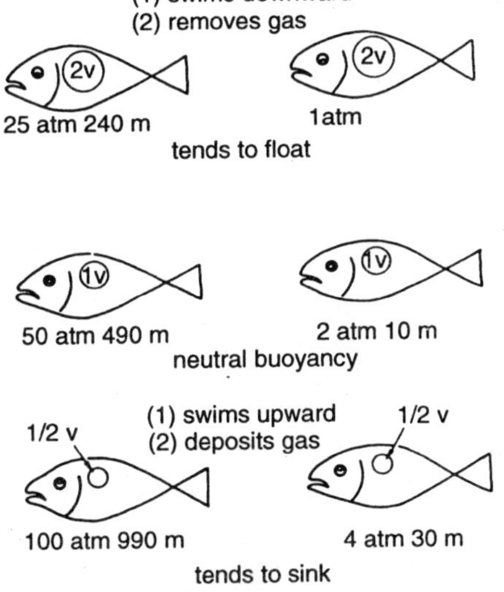

Figure 6.6 : The behaviour of the swim bladder volume of a fish during vertical displacements. The response of the fish is also indicated.

Fishes compensate for *transient changes* of pressure by swimming. Moreau showed that when a fish voluntarily changes depth, even by a very small distance, the volume of gas within the swim bladder simply changes in accordance with Boyle's law. Even the gurnards, which have particularly well-developed muscles around the bladder, do not use them primarily for controlling the volume but for making sounds.

Further research has added details to Moreau's conclusions.

Alexander showed that only in Cyprinidae does the swim bladder restrict, although passively, the expansion of the gas when the external pressure falls. For 22 other species he found that the swim bladder obeyed Boyle's law remarkably well. The sea horse, *Hippocampus brevirostris,* has been shown to use the muscles of its swim bladder wall to control not its density but its posture.

Moreau also demonstrated that in response to *persistent changes* in pressure the fish can restore the original bladder volume by changing the amount of gas in the bladder (the pressure of the gas is always equal to the external hydrostatic pressure). It thus became easy to see why codfish caught at 100-meter depth had very distended swim bladders when brought to the surface while cod caught in shallow waters had no such difficulties.

These early investigations thus drew up the contours of the hydrostatic function of the swim bladder. When a fish ascends the bladder expands and gas is removed; when it descends the bladder is compressed and gas is deposited. The crucial point is that the fish is able to vary its gas content in such a way that the gas volume is almost constant regardless of the hydrostatic pressure.

The Swim Bladder as a Float

In most common marine fishes the swim bladder occupies about 5% of the total volume. This brings the density of the fish very close to that of seawater, which is about 1.025. The density of muscle tissue is about 1.05 and that of the skeleton 2-3 times that of water. The density of fat, on the other hand, is about 0.9. In comparison the density of air at 1 atm is about 0.00125.

The efficiency of the swim bladder as a float is reduced with depth. Thus, while water and the organic components of animals are compressed to half the volume by 20,000 atm, gases obey Boyle's law fairly well. Thus their volume is halved when the pressure is doubled, for example, on going from the surface to 10-meter depth of the sea. According to Alexander the specific gravity of O_2 at 700 atm is 0.7, while the density of water is changed insignificantly. Thus even at 7000-meter depth a gas-filled swim bladder gives some buoyancy.

Energy Saved by Neutral Buoyancy

The possession of neutral buoyancy enables a fish to save energy in two ways: (1) It can remain motionless in midwater, and (2) the energy expenditure to swim horizontally is reduced.

(1) Since a fish without a bladder is about 5% heavier than water,

it will have to exert a force of 5% of its body weight downward on the water if it wishes to stay at one level. A few percent of the body weight may seem quite a small load, but a weight easily supported on land can only be sustained in water by continuous movement. Even an active pelagic fish very infrequently exerts a swimming force corresponding to more than 25% of its weight. To exert continuously a force of a few percent of its body weight seems therefore to be a formidable task.

(2) Alexander has made an interesting calculation of the amount of energy saved by the presence of a swim bladder during horizontal swimming. When a fish moves at a velocity of one body length per second, the power needed to overcome the sinking tendency is 60% of the total power of movement. When it moves at 10 lengths per second, only 5% of the total power is used to keep it at the same vertical position. At normal cruising speeds, which according to Bainbridge is 3-4 lengths/sec, about 20% of the total power is expended to overcome sinking. Since we do not know the efficiency of horizontal versus vertical movements, we can only guess that the energy partition must also be of this magnitude.

At present we are not in a position to make a quantitative comparison between the energy needed to stay at a certain depth by swimming as compared to the cost of maintaining a gas-filled swim bladder. It appears, however, that the swim bladder solution is by far the least expensive.

The Cost of Depositing Gas

Kanwisher and Ebeling investigated the metabolic cost connected with vertical migrations. On the assumption that the fish must keep a constant bladder volume during vertical migrations, the work performed by the bladder in migrating from pressure P_1 to P_2 is given by the formula:

$$W = 2.3vP_2 \log P_2/P_1$$

where v is the volume.

The accumulated work necessary to migrate from the surface to various depths. This work can be compared to the total metabolism of a fish. From data of Scholander and van Dam a 10-g fish should use about 0.4 ml of O_2 per hour under conditions of aerobic metabolism. This is equal to about 2 g cal. If one-third of this can be used by the swim bladder and the process of gas deposition is 25% efficient, then 0.15 g cal could be used per hour to deposit gas. On this assumption a fish would need 5 hr to go from the surface

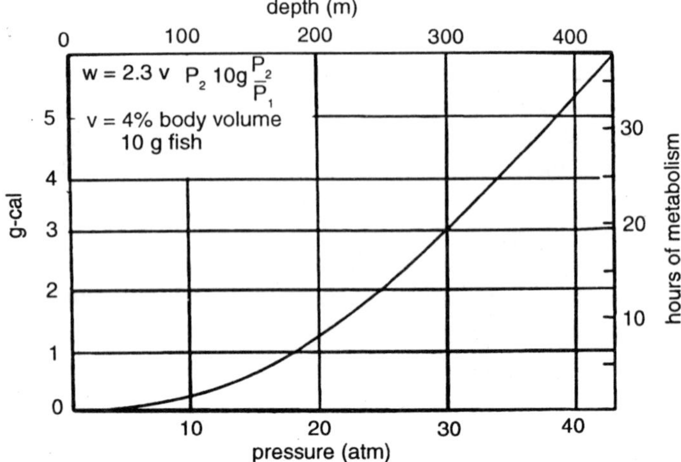

Figure 6.7 : The accumulated work necessary to migrate from the surface to various depths for a 10-g fish with a 0.4-ml swim bladder. The scale at the right indicates the number of hours necessary to attain neutral buoyancy at the various depths under condition specified in the text.

to a depth of 150 meters all the while being neutrally buoyant. Direct observations on the rate of gas deposition indicate, however, that an appreciably longer time is needed.

The Ability to Concentrate Gases—The Composition of the Swim Bladder Gas

Analysis of the swim bladder gas has been an important method for assessing the capacity of the bladder to concentrate gases. To take full advantage of such information two more parameters must be known. One is the total pressure at which the fish was caught, the other is at what pressure the swim bladder gas would have made the fish neutrally buoyant. The first of these values is used to calculate the partial pressure of each gas, the second gives the minimum pressure against which the fish is able to deposit gas.

Biot was the first to analyse the gas of the swim bladder. He constructed a gas analyser where he mixed the unknown gas with hydrogen, ignited the mixture, and weighed the resultant water. He was primarily interested in variations of the composition of the atmosphere; and possibly bored by its constancy, he introduced a sample of swim bladder gas into his analyser. Much to his delight, it is presumed, the gas analyser exploded. Biot reconstructed his gadget and confirmed the high O_2 content of the gas. More extensive analysis of swim bladder gases from deep sea fishes was first reported by Schloesing and Richard in 1898. A typical example of their results in a specimen caught at 900-meter depth was: oxygen 75.1%, nitrogen 20.5%, carbon dioxide 3.1%, and argon 0.4%. Scholander and van Dam studied gas taken from the swim bladders of fishes living at depths down to 1400 meters. The general picture emerging from these and other studies is that in fishes living near the surface the swim bladder contains a gas which is much like air, while at greater depths O_2 becomes an increasingly dominant component. In deep sea fishes the P_{O_2} and P_{N_2} in the bladder may be 100 and 25 atm, respectively, while the P_{O_2} and P_{N_2} of the ambient water, and consequently of the arterial blood, does not exceed 0.2 and 0.8 atm, respectively. The swim bladder has thus the ability to concentrate O_2 by a factor of at least 500 and N_2 by a factor of 30.

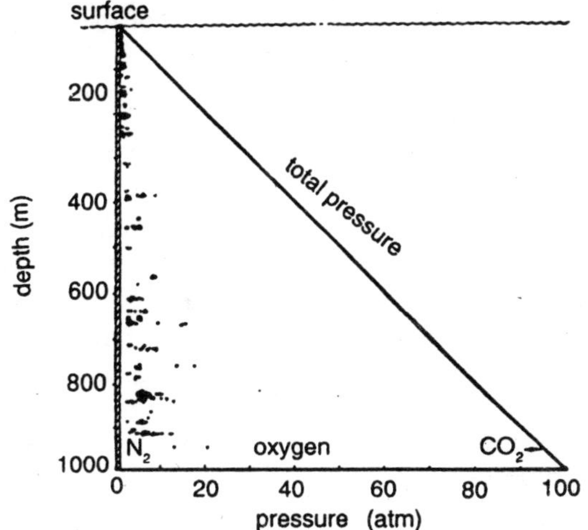

Figure 6.8 : The P_{N2} P_{O2} and P_{CO_2} in the swim bladder of fishes at increasing depth.

Almost a century after Biot's experiments, Hiifner published the observation that the swim bladder of whitefish, *Coregonus acronius*, from the bottom of Lake Constance at a depth of 60-80 meters contained 99% N_2. Accordingly, the P_{N_2} in the bladder at the depth at which the fish was caught must have been 7-9 atm. Similar findings on the swim bladder gas of several other species of freshwater physostomes from Lake Huron and Lake Michigan were later reported by Saunders, Tait, Scholander *et al.*, and Sundnes *et al.*. Sundnes added a significant piece of information when he observed that high P_{N_2} in several species of physostomes was found only in those fishes that had stayed for several weeks at the same depth. In the same species he found high P_{O_2} values in samples taken from fishes that had recently migrated to greater depths.

From the point of view of the mechanism of gas deposition the composition of freshly deposited gas has special interest. Unfortunately, such data are available only from experiments at surface pressures. Scholander covered the exposed gas gland of codfish and barracuda (physoclists) with a transparent material and analysed gas bubbles as they appeared. They contained 5-15% CO_2, 30-81% O_2, and 10-55% N_2. Wittenberg determined the compositions of gases in swim bladders from physostomes 12 hr to a week after they had been emptied. The collected gas always had a higher P_{O_2} and lower inert gases than the gas mixture with which the water was in equilibrium.

The information on the partial pressure of the gases in the swim bladder seems to indicate, therefore, that O_2 is the prominent swim bladder gas in both physoclists and in physostomes.

THE MECHANISMS OF GAS TRANSPORT

The Removal of Gas

The removal of gas occurs in two ways: direct release of gas into the water or reabsorption of gas into the bloodstream. Direct release of gas occurs in physostomes as a consequence of decreased hydrostatic pressure. This is well demonstrated when a herring filled purse seine is hauled up through the water. An ascending school of herring can also be identified on an echo sounder screen by a "cloud" above it. The cloud is reflections from rising gas bubbles, the school itself is detected by reflection from their bladders.

In physoclists, and also in physostomes exposed to mild reduction in external pressure, gas is removed by reabsorption into the blood circulating in the resorbent area. The mechanism in this case is identical to that in respiratory organs: Gases diffuse according to their partial pressure, and the amount of reabsorption is equal to the arteriovenous difference in gas content times the blood flow. Reabsorption is regulated partly by varying the proportion of this area exposed to the gas phase and partly by variation of the blood flow through the resorbent area. In a few physostome fishes (the common eel, *Anguilla vulgaris,* is an example) the pneumatic duct acts as an organ of gas reabsorption in addition to being a direct route for gas release.

The oxygen reabsorbed for hydrostatic purposes is most likely utilised metabolically. However, since the blood from the reabsorbent regionbe it a pneumatic duct or an oval drains directly to the gills, the possibility exists that some of the reabsorbed O_2, under certain conditions, may be lost to the ambient water.

The Maintenance of a Gas-Filled Swim Bladder

The swim bladder is in principle in the same situation as a gas bubble submerged in water which is equilibrated with air: The gas is compressed, and diffusion will ceaselessly cause reduction in the amount of gas. This is directly limited by a supposedly low gas permeability of the swim bladder wall. However, the most important mechanism to reduce gas loss is the barrier function created by arteriovenous exchange in the rete. Since this organ also plays the crucial role in the deposition of gas, we shall discuss the general features of countercurrent exchange.

Concurrent Exchange

Consider Figure 10. The arterial blood has a P_{O_2} of 0.2 atm, and the bladder is filled with gas containing O_2 at 100 atm. Let us assume that blood from the bladder enters the rete with a P_{O_2} of 100 atm. In the rete O_2 will therefore diffuse into the arterial blood, thereby reducing the venous P_{O_2} and increasing the arterial P_{O_2} in the two flow directions. Thus, as arterial blood enters the bladder, its P_{O_2} will approach that of the venous blood leaving the rete. In the same way the venous P_{O_2} will approach the arterial P_{O_2} at the heart pole of the rete. The closeness of approach (i.e., the efficiency of the rete as a gas exchanger) is determined by the area, the thickness, and the gas permeability of the membrane shared by arterial and venous blood, as well as by the length of time the two types of blood are in contact.

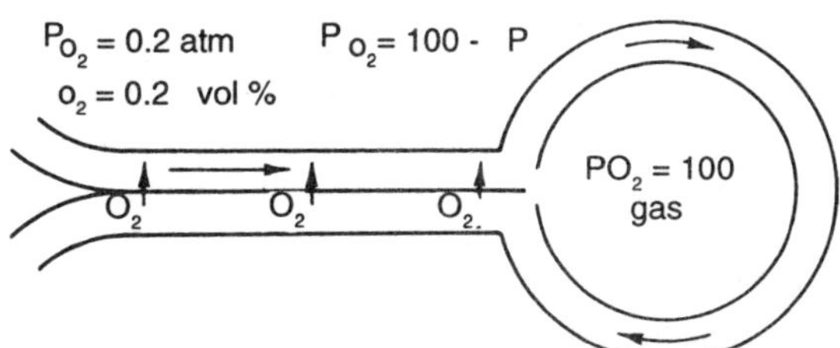

Figure 6.9 : The barrier function of the rete.

We should be aware of the importance of the number of capillaries in the rete. In a rete with many capillaries the intrarete diffusion will be large in comparison to diffusion between the rete and the surrounding tissue. In deep sea fishes with a P_{O_2} gradient of 1:1000 between the bladder and the tissue, this may be a factor of some concern. The number of rete capillaries will, on the other hand, not influence the efficiency of the rete as an exchanger but only its circulatory and thereby functional capacity.

Scholander developed a mathematical formula which expressed the efficiency of the rete as a barrier, its barrier function. He based his calculations on, anatomical dimensions from the rete and on reasonable values for velocity of blood flow. He found that the barrier function of the rete was so great that only when the P_{O_2} in the bladder exceeded 3000 atm (compared to an arterial P_{O_2} of 0.2 atm) would the O_2 loss via the bloodstream exceed that which one would accept as a reasonable amount of deposited gas. There is thus good reason to believe that the rete can reduce the gas leak from the bladder under natural conditions.

The Concentrating Ability of the Rete

Compared to the barrier function of the rete the multiplier function is distinguished by a reduction of the gas solubility, i.e., in the ratio of gas content to its partial pressure as blood circulates in the bladder. This results in an increasing P_{O_2} in the blood circulating in the bladder with consequent diffusion of O_2 into the bladder lumen. It also creates a higher P_{O_2} in prerete venous blood than in postrete arterial blood. This will cause diffusion of O_2 from venous to arterial blood of the rete and consequently an increased P_{O_2} in the arterial blood entering the bladder. When this blood in turn experiences a reduced solubility, the P_{02} will increase further and still more O_2 will enter the bladder. Thus the small increase in P_{O_2} caused by reduced solubility, the primary effect, will be accumulated by circulation through the rete. The process is cumulative and is termed "hairpin countercurrent multiplication" (ccm).

Haldane and, in particular, Koch were the first to suggest how the rete might function to concentrate O_2 and inert gases subsequent to release from the blood. The potential properties of such a system were, however, first clearly realised by Kuhn and Riiffel, who introduced to biology the concept of countercurrent multiplication. Hargitay and Kuhn developed an extensive model of such a mechanism in the case of urine concentration in the loop of Henle.

Kuhn and Kuhn expounded the cern of gases in the rete mathematically. [Helpful reviews are given by Kuhn *et al.*, Alexander, and Enns *et al.*.] They assumed the rete tissue to be *impermeable* to salts and acids and treated separately the importance of a salting out effect, the Bohr effect, and the Root effect on the concentrating ability. (The Bohr effect denotes the pH sensitive O_2 affinity of blood; the Root effect denotes the pH sensitive O_2 capacity.) When they inserted experimentally determined values for the permeability of O_2, blood flow, lactic acid production, and dimensions of the rete in the final formula they reached maximum values for P_{O_2} and P_{O_2} produced.

Table 6.1 : Estimated P_{N2} and P_{O2} in Atmospheres That Can Be Developed by Retia of Various Lengths.

	Length of rete (mm)		
	0	5	10
N_2 (by salting out) O_2	0.8	4	23
(1) By salting out; Bohr effect and Root effect	0.2	300	2000
(2) By salting out and Bohr effect only	0.2	60	300

These calculations indicate that a 1-cm long rete can produce gas pressures well above those ever recorded in deep sea fishes. Even without a Root effect the system can produce a P_{O_2} of 300 atm. The retia of *Anguilla* and of *Pornatomus* are about 1 cm long, but retia 2 cm long or more occur in benthic fish. Kuhn and Kuhn also showed that the rate of gas deposition depends on the extent of the reduction in solubility. Thus for O_2 it is higher when both the Bohr and the Root effect participates, than when only the salting out effect causes the primary reduction in solubility.

THE REALISATION OF COUNTERCURRENT MULTIPLICATION IN THE SWIM BLADDER

Much information has accumulated which supports the view that O_2 in particular, and probably also inert gases, are in fact concentrated by countercurrent multiplication of a primary gradient caused by acidification of the blood.

Release of O_2 from Blood

Hall found that pH of a dialysate of homogenate of gas glands from the yellow perch, *Perca flavescens,* during gas deposition was 6.4 against 7.1 for inactive glands. Ball *et al.* found that gas gland tissue from several species converts most of its glucose to lactic acid even in an atmosphere of pure 0_2. Fange found higher lactate content in homogenates of active glands than in those homogenates of passive ones in codfish. Steen found that the postrete venous blood contained more lactic acid than the prerete arterial blood and that the pH of the blood was drastically reduced as it circulated the secretory bladder. These results indicate that the blood is acidified by lactic acid as it circulates in the bladder.

The notion that this plays an essential role in the swim bladder function is supported through the demonstration by Root that fish blood apparently suffers not only a reduced O_2 affinity upon acidification but also a reduced O_2 capacity. This characteristic of fish blood is termed a "Root effect." Scholander and van Dam found a pronounced Root effect in the blood of several species of deep sea fishes where the hemoglobin at low pH attained full saturation only at P_0. between 20 and 130 atm.

The Root shift can occur in two directions in that pH can either increase or decrease. The phenomenon connected with acidification is termed a "Root off-shift," while the opposite is termed a "Root on-shift," "off" and "on" referring to the direction of 0_2 movement relative to Hb (or to O_2 affinity),

In some fishes a P_{O_2} of up to 140 atm did not increase the hemoglobin saturation above that recorded at 20 atm, indicating strongly that the acid blood suffers a permanently reduced O_2 capacity (Root effect). In response to reversal of the pH the blood regained normal O_2 capacity, thus showing that the Root effect is reversible. On the other hand, a Root effect is not found in fishes lacking a bladder.

The proposal that the O_2 present in the swim bladder is derived from molecular O_2 present in the water and transported in this form by Hb is directly supported by the experiments of Scholander *et al.*. Mass spectrometer analyses of gas from the bladder of codfish from aquaria to which $H_2{}^{18}0$ had been added revealed no trace of the heavy isotope. It has further been shown by tracer experiments that there is no exchange of 0 atoms between O_2 molecules as these pass from the water via the blood to the bladder during gas deposition in the toadfish, *Opsanus tau.*

Acidification alone is, however, not sufficient to explain the high P_{O_2} values found in deep sea fishes. Even if all the O_2 of fish blood were split off it would not raise the P_{O_2} to above some 5 atm.

Release of Inert Gases from Blood

As we have noticed in an earlier section the $_{P12}$ in the bladder may also exceed that in the water. The mechanism of inert gas deposition is thought to be based on release of some of the gas present in physical solution in the blood. Koch pointed out that an acid added to the blood would not only dissociate HbO_2, it would also increase the ionic strength of the blood and reduce its gas solubility. This type of gas release will of course affect all gases and may be brought about by any ion added to blood. It is commonly termed the "salting out" effect. If half of the dissolved gas were salted out, the partial pressure would double. Such an increase would, however, require an increase of ionic strength far above that which is compatible with normal physiology.

Multiplication by the Rete

The notion that the primary gradient caused by acidification is multiplied by ccm in the rete receives support from the very existence of the rete as well as from its unique architecture. This, together with the pH sensitivity of HbO_2 led Haldane to suggest that O_2 is deposited as a consequence to acidification and that the effect is possibly amplified by circulation through the rete.

The central position of the rete in the swim bladder function is clearly demonstrated by a comparison between the extent of its development and the depth at which the fish lives. Unfortunately, the only parameters of the rete structure that have been sufficiently studied are the length of the rete and the number of retial capillaries. Deep sea fishes have longer retia than shallow water forms. This indicates that the rete plays an essential role in gas deposition.

Before accepting the ccm explanation, three basic assumptions have to be discussed. (1) The effect of the added substance must be present at the pressure against which gas is to be deposited. (2) During gas deposition the gas pressure must be higher in postrete venous blood than in prerete arterial blood. (3) The rete must be impermeable to the substance causing the increase in gas pressure.

Pressure-Dependent Solubility

The most likely mechanism for a reduction in solubility is the addition of ions. The specific effect of H^+ ions on the O_2 capacity of blood (Root effect) has been mentioned earlier. In many fishes the effect does not persist to a P_{O_2} which is known to be present in their swim bladders. The effect of ions on the physical solubility will not, however, be influenced by the partial pressure of the gas.

It seems, therefore, that an acid like lactic acid will fulfill the desired requirements by salting out effect alone or, depending on the species, in combination with a Bohr or Root effect.

Arteriovernous Gas Gradient

The P_{O_2} in arterial and venous blood at both poles of the rete of the eel was estimated by Steen from O_2 content, pH, and hematocrit of appropriate blood samples. During established deposition of gas the postrete venous P_{O_2} was found to be *lower* than the prerete arterial. Similar results were obtained by Steen and Iversen who measured P_{O_2} polarographically.

Impermeability for the Rete to Acids

In circulating blood CO_2 will be in equilibrium with H^+ ions. Thus a change in pH will cause a change in P_{CO_2}. Carbon dioxide is known to pass biological membranes faster than O_2 (about 20 times). If this is the case also for rete tissue, then the ccm will be short-

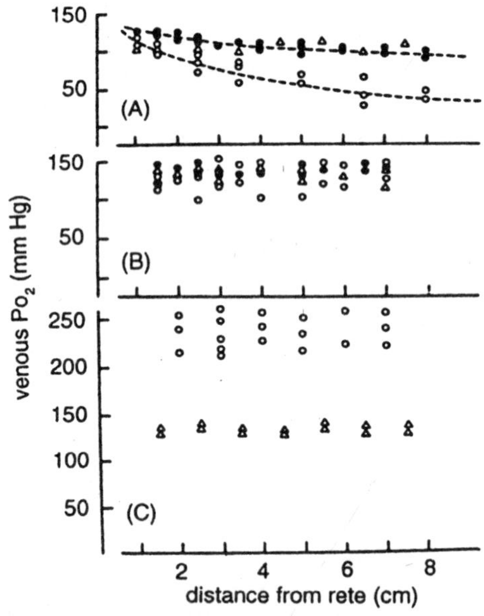

Figure 6.10 : (A) P_{O2} *of prerete arterial blood and of postrete venous blood from an actively secreting swim bladder. (B)* P_{O2} *of arterial (Δ) and venous blood (O) at various distances from the rete of a nonsecreting swim bladder. (C)* P_{O2} *of prerete arterial (Δ) and postrete (O) venous blood in experiments where countercurrent exchange had been abolished by appropriate cannulation.*

circuited. On the other hand, ions are known to penetrate slower than water (about 10 times). Thus lactic acid deposited in the swim bladder capillaries may be instrumental in a concentration mechanism based on salting out, but hardly on one based on its acid effects. Steen experimentally showed that the rete is in fact permeable to acids.

Thus a considerable pH change is detectable in both arterial and venous blood during circulation of the rete.

It thus appeared that the experimental evidence on two of the three crucial points clearly contradicted the requirements of the model. For a while it appeared that we had in the swim bladder a system which was perfectly engineered for gas concentration by ccm but that the fishes did not use it that way.

The anomalous evidence was based on measurements done on blood samples. Thus, transient changes in pH or P_{O_2} might have been over looked. Manwell urged students of the swim bladder to be aware of such transients. How right he was was demonstrated by Berg and Steen. In venous blood leaving the rete they recorded a lower P_{O_2} the farther away from the exit of the rete it was recorded. The half-time of the change was estimated to be 10-20 sec. This implies that the earlier estimates of venous P_{O_2} represented the stable end value and not the P_{O_2} actually present at the exit of the rete. In fact the P_{O_2} at that point seemed equal to or even slightly higher than the arterial P_{O_2} which was shown not to vary with distance from the rete.

The matter was further studied by measurements on the rate of the Root shifts in suspensions of eel red cells. At 23°C, the Root off-shift (acidification and subsequent splitting off of O_2) showed a half-time of 0.05 see, while the reverse process, the Root on-shift, had a $t_1/_0$ of 10-20 sec.

Present Concept of Countercurrent Multiplication in the Rete

Blood enters the rete with arterial pH and P_{O_2} values. As it flows through the rete it receives O_2 and acid (CO_2 and H^+) from the venous blood draining the bladder. The rate of the concomitant Root off-shift has a t1,o of 50 msec at 23°C.

Upon entering the bladder epithelium the arterial blood has thus become enriched in O_2 and acidified. During its further circulation through the bladder epithelium the blood receives still more lactic acid whereby its P_{O_2} is further increased causing O_2 diffusion into the bladder. The addition of acid probably also increases the osmotic pressure of the plasma, causing a reduction in the physical solubility of all gases.

When the blood returns to the rete it has a lower pH and O_2 content, but a higher P_{O_2} than when it left the rete to enter the bladder. It may also have a lower physical gas solubility. As this blood flows through the rete, O_2 will diffuse according to the gradient in P_{O_2} from venous to arterial blood. At the same time, lactic acid and CO_2 pass in the same direction thus causing increasing pH in venous blood. The concomitant decrease of venous P_{O_2} (Root on-shift) has a $t_{1/2}$ of 10-20 sec. The degree to which the venous P_{O_2} changes are completed when the blood leaves the rete will depend on the rate of blood flow. Final equilibrium between H^+ ions, Hb, and O_2 is established as the blood flows away from the rete.

Thus, as suggested by the Kuhns, O_2 is concentrated by countercurrent multiplication of a P_{O_2} gradient caused by acid; however, this gradient is not based on impermeability of the rete to acid but instead on the slow response of the O_2 capacity of blood to decreasing H^+ ion concentration. The slow Root on-shift thus explains how the rete can work in spite of the fact that it is permeable to acid.

The rate constants hereby introduced into the system will also make the system dependent upon blood flow in another and biologically more attractive manner than previously assumed. The Kuhn model implies that at a low flow rate the extraction of O_2 from the blood is maximum. Introduction of rate constants indicates that when the flow is low the arteriovenous acid flux will have sufficient time to influence the arteriovenous P_{O_2} gradient and thus reduce the degree of extraction. It is thus quite possible to circulate a bladder at a low perfusion rate without significant deposition of O_2. The full consequence of these kinetic parameters will be evident only when incorporated into a mathematical model of the countercurrent multiplying system.

It was shown by Fange that acetozolamide (Diamox), which inhibits the catalytic action of carbonic anhydrase on the reaction of H_2O with CO_2 will abolish gas deposition. In a recent investigation, Forster and Steen showed that this inhibitor lengthened the half-time of the Root off-shift from a normal value of 50 msec at 20°C to at least 30 sec. It is very likely that this explains Fange's observation. The acid secreted by the bladder epithelium would act too slowly to cause significant O_2 release from the blood during the period of time it circulates in the swim bladder.

The model for inert gas concentrations based on salting out effect should also take into account the diffusion constants of inert gases relative to that of water and ions. The half-time for water penetration of eel red cells during osmotic swelling has been measured by Blum (personal communication) to be 2-3 min, as compared to less than 20 msec for O_2 diffusion out of the red cell. N_2 will diffuse through water roughly at half the speed of O_2 and is thus apparently so much faster than H_2O diffusion that countercurrent multiplication of inert gases via the salting out effect can take place.

CONCLUSION

At present it seems feasible to explain the concentration of O_2 and of inert gases in fishes with a rete by countercurrent multiplication of a primary effect caused by lactic acid. Other ions, yet unidentified, may possibly be involved.

Gas Deposition in Fishes with Poorly Developed Rete

The situation becomes more difficult when we attempt to explain the high concentration of inert gases found in many physostomes, the above-mentioned whitefish of Hiifner is an example. The swim bladder of these fishes possesses only a very poorly developed rete; in many cases, it consists of only a few arteries and veins. One possibility is that inert gases in these fishes are concentrated by an entirely different process. However, Wittenberg showed that even in these fishes the freshly deposited gas contained predominantly O_2. This indicates that the final high tension of inert gases may result from secondary reabsorption of O_2. The presence of Hb would certainly increase the rate of O_2 reabsorption above that of inert gases. It is most likely, as concluded by Wittenberg and substantiated by Sundnes, that the high partial pressures of inert gases found in these fishes result from the interaction between a countercurrent mechanism of lower capacity than that found in physoclists and a reabsorptive process with a high capacity.

Gas deposition in the herring, which lacks a rete altogether, has not been investigated. Most likely it depends on acidification alone. Variation in gas composition will be accomplished by the interplay between the rate of deposition and that of reabsorption. The problem is obviously in need of experimental data.

NERVOUS CONTROL OF THE HYDROSTATIC FUNCTION OF THE SWIM BLADDER

The volume of the swim bladder may be considered to be steadily controlled by two reflex mechanisms: an inflatory reflex (gas deposition) and a deflatory reflex (gas reabsorption). Both these reflexes seem to be complex processes involving reactions of several categories of effectors in the swim bladder.

Treatments that increase the specific gravity of the fish act as stimuli initiating the inflatory reflex. Thus gas deposition (inflation) can be produced experimentally by an increase of the external hydrostatic pressure, by removal of a part of its gas content, or by attachment of small weights to the fish. All these treatments reduce the degree of stretch in the bladder wall.

Treatments that lower the specific gravity of the fish, or increase the stretch in the bladder wall, act as stimuli for the deflatory reflex. Thus von Ledebur reports that reabsorption of gases can be provoked by a decrease of the hydrostatic pressure of the surrounding water, by attachment of air-filled balloons to the fish, or by injection of gases into the swim bladder. It is also known that during asphyxiation oxygen is reabsorbed from the bladder.

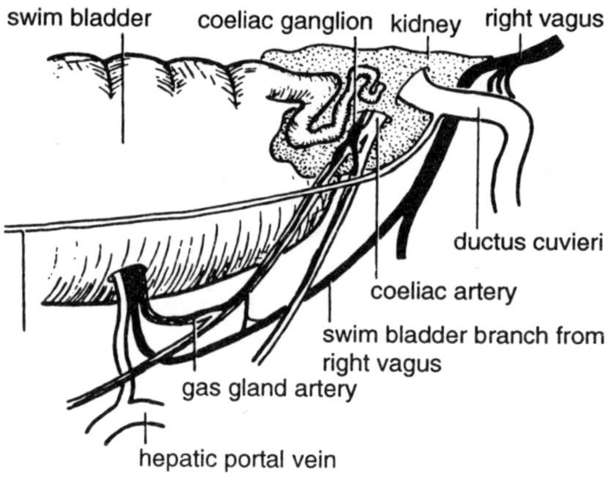

Figure 6.11 : Blood vessels and nerves to the swim bladder of the cod.

The swim bladder is innervated by branches from the vagi and from coeliac ganglia. Nerve endings have been observed in the reabsorbent region, the oval the rete, and in the secretory epithelium. Near the heart pole of the rete there are numerous large ganglion cells. The muscular layer of the swim bladder wall is also well supplied by nerves.

The functioning of this nerve supply is not yet clearly understood. For example, we do not know what part of the swim bladder function is nervously regulated. Most likely the quantitative extent of gas reabsorption and gas deposition is accomplished by regulation of the blood flow to the appropriate areas. Gas reabsorption, being a pure diffusion process, will increase when the blood flow through the reabsorbent area increases. Deposition of gas, being dependent on interaction between acid deposition and circulation rate, appears at first sight more complex. However, as shown by Ball *et al.*, the production of lactic acid decreases with increasing H^+ concentration. This implies that lactic acid production will be increased at increased blood flow, since under these conditions H^+ ions are more quickly removed. In addition, the retention of H^+ ions in the bladder circulation will increase at low flow owing to enhanced arteriovenous exchange in the rete.

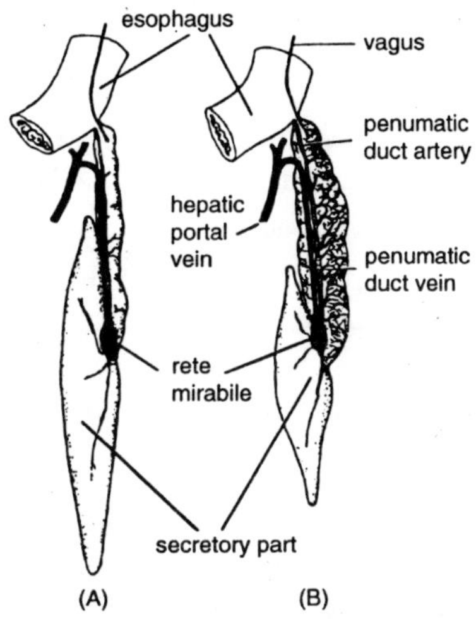

Figure 6.12 : (A) Normal appearance of the eel swim bladder. (B) Its appearance after stimulation for 5 min of the left vagus. The secretory part is contracted and the resorbent part relaxed. From Fange.

Fange showed that electric stimulation of the vagus will relax the resorbent areas or cause an opening of the oval. Adrenaline had the same effect on the resorbent part but caused the secretory bladder to contract. Acetylcholine has no effect on the secretory bladder, but it has a contracting influence on the resorbent bladder.

Stray-Pedersen has shown that the blood flow through the secretory bladder of the eel is especially reduced by adrenaline and to some extent by acetylcholine. The vascular resistance of the rete was unaffected by acetylcholine, but it increased on adrenaline injections. The reabsorbent bladder showed increased circulation when adrenaline was added to the perfusate.

It may appear, therefore, that reabsorption of gas is stimulated by catecholamines, while deposition is increased by the absence of such stimuli. It is difficult to find support for the presence of "secretory cholinergic stimuli."

7

AUTONOMIC NERVOUS SYSTEM

The current revival of interest in the general physiology of fishes has emphasised our ignorance of autonomic functions in these animals. Since Nicol reviewed the anatomy and function of the autonomic nervous system of fish in 1952, there have been many new investigations; these have often been concerned with details of the innervation of systems already examined in some detail, namely, the heart, the gastrointestinal tract musculature, and chromatophores. It is rather surprising that there has not been greater interest in the innervation of, for example, the vasculature, the internal genitalia, and the digestive glands. It is possible that this will be the last of the old-style reviews of this subject, for new techniques, including the now widespread use of delicate mechanoelectric transducer recording systems and the introduction of the histochemical technique for visualising tissue catecholamines, are already starting to revise our opinions on the autonomic nervous system of fishes.

CYCLOSTOMES

Cranial Autonomic Nerves

The oculomotor nerve is absent in hagfish, in which the lateral eyes are degenerate. The eyes of lampreys are also small, but oculomotor nerves are found. Anatomical evidence for an autonomic component of the oculomotor in lampreys is controversial, and there would appear to be no function for such a component since the eyes are probably completely devoid of intrinsic musculature.

An autonomic component of the facial nerve has been claimed for *Myxine,* largely on the basis of brainstem analysis. Lindstrom has suggested that the autonomic fibers innervate the lingual artery, but he does not exclude the possibility that the fibers are afferent. No physiological investigations of this nerve have been made.

In both hagfish and lampreys, the two vagus nerve trunks unite to form a cardiac plexus, lying on the gut wall, containing numerous nerve cell bodies and a few ganglionic aggregations. From this plexus, autonomic fibers are distributed to the region of the heart, major veins, gall bladder, and, via an unpaired nerve trunk, to the entire length of the intestine. The intestinal wall contains a nervous plexus with numerous cell bodies, some of which may be sensory neurons. The neurons become sparse in the posterior region of the gut, although the fiber plexus persists. Many of these enteric neurons are probably innervated by vagal fibers.

Of the vagal innervation systems, that of the heart has been studied in greatest detail. Greene first reported that the heart of the hagfish *Polistotrema (=Bdellostoma) stouti* was not affected by stimulation of the vagus nerve, or indeed of any other nerve. This was confirmed for the same species by Carlson and for another hagfish, *Myxine glutinosa,* by Augustinsson *et al.*. In agreement with this observation, no nerve fibers were found during an electron microscopic investigation of *Myxine* heart. Normal hearts of both species are virtually unaffected by acetylcholine and by catecholamines. In contrast, the lamprey heart is innervated by the vagus nerves. The main response to stimulation of the medulla oblongata is an acceleration of the heart followed by a period of deceleration after periods of intense stimulation, although a period of inhibition preceding the acceleration has been reported. The heart of *Lampetra* does not contain adrenergic nerve fibers demonstrable by fluorescence histochemistry, and catecholamines are variously stated to cause a slight depression or an acceleration and augmentation of the cardiac beat. It would therefore seem that the vagal innervation is cholinergic, since acetylcholine accelerates the heartbeat, at the same time depressing the force of contractions. Similar actions are exerted by other choline esters, including succinylcholine, and nicotine, but not by muscarine; the response to acetylcholine is inhibited by the nicotinic blocking drugs hexamethonium and tubocurarine but is unaffected by atropine. The cholinergic receptors are therefore of the nicotinic type. Because of this, Augustinsson *et al.* suggested that acetylcholine was acting on ganglion cells in the heart. It seems a simpler view that the lamprey heart muscle has a closer affinity to the skeletal muscle than to the cardiac muscle of higher vertebrates and that acetylcholine is acting directly on nicotinic receptors in the heart muscle itself. In this case, one would expect treatment with curare to prevent vagal transmission to the heart, but Zwaardemaker observed cardiac responses to medulla oblongata stimulation in lampreys immobilised with curare. Perhaps the cardiac cholinergic receptors are relatively resistant to blockade by curare. The vagus does not appear to innervate the heart in larval lampreys.

Stimulation of either vagal trunk in *Myxine* causes a contraction of the strongly muscular gallbladder. This pathway involves a ganglionic synapse in the hepatic plexus since topical application of nicotine to the plexus abolishes the responses. The terminal neurons are probably cholinergic, for acetylcholine causes a contraction of the gallbladder, whereas epinephrine appears to relax it.

The vagal innervation of the intestine is poorly understood. The muscularis of the gut wall is poorly developed, but it seems that all regions of the intestine are contracted by acetylcholine and relaxed by cate cholamines. In addition, the isolated intestine is reported to undergo peristaltic movements. It is therefore clear that the gut musculature is functional. However, Fange and Fange and Johnels could not detect any intestinal response to vagal stimulation in *Myxine*. Previously, Patterson and Fair had found that stimulation of the vagus in *Bdellostoma* caused a very slight relaxation of the gut, but their records were obtained *in situ,* and in view of the thinness of the gut wall the response could have been artifactual. On the other hand, Fane and Johnels pointed out that they had allowed the animal to warm up to room temperature (18°C), and they suggested that the vagus might be ineffective at this high temperature. The only independent support for an inhibitory vagal innervation of the gut is the observation that nicotine causes a relaxation of previously contracted *Mijxine* intestine *in vitro,* which may indicate the presence of inhibitory neurons in the gut wall. It has been postulated elsewhere that the vagus nerve is primitively inhibitory to the vertebrate gastrointestinal tract and that the inhibitory nerve fibers are of a nonadrenergic type recently demonstrated in the mammalian gastric vagus.

The possibility has been raised that the vagus nerves contain adrenergic nerve fibers since the vagus of *Myxine* contains considerable concentrations of catecholamines, predominantly norepinephrine.

Adrenergic fibers have not been found in the cranial autonomic outflow of any higher vertebrate, but this does not necessarily mean that they will be absent in cyclostomes. Alternatively, sympathetic adrenergic fibers could enter the vagi from the spinal outflow, as claimed by Marcus, since the vagi run extremely close to the mixed ventral rami of the anterior spinal nerves. Finally, the catecholamines could be stored in chromaffin cells; the vagus contains many ganglion cells, and one must remember that the "ganglion cells" found in cyclostome hearts by Augustinsson *et al.* were later shown to be specialised chromaffin cells. A fluorescence histochemical study of this problem would be most welcome.

Spinal Autonomic Nerves

There is still much confusion about the distribution of autonomic nerve fibers from the spinal cord. Some of this confusion arises from the fact that many peripheral nerve cell bodies have been observed

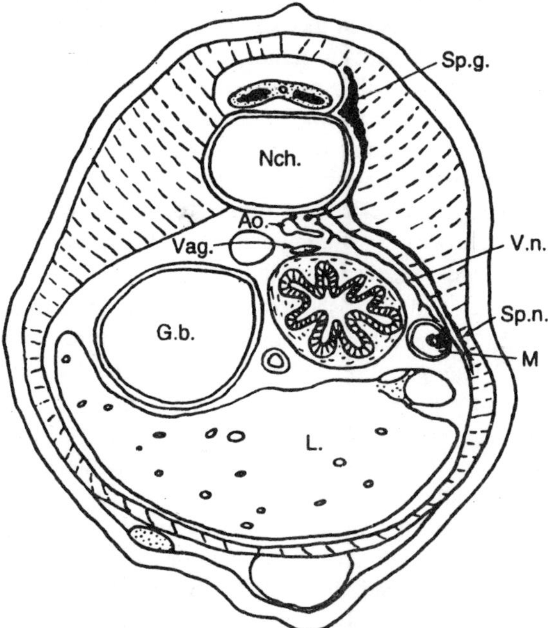

Figure 7.1 : Spinal autonomic fibers in Myxine. Diagrammatic illustration of the course of the visceral (spinal sympathetic) fibers as seen in a projection of a transverse section. Longitudinal fibers below the notochord indicated by two dots. An., dorsal aorta, at the beginning of a segmental artery; G.b., gallbladder; L., liver; Nl, mesonephros; Nch., notochord; Sp.g., spinal ganglion; Sp.n., spinal nerve; Vag., vagus nerve; V.n., visceral nerve.

in cyclostomes, but it is not always clear which of these neurons are autonomic and which sensory. In addition, there are reports of fibers leaving the spinal cord of *Lampetra* in both dorsal and ventral spinal nerves and running directly, without an intervening ganglionic synapse, to visceral structure. Whether all or any of these fibers are autonomic, i.e., efferent, is unknown and will probably remain so until physiological observations have been made.

There is, however, general agreement that there are neither organised sympathetic chains nor strictly segmental sympathetic ganglia in cyclostomes. There are collections of ganglion cells along the cardinal veins in the abdominal cavity of lampreys which may represent sympathetic ganglia. In *Lampetra,* visceral branches of the spinal nerves run directly to the region of blood vessels, the kidneys, and the gonads. In posterior regions there is a spinal outflow to the cloaca, rectum, and ureters, with many ganglion cells in the pathway. A similar segmental visceral outflow, possibly vasomotor, has been demonstrated in *Myxine*, but in this animal the outflow to the posterior gut appears to be mainly somatic motor, supplying the striated muscles of the cloacal sphincter, with a sensory component. It is highly likely that there is no spinal autonomic innervation of the gut in hagfish but that the smooth muscles of at least the posterior intestine of lampreys is innervated from the spinal cord. Support for a vasomotor function of the sympathetic has come from the observation of Leont'eva, who showed histochemically that there is an adrenergic innervation of blood vessels in cyclostomes.

Both lampreys and hagfish possess a rich subcutaneous nerve plexus containing uni- and bipolar nerve cells, a feature unique among the vertebrates. The ganglion cells are innervated by spinal nerve fibers and are almost certainly autonomic. A number of functions could be subserved by the fibers. Bone has found that the plexus is particularly concentrated at the orifices of the slime glands in *Myxine,* associated with bundles of smooth muscle which form sphincters for the glands. In addition, the plexus is continuous about the walls of the gland and may regulate the activity of smooth muscle cells around the gland. Bone has also noted that cell bodies in the plexus often lie beside the small blood vessels of the subcutis, and he has suggested that the plexus may control blood flow in the subcutaneous blood sinuses. One further function may be proposed for the subcutaneous plexus, that of chromatophore control. Color changes in most cyclostomes appear to be extremely slow if they occur at all. However, Wild has reported that *Petromyzon marinus* shows dramatic and rapid color changes, indicating a nervous control of the chromatophores in this species, presumably via the plexus.

GNATHOSTOMATOUS FISH

Physiological observations on the autonomic nervous system of gnathostome fishes have been virtually restricted to selachians and teleosts. Although there are conspicuous anatomical differences between the autonomic systems of these two groups, especially in the arrangement of the spinal outflow, it seems likely that there are no gross physiological differences.

Cranial Autonomic Nerves

Autonomic fibers leave the brainstem of selachians in the oculomotor, facial, glossopharyngeal, and vagus nerves. In teleosts, the outflow is restricted to the oculomotor and vagus nerves. There have been no physiological investigations of the autonomic functions mediated by the selachian VIIth and IXth nerves, but it has been suggested that both outflows provide vasomotor fibers to the pharyngeal region. Presumably this function has been taken over completely by the vagus in teleosts.

Oculomotor Nerve

Oculomotor nerves are absent from forms with reduced eyes, but where present they supply preganglionic fibers to the ciliary ganglion, from which postganglionic fibers pass to the eyeball. Stimulation of the oculomotor nerves causes pupillary dilatation in the teleost *Uranoscopus scaber* and in one selachian, *Scyllium,* although this was not seen in two others. This response is in marked contrast to the pupillary constriction caused by oculomotor nerve stimulation in mammals. There is no direct evidence to show whether the response represents an inhibition of activity in the sphincter or an excitation of the dilator muscle. In *Uranoscopus,* both epinephrine and acetylcholine can cause both dilatation and constriction of the pupil; in the selachians studied both epinephrine and acetylcholinc caused predominantly dilator reactions, and it has been found that acetylcholine causes contraction of the isolated iris dilator muscle of *Scyllium*. Atropine inhibits constrictions of *Uranoscopus pupil* caused both by sympathetic nerve stimulation and by acetylcholine, but the effects of blocking drugs on responses to oculomotor nerve stimulation have not been tested. The most simple, but not necessarily the correct interpretation of these limited observations, is that both the sympathetic and the oculomotor are cholinergic and that the sympathetic provides

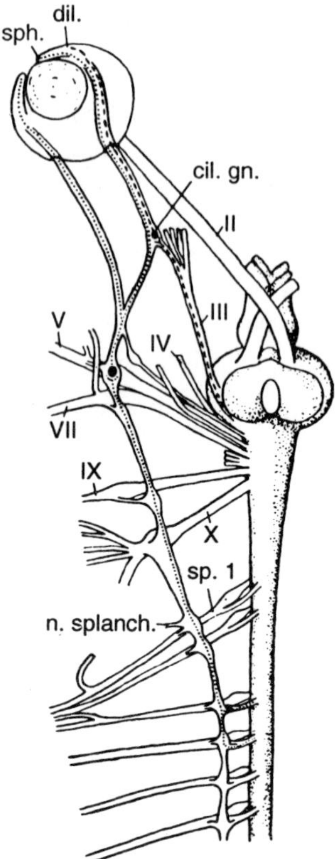

Figure 7.2 : Diagram of a ventral view of the anterior sympathetic system in Uranoscopus, showing the sympathetic and parasympathetic innervation of the iris. cil. gn., ciliary ganglion; dil., pupil dilator muscle; n. splanch., anterior splanchnic nerve; sph., pupil sphincter muscle; sp. 1, first spinal nerve; cranial nerves numbered II, III, IV, V, VII, IX, and X. (. . .) Sympathetic pathways and (—) parasympathetic pathways; positions of ganglionic synapses shown by filled circles.

an excitatory innervation to the sphincter (in teleosts only) while oculomotor stimulation excites the dilator muscle.

Vagus Nerve

The posttrematic rami of the vagus nerve in teleosts (as well as of the glossopharyngeal nerve in selachians) almost certainly innervate the branchial vascular bed. Direct evidence of efferent innervation is wanting, but elevations of branchial vascular resistance are known to occur during, for instance, periods of anoxia. Determination of whether this phenomenon is nerve mediated is made difficult by indications that the gill vasculature can constrict in direct response to the anoxic conditions. It is well established that catecholamines dilate and acetylcholine constricts the branchial vasculature in teleosts, although no vasomotor responses could be obtained from the isolated gills of the selachian *Squalus acanthias*. Similar vasoconstrictor actions of acetylcholine have been observed in the amphibian pulmonary vascular bed, a system derived from branchial arch vasculature. It is then reasonable to suggest that the vagi provide cholinergic vasoconstrictor fibers to the gills, if only because of the analogy with the established cholinergic vasoconstriction mediated by the vagi in amphibian lungs.

The visceral vagal rami distribute efferent fibers to the heart, the stomach, and usually no more than the most oral region of the intestine, the swim bladder in teleosts, and possibly to accessory digestive organs and to blood vessels in this region. The innervation of the heart has been reviewed by Nicol and more recently by Johansen and Martin. The evidence for a cholinergic vagal supply to the heart, causing a reduction in the rate of beating, is so overwhelmingly strong that no further comment is needed. The consensus has been that the vagi do not contain cardiac augmentor or accelerator fibers, whether of sympathetic or parasympathetic origin, and it is well established that there are no distinct sympathetic nerves innervating the heart directly. Jullien and Ripplinger found that after degenerative sections made high in the vagal trunk stimulation of the vagal cardiac branches still caused negative chronotropic and inotropic responses in teleosts, indicating that the ganglionic synapses in the vagal pathway occur within the vagal trunk itself. They also observed that the heart rate was slow and that arrhythmias were common in bivagotomised tench and that the heart returned to a fast, regular beat after atropine treatment. They suggested that the vagal postganglionic cell bodies are tonically active, holding the heart muscle under an inhibitory drive, and that the activity of the neurons is normally inhibited by fibers from the central nervous system running in the vagi, a suggestion which deserves further consideration.

A further observation made by Jullien and Ripplinger is also worth reconsidering. In a series of papers around (summarised in Jullien and Ripplinger), they provided evidence for a unique system of noncholinergic "negative tonotropic" nerve fibers in the vagal supply to the teleost heart. The crux of their argument is that in nonbeating hearts, e.g., after treatment with acetylcholine, stimulation of the vagus causes elongation of the heart; on the other hand, in hearts treated with atropine, stimulation of the vagi causes elongation of the heart when all signs of negative inotropic and chronotropic responses have vanished. Since their recordings have been made by attaching the heart to a kymograph lever *in situ,* leaving the heart attached to the body by the large veins alone, one might suggest that the " tonotropic' responses represent inhibition of venous smooth muscle tone. This suggestion is striking enough but is perhaps more acceptable than the alternative that the cardiac muscle cells are in a condition of tonus which is not related to cardiac action potential activity.

The vagal innervation of the gut of teleosts and selachians has been reviewed by Nicol, Barrington, and Campbell and Burnstock. All reviewers agree that stimulation of the vagi causes contraction of the stomach, but not of the intestine, in those animals possessing a stomach. When there is no stomach, the vagi may cause contraction of part or all of the intestine. However, cautious interpretation of this result is needed for two reasons. First, part or all of the responses observed might result from stimulation of sympathetic nerve fibers running with the vagus, at least in teleosts. Young described connections between the cranial sympathetic chain and the vagus nerve in *Uranoscopus*, and Fange has already raised this question with respect to the innervation of the swim bladder. It is pertinent to raise this possibility since all of the excitatory nerve fibers in the vagosympathetic trunk supplying the gastric musculature of an amphibian, *Bufomarinus*, are of sympathetic origin, the vagus supplying only nonadrenergic inhibitory fibers to the stomach. The presence of these nonadrenergic inhibitory fibers in the gastric vagi of both mammals and amphibians provides the second reason for caution, for it is not yet clear whether such fibers exist in fish. Campbell and Burnstock have pointed out that many of the excitatory responses to vagal stimulation recorded by earlier workers on both selachian and teleostean gut do not start until some time after the period of stimulation of the nerves is over. In this respect the contractions appear more like the "rebound" or recovery contractions which follow responses to inhibitory nerve stimulation in mammalian gut than like the virtually immediate primary contractions caused by stimulation of the excitatory innervation. In other words, there is still considerable doubt as to whether the contractions seen in fish stomach preparations following vagus nerve stimulation are mediated by excitatory or inhibitory nerves. The absence of records of actual inhibition coulα be easily explained under the conditions of experimentation for the nerves have been stimulated for relatively short periods, the spontaneous contractions occur at a very low rate, if at all, and the musculature has little or no tonus. Pharmacological investigations have been of little help in answering this question so far. For instance, vagal excitation of the stomach is not inhibited by atropine in the brown trout or in *Raja*, whereas vagal excitation of the smooth muscle of the intestine of the tench is prevented by atropine; in all three preparations, atropine prevents the excitatory action of acetylcholine. Only further experimentation can settle this question.

There have been no further investigations of neural control of gastric, hepatic, or pancreatic secretion in any fish since the subject was reviewed by Barrington in 1957. One can only agree with Barrington that there is no evidence for such a nervous control, while remarking that there have been too few investigations to allow any conclusion to be reached.

The processes causing filling and emptying of the swim bladder of teleosts are complex. At least three processes seem to be implicated in determining whether a net secretion or absorption of gases occurs. First, the relative amounts of secretory and absorptive epithelium exposed to the lumen of the swim bladder can be varied by differential contraction of the muscularis mucosae or by variable closure of the oval. Second, the rates of blood perfusion of the secretory epithelium, and therefore of the retia mirabilia, and of the resorptive epithelium are probably independently controlled. Third, the activity of the secretory cells themselves, which appear to act by adding metabolites to the blood leaving the secretory epithelium, is probably under nervous control.

Bohr first showed that following section of the vagi, experimentally deflated swim bladders were no longer refilled. This observation shows that the vagi contain nerve fibers which are in some way responsible for secretion into the swim bladder. However, Fange was unable to show any filling responses to vagal stimulation. In fact, the only clear response to vagal stimulation that he observed was an increase in the relative area of resorptive epithelium exposed to the enclosed gases brought about in a number of ways by contraction of the muscularis mucosae, and a vasodilation in the resorptive epithelium. This response would favor emptying of the swim bladder, and it probably masked completely those effects of vagal stimulation favoring filling. Fange showed that these "emptying" responses of the muscularis mucosae and of the vasculature were probably mediated by adrenergic nerves which, he suggested, contracted the muscularis mucosae of the secretory portion but relaxed that of the resorptive portion of the swim bladder and dilated the blood vessels in the resorptive region. In view of their adrenergic nature, Fange postulated that these fibers were sympathetic in origin, joining the vagus intracranially. In the absence of conflicting evidence, this is still the safest interpretation. Fange's pharmacological arguments that the nerves are adrenergic have received support from studies using the histochemical technique for demonstrating tissue monoamines. Fahlen *et al.*, using this technique, have shown that there is an adrenergic (norepinephrine) innervation of blood vessels throughout the swim bladder and of the muscularis mucosae only in the secretory region in cod and trout; Fahlen found that the muscularis of only the postulated resorptive region in *Argentina* was adrenergically innervated. Many or all of these adrenergic neurons have their cell bodies in or near the swim bladder, a fact which would explain why vagotomy does not alter catecholamine levels in this organ .

It is highly likely that the vagal nerve fibers which cause secretion are cholinergic, for Fange (1.953) has shown that atropine treatment, like vagotomy, inhibits gas secretion into a deflated swim bladder. The problem is which structure or structures are being affected by the cholinergic fibers. Fange has suggested that the vagi provide cholinergic fibers to contract the muscularis mucosae of resorptive regions, a response which would tend to cause filling by decreasing the area of resorptive epithelium exposed. He reached this conclusion because he found that isolated preparations of the tissue were contracted by acetylcholine and relaxed by atropine. However, he did not observe muscular contractions of this type in response to vagal stimulation, nor did atropine treatment affect the position of the swim bladder diaphragm in *Ctenolabrus*. Fange has provided evidence for vagal cholinergic fibers causing vasodilatation in the secretory portion of the swim bladder. He noted that during secretion there was a marked vasodilatation in the gas gland, whereas after vagotomy, and especially after atropine treatment, there was considerable vasoconstriction in the gland. But Fange rightly raised the still unresolved question of whether the vasodilatation occurring during activity is a primary effect caused by specifically vasodilator nerves or a secondary effect caused by the release of metabolites from cholinergically activated secretory cells.

Spinal Autonomic Nerves

The sympathetic nervous system of selachians consists of a series of paravertebral ganglia, one or more occurring per spinal nerve, linked together at most by a loose plexus of nerve bundles. There is no compact sympathetic chain. The ganglia are connected to the spinal nerves by white rami communicantes. Young claims that gray rami are absent, i.e., that there is no entry of postganglionic fibers into the spinal nerves to reach dermal structures, but there is some evidence for sympathetic nervous control of dermal melanophores. The ganglia extend no further caudally than the posterior end of the mesonephros in adults. There are no cranial sympathetic ganglia nor is there any semblance of a sympathetic chain entering the head region; however, this does not preclude a sympathetic innervation of cranial structures via perivascular nerve plexuses. There are no prevertebral ganglia, and fibers from the paravertebral ganglia are collected into anterior, middle, and posterior splanchnic nerves extending to the alimentary canal via plexuses on the arterial supply. Other fibers run directly to the genital and urinary ducts.

In teleosts, the sympathetic system has a well-defined structure resembling that found in tetrapods. Paravertebral sympathetic ganglia, usually two per spinal segment, arc linked together regularly to form two sympathetic chains which fuse to a greater or lesser extent in different species. The sympathetic chains are connected to spinal nerves by both gray and white rami communicantes. The chains extend caudally to the end of the tail and also extend rostrally into the cranium where connections are usually made with the IIIrd, Vth, VIIth, IXth, and Xth cranial nerves. The gastrointestinal tract is mainly innervated by an anterior splanchnic nerve, leaving the right sympathetic chain, but a posterior splanchnic nerve has been found in trout. Genital and vesical nerves leave the posterior regions of the chains, supplying the urinary andd genital ducts via periarterial nerve plexuses.

A sympathetic innervation of the iris occurs in teleosts, but it has never been observed in selachians. The fibers involved leave the spinal cord in the anterior spinal nerves and run rostrally up the chain to form synapses in the sympathetic ganglion in the trigeminal nerve. Stimulation of the sympathetic chain causes pupilloconstriction. The response is prevented by treatment with atropine, indicating that the nerves are cholinergic. The sympathetic also appears to provide a motor innervation to the striated extraocular muscles in teleosts because sympathetic section causes abnormal protrusion of the eyeball. There is no evidence concerning the transmitter substance mediating this effect, but it has been found that sympathetic adrenergic fibers can cause contraction of the superior rectus muscle in a mammal.

It has long been denied that there is a sympathetic innervation of the heart in fish. However, fluorescence histochemical studies of the heart of teleosts have now shown that in spite of one report to the contrary there is an adrenergic innervation of the cardiac chambers. Nerve fibers containing small granular vesicles, typical of adrenergic nerves in other animals, have been seen during an electron microscopic investigation of the trout heart. The adrenergic nerve fibers appear to come both from the vagi and along the coronary arteries and occur most densely in the sinus venosus and auricle. Gannon and Burnstock have confirmed Kulaev's report that stimulation of the vagi can cause ardioacceleration in teleosts. They have also shown that this response is prevented by treatment with the adrenergic neuron-blocking drugs, bretylium and guanethidine, or with an epinephrine-blocking drug, propranolol, a congener of which inhibits cardiac responses to catecholamines in plaice. As in the case of the vagal adrenergic innervation of the swim bladder, it is probably safest to assume that these adrenergic cardioaccelerator nerve fibers enter the vagi from the sympathetic chain via the intracranial connections. In selachians, considering the absence of sympathetic nerves running directly to the heart and the absence of intracranial connections between the sympathetic and the vagus, it seems unlikely that a sympathetic adrenergic innervation of the heart will be found.

In both selachians and teleosts we are temptingly close to being able to say that there is sympathetic nervous control of the vascular system. On the one hand are reports that catecholamines injected into the circulation cause pressor responses in selachians and teleosts. On the other hand are the many reports of dense nervous plexuses investing blood vessels and, recently, a number of histochemical studies which show that there are adrenergic nerve fibers in some of these plexuses, at least in teleosts. Kirby and Burnstock have shown that adrenergic and cholinergic nerves, both mediating contraction of isolated spiral strips, occur in the ventral aorta of trout and eels. And yet there is still no report of a direct effect of autonomic nerve stimulation on the resistance of any vascular bed in fishes. What little evidence there is goes against belief in a sympathetic vasomotor tone. Wilber and Sudak found that there were no compensatory cardiovascular responses to hemorrhage in *Mustelus* and *Squalus*, while Burger and Bradley showed that a ganglionblocking drug, tetraethylammonium, and an epinephrine-blocking drug, dibenamine, did not cause any alteration of ventral or dorsal aortic pressure in *Squalus*. Randall and Stevens found that the epinephrine-blocker, phenoxybenzamine, caused a fall of only 1-2 mm Hg in dorsal aortic pressure in coho and sockeye salmon. At the moment it seems likely that if there is a sympathetic vasoconstrictor innervation of systemic vascular beds, it is used to regulate blood flow to individual organs rather than to sustain a certain level of blood pressure. Any generalised control of systemic vascular resistance is likely to be mediated by circulating epinephrine from chromaffin cells.

The sympathetic innervation of the gastrointestinal tract has been reviewed recently. Interpretation of the available results is again made difficult by the possibility that the observed excitatory responses to sympathetic nerve stimulation are in reality rebound contractions caused by the stimulation of inhibitory nerves. However, these authors concluded that a number of workers had shown true primary contractions mediated by sympathetic nerves. Other excitatory responses could be interpreted as rebound contractions, and there is one clear example of an inhibitory response to ' posterior splanchnic nerve stimulation in trout. In summary, these authors suggested that the sympathetic supply to all regions of the gut in both selachians and teleosts contains a mixture of excitatory and inhibitory nerve fibers. One may suggest that the excitatory nerve fibers are cholincrgic, for at least some of the excitatory responses are abolished by atropine treatment. The inhibitory nerve fibers may be adrenergic since catecholamines cause relaxation of the gut of a number of species and an adrenergic innervation of gastrointestinal muscle has been demonstrated histochemically in trout, eel, and tench, but direct evidence is lacking. In addition to the adrenergic, presumably sympathetic, nerves reaching the swim bladder in the vagus trunk of teleosts, there is a direct sympathetic innervation by a branch of the splanchnicnerve. Fange showed that if this branch is cut there is a slight rise in the oxygen content of the swim bladder gases. Fange suggested that this might occur because of a loss of sympathetic vasoconstrictor tone in the gas gland. However, Fahlen *et al.* have shown that adrenergic nerve terminals, deriving from cell bodies in the coeliac sympathetic ganglion, form pericellular networks about the adrenergic neurons of the gas gland ganglion in cod. This raises the possibility that bysectioning the splanchnic nerve one removes an inhibitory influence on synaptic transmission in the gas gland ganglion, thus enhancing a secretory activity being mediated by that neural pathway.

A few passing observations have been made on the sympathetic innervation of the urinogenital system. Stimulation of the sympathetic ganglia in selachians causes contractions of the oviducts, as does the application of acetylcholine. Similar responses of the ovaries occur in the teleosts *Lophius* and *Uranoscopus*. In addition, stimulation of the sympathetic innervation of the urinary bladder in these teleosts causes a contraction and the expulsion of urine, a response also mimicked by acetylcholine. Young found that the excitatory responses of the bladder to nerve stimulation were reduced by atropine, suggesting that the nerves are cholinergic.

The number of investigations of nervous control of dermal chromatophores in fish is so great that no detailed discussion will be

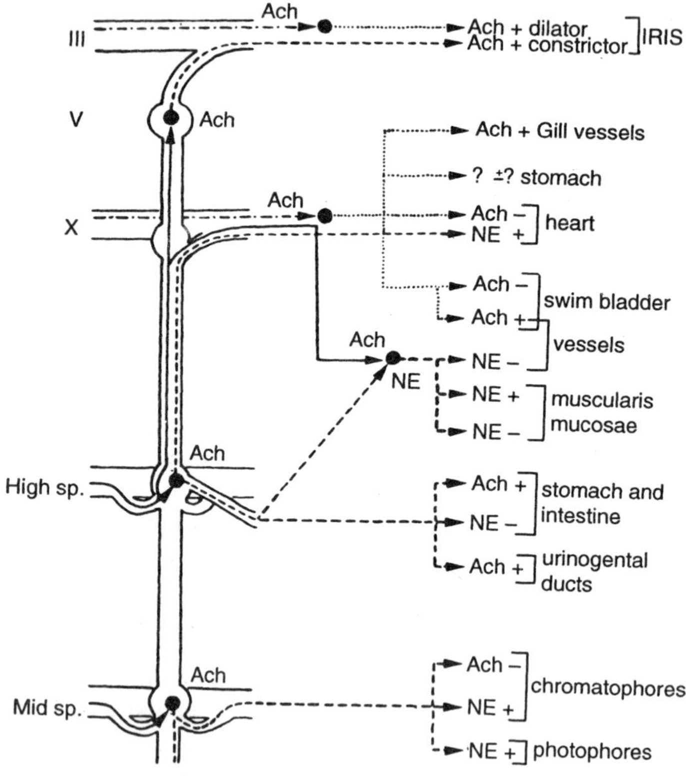

Figure 7.3: Diagram to illustrate some principles of the arrangement of the autonomic nervous system in teleosts, showing the autonomic outflow in the oculomotor (III), vagus (X), anterior spinal (high sp.), and midspinal (mid sp.) nerves. Nerve types shown as (—) preganglionic sympathetic, (—) postganglionic sympathetic, (-) preganglionic parasympathetic, (. . .) postganglionic parasympathetic. The filled circles represent ganglionic synapses. Transmitter substances shown as Ach (acetylcholine) and NE (norepinephrine). Effects of the nerves shown as (excitation of muscle and gland, concentration of chromatophores) and -.

given here. The main conclusion reached has been that the melanophores in many if not all teleosts are under the control of sympathetic adrenergic nerve fibers, stimulation of which causes concentration of the melanophore pigment. A comparable sympathetic adrenergic innervation of dermal photophores has been postulated for the teleost *Porichthys*. Although it has been suggested that there are cholinergic sympathetic fibers which mediate pigment dispersal in melanophores, it seems to this reviewer that there is no definitive evidence for their existence. In selachians, color changes are generally slow or absent, but evidence has been produced for melanophore-concentrating nerves in *Mustelus*. In a few teleosts, catecholamines cause dispersal rather than the normal concentration of melanophore pigments, and it is not clear how this will affect the chromatophore innervation. There is very little information concerning the innervation of other chromatophore types (xanthophores and erythrophores), although it is well known that they may, for example, disperse under conditions which induce melanophore concentration. Virtually nothing is known of the control of internal melanophores. Perhaps the most striking feature of chromatophore control in fish is that the animals can not only match the shade of the background but can also match its pattern, an ability which is highly developed in various flatfish. The presence of such responses indicates that the sympathetic system can be used to pick out single or, at most, small groups of chromatophores as required, showing a fineness of control which is not usually attributed to the sympathetic system.

There is no clear evidence for the existence of a "spinal parasympathetic" autonomic outflow in fish. But Friedman has shown that stimulation of posterior spinal nerves in skates causes erection of the claspers by both muscular contraction and vasodilatation and elicits secretion from the clasper gland. As Nicol has remarked, these phenomena resemble effects of stimulation of the sacral parasympathetic system in mammals.

8

REPRODUCTION

Reproduction is the process by which species are perpetuated and by which, in combination with genetic change, characteristics for new species first appear.

TYPES OF REPRODUCTION

At least three types of reproduction are possible: bisexual, hermaphroditic, and parthenogenetic. In bisexual reproduction, which is the prevalent kind, sperms and eggs develop in separate male and female individuals. In hermaphroditism (a type of intersexuality), both sexes are in one individual and, as among certain serranids and a dozen or more other families, self-fertilisation or true functional hermaphroditism exists. Such synchronous hermaphroditism, from an evolutionary point of view, could be the most advantageous form of reproduction.

Hermaphroditic sex glands are known for several species, including some trout relatives (salmonoids), perches *(Perca),* walleyes *(Stizostedi-qn),* darters *(Etheostoma),* and some of the black basses *(Micropterus).* Some seabasses (Serranidae) are protandric hermaphrodites, being males at first, females later.

Parthenogenesis is the development of young without fertilisation, and a condition that has been called parthenogenesis (more properly gynogenesis) occurs in a tropical fish, the live-bearing Amazon molly *(Poecilia formosa),* and is also known in *Poeciliopsis.* Mating with a male is required, but the sperm serves only one of its two functions, that of inciting the egg to develop; it does not take any part in heredity. The resultant young are always females, with no trace of paternal characters.

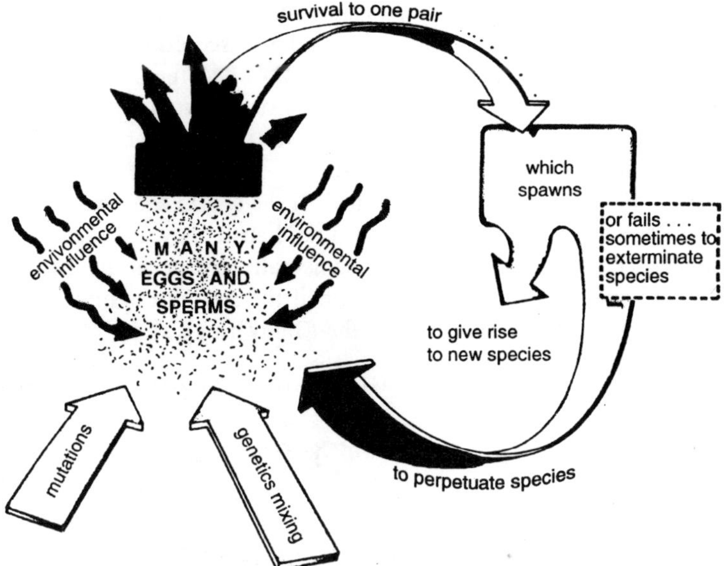

Figure 8.1 : Role of reproduction in the survival and evolution of species.

THE REPRODUCTIVE SYSTEM

The reproductive function in fishes is primarily the job of the reproductive system. Previously described are the components of the system, the sex glands or gonads, ovaries in the female and testes in the male, and their ducts. The endocrine system plays an important regulatory role in reproduction.

SPERMATOZOA AND THEIR FORMATION

The process by which sperm cells are formed in the testes is called spermatogenesis or spermiogenesis. It is in this process that the materials determining the characteristics which the offspring will inherit from its parents are readied for union with like materials in the egg cell. Not only do the spermatozoa or sperms of different fishes contain different hereditary materials, they differ in shape as well. It is also in this process that sperm cells gain a whip-like tail which makes them motile and with which they make their way ultimately to and into eggs for fertilisation. Furthermore, during development spermatozoa attain the differences in shape that they exhibit as species characteristics. In order to insure fertilisation, each male fish, like the male of other vertebrate species, produces huge numbers of sperm cells so tiny that a million of them can be held in a droplet. The spermatozoa, plus secretions of the sperm ducts, compose the milt that the male fish exudes at spawning (like the semen of man). The sperms are inactive and immobile until the secretion of the sperm ducts has been encountered. Once they have become active and have started to move by the whipping motions of their tails they will soon perish unless an egg is encountered for fertilisation. The life span of sperms varies considerably according to species and according to substrate into which they are deposited. If the sperm cells are deposited in water, they live a much shorter time than if they are deposited in a female. If water has nearly the same salt content as the fish body fluid, the sperm cells live longer than if the medium is either higher or lower in effective salt content than body fluids.

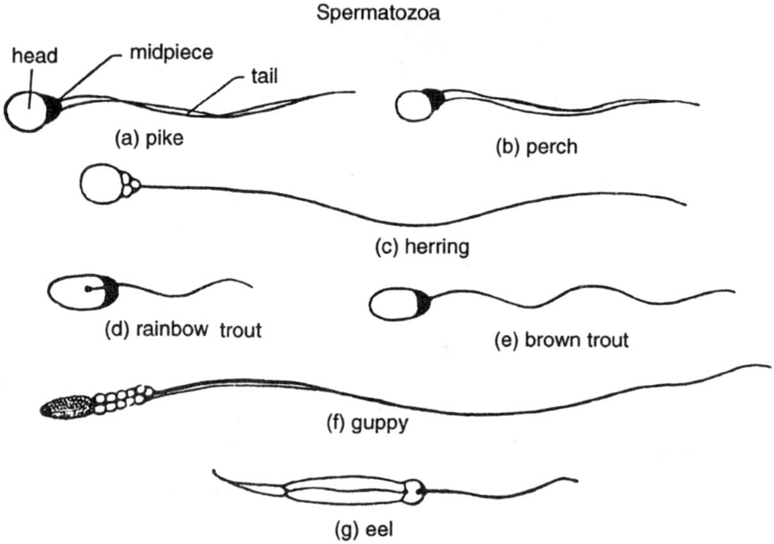

Spermatozoa

head

midpiece

tail

(a) pike

(b) perch

(c) herring

(d) rainbow trout

(e) brown trout

(f) guppy

(g) eel

Figure 8.2a : Variations in shape among spermatozoa (greatly enlarged) of certain bony fishes: (a) northern pike (Esox lucius); (b) European perch (Perca fluviatilis); (c) Atlantic herring (Clupea harengus); (d) rainbow trout (Salmo gairdneri); (e) brown trout (Salmo trutta); (f) guppy (Lebistes reticulatus); and (g) European eel (Anguilla anguilla).

Viability of sperm cells is also affected by temperature; in general, they live longer in lower temperatures than in higher ones and, just as human sperms may be stored under refrigeration for artificial insemination, it is possible to store the milt of fishes. The spermatozoa of fishes, like those of man, bull, fowl, and frog, have been successfully preserved by freezing-for example, carp *(Cyprinus)*, Atlantic herring *(Clupea)*, and salmon *(Oncorhynchus)* sperms have remained viable after freezing.

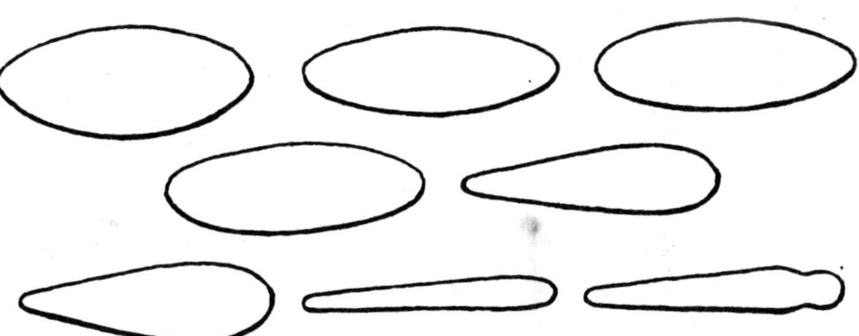

Figure 8.2b : Nonspherical fish eggs (greatly enlarged) as shown in selected gobies. (Gobiidae).

Eggs and Their Formation

Oogenesis is the process of egg development in ovaries parallel to that of sperm manufacture in the testes. The result is not only the largest of the cells to be found in fishes (mostly from 0.5 to 5.0 millimeters in diameter, with few to golf-ball size and larger) but also a cell endowed with the peculiar quality to produce after fertilisation a new individual of the species. During oogenesis the surrounding epithelial cells (granulosa) provide the developing egg cell with relatively large quantities of stored food material in the form of granular yolk (protein) and fat commonly in the form of oil droplets.

Fecundity is a general term used to describe the number of eggs produced. The fecundity of an individual female varies according to many factors including her age, size, species, and conditions (food availability, water temperature, season).

Fecundity appears to bear some broad relationship to the care or nurture accorded to the eggs. Thus, the ocean sunfish *(Mola mola)* may produce huge numbers of eggs but they are merely broadcast in the open sea water by the female while swimming with males. A 54-pound fish of this kind is said to produce 28,000,000 or more eggs in a single season. The cod *(Gadus)*, which is also a pelagic open-water spawner, produces as many as 9,000,000 eggs per female per season. Such species accord no care to their eggs, other than to emit them in the vicinity or in the company of males depositing sperms. Compare with this the sticklebacks *(Culaea* and *Gasterosteus)* which make elaborate nests and then provide parental care to the developing fish. Here the usual number of eggs ranges between thirty and a hundred. In guppies *(Lebistes)*, which bear their young alive and thus

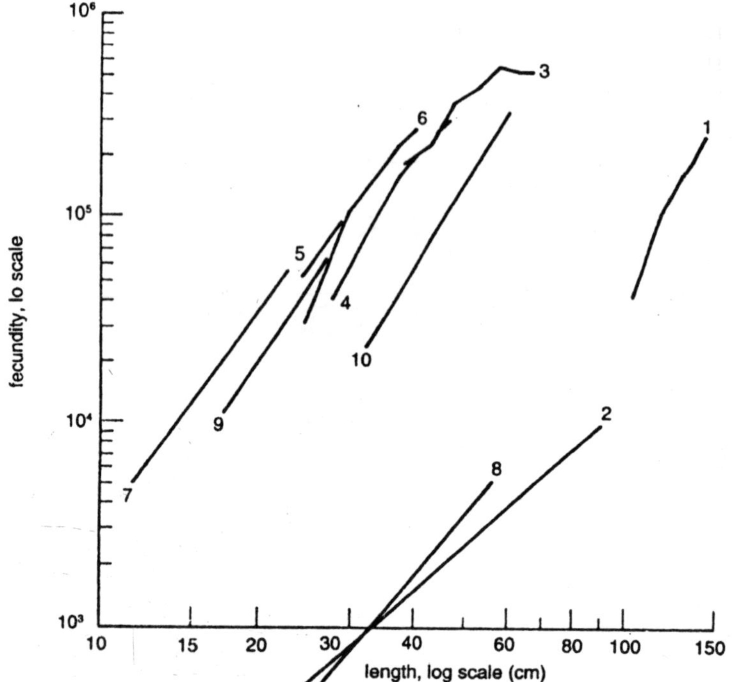

Figure 8.3 : The relationship between fecundity and length within a species. 1. Acipenser stellatus; 2. Salmo salar; 3. Cyprinus carpio; 4. Pleuronectes platessa (Clyde); 5. Clupea harengus (northern North Sea); 6. Melanogramus aeglefinus; 7. Osmerus eperlanus; 8. Salvelinus fontinalis; 9. Sardinops caerulea; and 10. Sebastes marinus (Faroe Island).

accord an extreme amount of care (that is, internal incubation), the usual brood number is less than two dozen. Some fishes even develop at one time, only a single egg, or one in each ovary.

The eggs of most fishes have shells secreted around them. The egg cell itself has only a tender membrane when ready to leave the ovary. As the eggs of oviparous species pass through the oviducts, they go through glands that secrete the shells about them. In fishes that accord care to their young, either by having the parents stay on guard or by burying the eggs in gravel, the shells are often more tender and less limy or horny than they are in eggs exposed to waves, currents, or other hazards. In many skates and some sharks, for example, the shells are horny-hard and almost fingernail-like in strength. Such eggs resist damage, even drying, to such an extent that the eggs, when washed up on shore, may lie there for some time and still hatch when returned to the sea. An egg of the little skate *(Raja erinacea)* was once mailed without padding from the New Jersey coast to our Ann Arbor laboratories in a common letter envelope. It was out of the water in transit for several days and traveled some seven hundred miles. Yet it hatched successfully and became a pet for some months.

Fish eggs have many other interesting characteristics than their covering. Some are buoyant (pelagic) and have a specific gravity about like that of fresh water. Although most marine food fishes, for example, cod *(Gadus),* have buoyant eggs, some marine species, for example, herring *(Clupea harengus),* have nonbuoyant eggs that are attached to the substrate. Most stream fishes have eggs that are heavy and sink (demersal), with a specific gravity greater than that of fresh water. Perhaps this is an adaptation that insures these eggs against being swept downstream at the time of deposition, which may take place in gravel beds in rather swift current in sites such as those chosen by stream chars *(Salvelinus),* the trouts and Atlantic salmon of the genus *Salmo,* and the species of Pacific salmons *(Oncorhynchus).*

Ova also vary as to adhesive stickiness. Most buoyant eggs are nonadhesive and do not normally stick to anything for holdfast purposes. Most eggs that sink, however, are at least temporarily adhesive. Trout and salmon (Salmoninae) eggs are temporarily adhesive during the process of water hardening in the early minutes after deposition while they are absorbing water at a great rate. One can almost visualise the eggs as "sucking in" water at such a rate that they stick to things, things stick to them, or they adhere to one another. In this process, the eggs of stream-spawning trouts and salmons pick up little particles of sand that help sink them to the bottom and anchor them there for covering and development. The stickiness of the eggs of other fishes such as the walleye *(Stizostedion)* serves the purpose of keeping the eggs from settling deep into bottom materials and becoming suffocated. Sticky, demersal eggs adhere to sticks and debris on the bottom without falling down into the deeper, softer materials that may be present. Adhesiveness can be due temporarily to the process of water hardening, or more permanently to a pasty, mucoid deposit on the egg that remains sticky throughout the whole developmental period.

Eggs also vary as to shape, but those of most species of fish are spherical. The spherical shape seems to be the natural, basic contour of most animal eggs. In many anchovies of the herringlike family Engraulidae, the outline of the egg is elliptical, occasionally spherical.

But in the minnow genus *Rhodeus,* the European bitterling, and in various other fishes including the cichlids, the eggs are oval in outline (ovoid). In the gobies they range from ovoid to teardrop-shape.

The eggs of fishes also differ considerably according to the appendages that they bear. In the hagfish *(Myxine),* for example, the egg is an elongate ovoid, with a number of tendrils at each end. Apparently, these tendrils anchor the eggs to vegetation or other material. In the American smelt *(Osmerus mordax),* the egg has a low stalk which is adhesive and becomes attached to the stony bottoms of streams where the species most often spawns. In the brook silverside *(Labidesthes sicculus),* there is a single elongate filament that serves first for temporary flotation and then for attachment. The eggs of some skates (Rajidae) and cat sharks (Scyliorhinidae) have tendrils that come out of each of the four corners of the more or less box-shaped case. Most flyingfishes (Exocoetidae) have eggs with numerous filaments, much like long hairs, that come out from the surface of the spherical egg. However, the eggs of a few flyingfishes have no filaments, some have the filaments confined to the two poles of the egg, and others, to one pole.

Quite as spermatozoa exhibit different viability in different situations, so do eggs have various survival features. Interestingly, not all mature eggs are laid in any one period of spawning. Some remain in the ovary and are normally resorbed. It would appear, however, that when disproportionately large quantities of eggs are not shed, the ovary is unable to resorb them. The individual may then become egg-bound, and the mouth of the oviduct plugged by a mass of fibrous tissue. Future passage of eggs may be barred.

SEX DIFFERENCES

Anyone can tell a male from a female fish when milt or eggs are running. A light press on the belly of most ripe fishes will bring the whitish milt, or the eggs, into view near the anus. Both humans and fish, however, also use features other than these to tell the sexes apart. The characteristics of sexual difference or sexual dimorphism that enable identification of the sexes are classed as primary and secondary. Primary sexual characters are those that are concerned actually with the reproductive process. Testes and their ducts in the male, and ovaries and their ducts in the female constitute primary sexual characters. The primary sexual characters often require dissection for their discernment, which makes the secondary sexual characters often more useful, although sometimes not so positive.

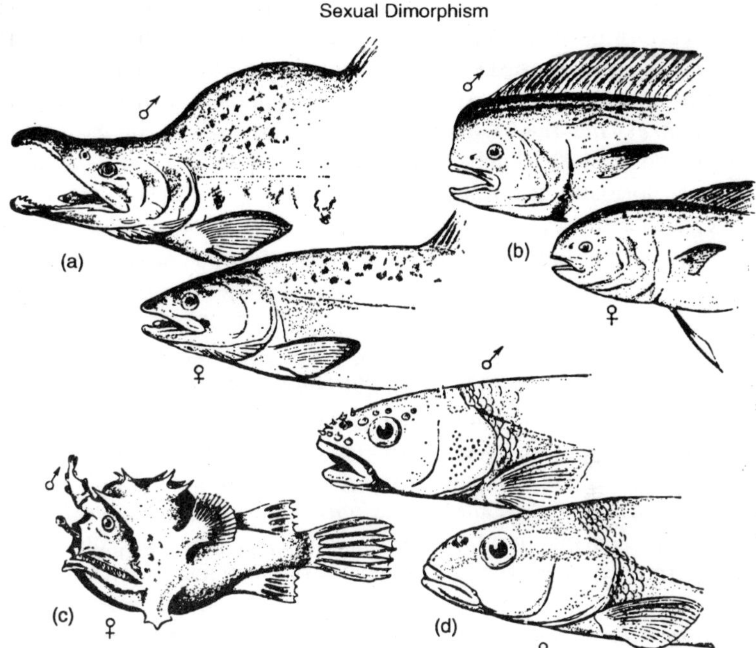

Figure 8.4 : Sexual dimorphism among fishes as shown by: (a) humped back and hooked jaws of the male in the pink salmon (Oncorhynchus gorbuscha); (b) domed forehead and anterior position of the dorsal fin in the dolphin (Coryphaena hippurus); (c) parasitic male of the deepsea anglerfish (Photocorynus spiniceps); (d) nhptial tubercles on the snout and forehead of the creek chub (Semotilus atromaculatus).

Secondary sexual characters themselves are really of two kinds: those which have no primary relationship with the reproductive act at all, and those which are definitely accessory to spawning. Some outstanding points of difference of a dimorphic nature between the sexes follow.

A genital papilla marks the male of such fishes as lampreys (Petromyzonidae), Johnny darters *(Etheostoma nigrum),* and white bass *(Morone chrysops).* The papilla is a veritable penis in one group of sculpins, the Oligocottinae. Body shape is an important secondary sexual character. Ordinarily the female is much more pot-bellied than the male, particularly when she is ripe or nearripe for spawning, because of the relatively larger quantity of sexual products which she contains. Pearl organs or nuptial tubercles appear on the males of many fishes to mark the sex; examples are the smelt *(Osmerus),* and most species of minnows (Cyprinidae) and suckers (Catostomidae). These tubercles are little horny excrescences that become evident just before the spawning season and disappear shortly after, under the influence of hormonal secretions. The fins often provide characteristics distinctive of the males; on the average they are larger than in females. Besides, in one North American fish, the fantail darter *(Etheostoma flabellare),* the tips of the spines of the dorsal fin of the breeding male end in fleshy knobs, giving them a clubbed shape, whereas the female normally has ordinary rays. In some other species males have over-all fleshy rays (bluntnose minnow, *Pimephales notatus),* whereas the rays in the fins of the female are just ordinary, fine ones. In some fishes, the caudal fin may show sexual dimorphism; for example, the lower lobe is greatly extended in the males of the swordtail *(Xiphophorus helleri)* of the aquarium fancier, and is somewhat enlarged in the white sucker *(Catostomus commersoni).*

An accessory sexual character marks the males of several species in which the anal fin becomes enlarged into an intromittent, copulatory organ. This organ, designated the gonopodium, occurs in such fishes as the mosquitofish *(Gambusia affinis),* the guppy *(Lebistes),* and in other livebearing topminnows (Poeciliidae). The anal fin in some viviparous perches (Embiotocidae) bears glandular structures that are of some use in intromission. In the male of the white sucker *(Catostomus commersoni),* the enlarged anal fin serves to convey the watery milt into the bottom of the currentplagued nest in streams, thereby improving the chances for contact between the sperms and the eggs. In the clasping cyprinodont *(Xenodexia)* the modified pectoral fin or heterochir is evidently used in mating for holding the gonopodium in position for insertion into the oviduct of the female. The pelvic fins are variously modified in the sharks and their relatives (Elasmobranchii) as intromittent structures, the myxopterygia, that help to insure internal fertilisation which is widespread in this group of fishes.

Obviously, colouration in fishes often serves as a mark of sexual distinction and recognition; it is termed sexual dichromatism. In general the males are brighter or more intense in colour than the females. This is also true among the birds but, as among the birds, there are some exceptions to the rule. An outstanding example of sexual dichromatism in the central North American fish fauna is the orangespotted sunfish *(Lepomis humilis).* In this sunfish the male has more numerous and brighter orange spots on the body than does the female. In the bowfin, *Amia calva,* a dark eye-spot develops in the tail region in the young of both males and females but becomes diluted and subdued in the adult females whereas in the males it becomes intensified and develops a brightly coloured ocellation (eye-like ring) around it. Wrasses (Labridae) and parrotfishes (Scaridae) in the marine realm are prime examples of sexual dichromatism and in some cases also of dimorphism.

Intromittent Organs

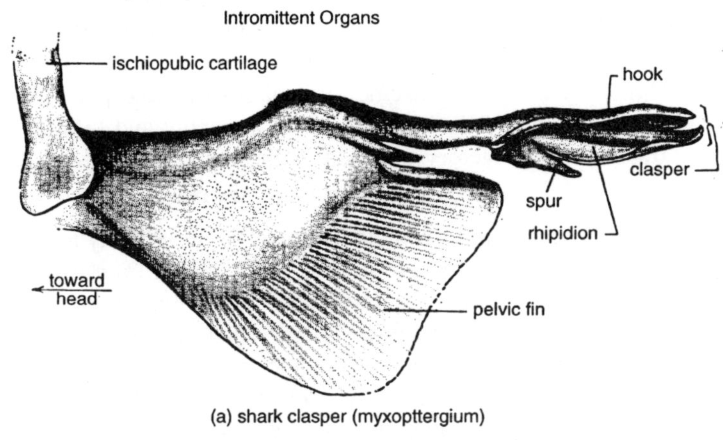

(a) shark clasper (myxopttergium)

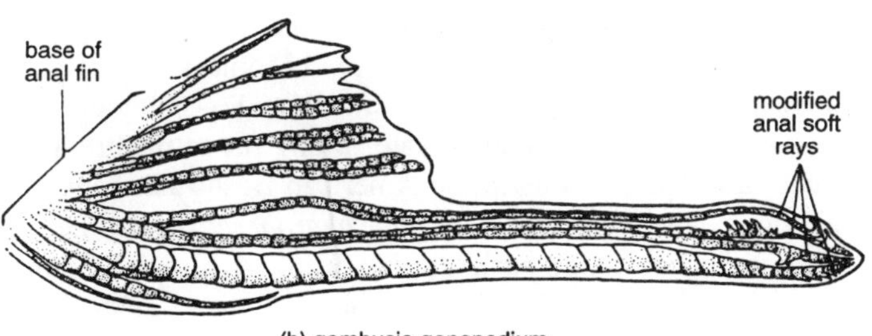

(b) gambusia gonopodium

Figure 8.5 : Male intromittent organs of two fishes: (a) left pelvic fin of a horn shark (Heterodontus francisci) showing its modification into a myxopterygium, ventral view; (b) anal fin modified into a gonopodium in the mosquitofish (Gambusia affinis), lateral view.

Several different head characteristics also serve to distinguish the sexes among fishes. In the chimaeras (Chimaeridae), the male develops a spiny, stout, retractile knob, the frontal clasper, on the upper part of the head; a similar structure characterises the forehead of the forehead brooders. In the salmons and the trouts (Salmoninae), the breeding males typically develop a knobby hook or kype near the tips of both the upper and lower jaw.

A few accessory reproductive structures among females serve as sexual characteristics. An outstanding example is the egg-laying tube or ovipositor in the females of the European bitterling, *Rhodeus amarus,* and an Asiatic lumpsucker, *Careproctus.* In a few species of fishes there have been discovered differences in fin-ray and vertebral counts between males and females. Neoteny, the attainment of sexual maturity by larval individuals, is a sexually dimorphic feature of males of the deepsea stalkeyed fishes *(Idiacanthus)* and among certain European lampreys (Petromyzonidae).

Probably the most extreme form of sexual dimorphism occurs in the deepsea angler fish where the male is minute and little more than a parasitic sperm factory attached to the much larger female.

SEXUAL MATURITY

Several factors influence the first attainment of ability to reproduce, or sexual maturity, in fishes. Among these are differences in species, in age and size, and in individual physiology. In general, species of small maximum size and short life span mature at younger ages than do kinds of large maximum size. Some fishes are sexually mature at birth; males of the reef perch *(Micrometrus aurora)* and of the dwarf perch *(M. minimus)* spawn shortly after birth. Although the females of these species receive sperm soon after they are born, they do not bear young until a year later. Other fishes, such as the guppy *(Lebistes)* and the mosquitofish *(Gambusia affinis),* become sexually mature at ages less than a year and lengths of an inch or less, although they are not born (or hatched) in an adult state. Many fishes mature at the age of 1 year. Examples of this in the American freshwater fauna include the inch-long least darter *(Etheostoma microperca)* and other species less than 6 inches long, such as the brook silverside *(Labidesthes sicculus);* a marine example, of which there are many, is the California grunion *(Leuresthes tenuis).* Many species attain sexual maturity for the first time at ages of 2 to 5 years, with lengths from 3 to 12 inches or more. Among these are trouts *(Salmo),* black basses *(Micropterus),* and sunfishes *(Lepomis).* Eels *(Anguilla)* become sexually mature at 10 to 14 years of age and 2 feet or more in length, and sturgeons *(Acipenser)* at 15 years or more, with lengths of about 3 feet (1 meter) and more. Both age and associated size are thus significant factors in the determination of adulthood.

Once a fish is sexually mature, the sex products must still become ripe and the reproductive act must take place. Many factors interplay to bring about these events. The forces at work can be grouped roughly into those that arise within the fish and those that are in the surroundings (extrinsic). The individual fish has essentially no control over either set. Among the intrinsic factors determining the attainment of sexual maturity and subsequent breedings and ripenings are the kind of fish and its heredity, foods chosen, and finally the whole physiological makeup of the individual, which of course, would be very difficult to separate from Its heredity. Within a species, sex of the individual may determine importantly how it reacts to the factors that set up the drive to spawn. In many kinds of fishes, the males mature and ripen earlier than the females, either in the entire life-cycle or in the course of a single spawning season. For example, the sex glands of some yearling males of certain Pacific salmons *(Oncorhynchus)* may ripen, and the individuals may spawn, although the species normally first spawns 2 or 3 years later. Some of the early-ripening males, the so-called grilse, which often do mate with females, ripen a year earlier than the females of the same age. In many freshwater fishes like trouts *(Salmo),* suckers *(Catostomus),* and creek chubs *(Semotilus),* the males arrive first on the spawning grounds, which for the kinds listed are the riffles of streams for the stream-spawners. Here the male starts building the nest or just awaits the arrival of the females. This early ripening of males has been held to be due to their higher rate of metabolism, perhaps through endocrine influence in response to genetic and environmental forces.

The internal physiological rhythm of gonadal maturation, involving predominantly pituitary-gonadal interactions, is adjusted to insure that breeding will occur at a time when environmental conditions are most favorable for survival of the offspring. Changes in the duration of daylight (photoperiod) and the temperature affect the rhythm of gonadal maturation and spawning of fish in the Temperate Zone and higher latitudes. Near the equator where changing day length could not possibly account for the many examples of strictly seasonal reproduction, reproductive activity in many tropical fishes often coincides with extensive flooding during the rainy season. Other extrinsic factors that may affect reproduction are presence of the opposite sex, current, tide, stage of the moon, and the presence of Spawning facilities. In the bitterling, *Rhodeus amarus,* it seems that the presence of a freshwater mussel, into which the eggs are to be laid, stimulates ripening. And the currents from the exhalent syphon of the mussel finally stimulate egg laying by the fish.

REPRODUCTIVE CYCLES

Reproductive behaviour in most animals is cyclic-more or less regularly periodic; this is so for nearly all fishes. The reproductive act in some fishes occurs only once in a very short lifetime; an example of this is the brook silverside *(Labidesthes sicculus)*. In still other fishes, it occurs once in a moderately long life span: in the Pacific salmons *(Oncorhynchus)*, once in 2 to 5 or more years; in the sea lamprey *(Petromyzon marinus)*, once in 5 or 6 years; and in the freshwater eels *(Anguilla)*, once in 10 to 14 years. Most fishes, however, have a yearly cycle of reproduction, and, once they have begun it, they follow it until they die. Several other species spawn more than once in a year and more or less continually. The guppy *(Lebistes)* may bring out broods at approximately 4-week intervals; also, broods may be several per year in certain tilapias *(Tilapia)*. Around Hawaii, the threadfin *(Polydactilus sex filis)* spawns once a month between May and October. The grunion *(Leuresthes)* of the California coast may spawn every couple of weeks in season during the spawning period until it is spawned out. In all of breeding the objective is to bring about union of the sperm and egg. Other adaptations then lead to the survival of the fertilised egg.

BREEDING

Fish have developed many different ways for gaining nearness of sperms and eggs to each other in order to facilitate and insure their union. Outstanding among these ways are the nicely timed relations of the various reproductive functions. In the prolific open-water spawners, for example, the millions of eggs do not all mature at one time; just enough mature in each batch so that at the final stage of rapid growth of the eggs, the female is not too greatly distended with the enlarged sex products. If all ripened simultaneously, they would occupy a space much larger than the parent fish, so batches are timed in ripening to accommodate the parent. Many fishes have a very short breeding season, occurring only once a year as we saw, yet the males and females are in fully ripe condition at the same instant and both then are capable of exhibiting their complex breeding reactions. The secondary sexual characteristics and accessory structures which are used in courtship, clasping, or intromission develop simultaneously with the readying of the primary sexual products.

Perhaps the most famous case of accurately timed relationships in reproduction is in the palolo worm which swarms to the surface of the open seas once a year at a certain phase of the moon to breed. A few fishes show similar marvelously timed relationships for breeding. The best known of these is the grunion *(Leuresthes tenuis)* of the California coast. This fish spawns just after the turn of high tide at certain times of the year. It spawns almost literally out of water, as far up on the beach as the largest waves will carry it at this time of high tide. It deposits eggs and sperm in pockets in the wet sand at just such a time and in just such a position that the eggs are not likely to be washed out by waves until two weeks or a month later at the time of the next high tide. At this time the waves come up on the sand again and when they hit the places where the nests are deposited and stir up the sand, the young hatch almost instantly and go out with these waves before the tide can recede and make it impossible for them to do so. This is certainly a remarkable instance of timed reproduction as well as development and hatching. Regarding timed relationships, too, we should think that somewhere in a continent such as North America, some fishes are spawning at almost any time, because of the influence of latitude and altitude, as well as individuality of species and races or varieties that compose them.

Most fish species have definite seasons for spawning as a part of their timed reproductive relationships, and are generally grouped as follows. Warmwater fishes are summer spawners and coldwater fishes, fall and winter spawners. Species tolerating intermediate temperatures are generally spring spawners. Some tropical species spawn the year around. Fixed spawning seasons are roughly correlated with the developmental period that the fish require. Warmwater fishes (such as the black basses and sunfishes, Centrarchidae) use only a few days to hatch and emerge into an environment generally favorable to growth and survival. Several months may be required for a char *(Salvelinus)* or a whitefish *(Coregonus)*. Some trouts *(Salmo),* at a mean temperature of 50°F (10°C), may average around *50* days for their developmental period. The whitefish *(Coregonus clupea f ormis),* at a temperature in the low 30°s F (1°C), may take as much as 130 to 150 days for development to hatching. Both seem to be timed to bring the young forth into conditions favorable to them. Timing of reproduction also makes it possible for more than one species to use the same breeding grounds in a calendar year. Thus in American streams that both occupy, spring-spawning suckers *(Catostomus)* may use the same nest (redd) sites as fall-spawning brook trout *(Salvelinus fontinalis)*.

Timed relationships, no matter how good, would not be very useful to a fish if it could not recognise a mate of the opposite sex when present. When a scientist is sexing fish he uses such characters as structure, form, and colour as described previously. With live fish, he may also use behaviour. Fish themselves are also aware of form, colour, and behaviour when recognising their mates. An outstanding experimental study of the value of these attributes for sexual recognition to fish is that by Tinbergen, who has demonstrated all of these points by observing sticklebacks *(Gasterosteus)* in aquaria.

In addition to the evolution of timed relationships, and abilities in sexual recognition, there have developed among fishes several useful devices for insuring either external or internal fertilisation. For external fertilisation, proximity of two individuals of the opposite sex for spawning is the most common means employed. Actual pairing and some form of holding (amplexus) is sometimes used as a special development of proximity. In pairing, some fishes come side by side in actual contact and simultaneously emit eggs and sperms and in other instances the male twists his body around that of the female, in a semicircle, or even in a corkscrew spiral for a fish with a much-elongated body such as a lamprey (Petromyzonidae).

The parasitic dwarf male of deepsea anglerfishes *(Photocorynus)* adheres to the skin of the female by its mouth. Here the male has solely a reproductive function, and has virtually lost the digestive tract as a result of his parasitic adaptation. Concentration of eggs in masses is a feature that insures external fertilisation. We see eggs roped together with watery jelly in perches. Eggs are massed underneath stones in such fishes as the Johnny darter and the bluntnose and fathead minnows. The marine scorpionfishes *(Scorpaena)* make eggs into a two-lobed balloon. There are relationships here with the features of viability and specific gravity of sperms and eggs and also of their abundance. Mass aggregations for spawning are also a means employed for insuring external fertilisation.

For internal fertilisation, several devices have evolved in fishes. Most common among them is the placement of sperm by the male into the reproductive tract of the female in the process of intromission, for which important adaptations include: special modifications (Figure 10.5) of the pelvic fins (as in the shark group, Elasmobranchii), and of the anal fin (as in the topminnowlivebearer group, Poeciliidae); and the development of particular genital organs in the region of the genital pore (as in the blind cavefishes, Amblyopsidae) are important. The function of all of these structures is to bring the sperms into the oviduct of the female. The normal composition of the sperm mass in these forms is fluid milt. In at least one fish, *Horaichthys,* of India, a further step has been taken to insure the transfer of a mass of sperm. The spermatozoa are packaged by the male into a tight mass similar to the spermatophore that is employed among salamanders for getting packages of sperm from the male to the female. The sperm ball, which carries an arrow-tipped , stalk, is impaled on the skin of the female near the opening of the oviduct, in a position ready to insure fertilisation as the female passes one egg at a time. The basking shark, *Cetorhinus maxim us,* also has spermatophores.

Still another manner in which the proximity of eggs and sperms is guaran*teed is in* the stimulus provided by one sex to the other to release the sex products at the right time. This is brought about in many fishes by definite courtship behaviour patterns, in which sound emission and pheromones play a role and the male may swim circles around the female or prod her, bunt her, rub her, fan her, herd her, or do most anything to signify that "now is the time." Interestingly, fish with the best-defined and strongest courtships appear to have produced smaller numbers of eggs than those in which there -is little or none. Furthermore, in fishes in which the males are very active courters they usually assume the task of caring for the eggs. In addition to the foregoing, reproduction in fishes is also affected by such obvious conditions as water levels, substrate or bottom type, weather, natural and manmade barriers, predation, food supply, and pollution.

Lest the reader assume that breeding habits of fishes are well known, it must be pointed out here that information is lacking or incomplete on most of the 20,000 species.

CARE OF EGGS AND YOUNG

The effective production of eggs and sperms alone would not be enough to bring about survival of species. Whereas many fishes, especially marine ones, rely simply on "safety in numbers," there have evolved various means for affording care to fertilised eggs and young by one or both sexes. One of the means evolved has been to get the eggs into the right place-the place for which their developmental machinery is suited. Thus, such anadromous fishes as the sea lamprey *(Petromyzon marinus),* sea-run sturgeons *(Acipenser),* and salmons *(Salmo, Oncorhynchus)* ascend freshwater streams to spawn. Many adult salmon home to the very streams in which they hatched and spent their youth, after travels of thousands of miles over periods of 5 years or more. The freshwater eels *(Anguilla)* have a catadromous habit, descending into the ocean to bring the eggs into the salt-water habitat to which they are bound for development.

Once on the spawning grounds, the selection of the spawning site takes place. This process appears to be a particularly careful one in

Egg Fate

Laid at Random

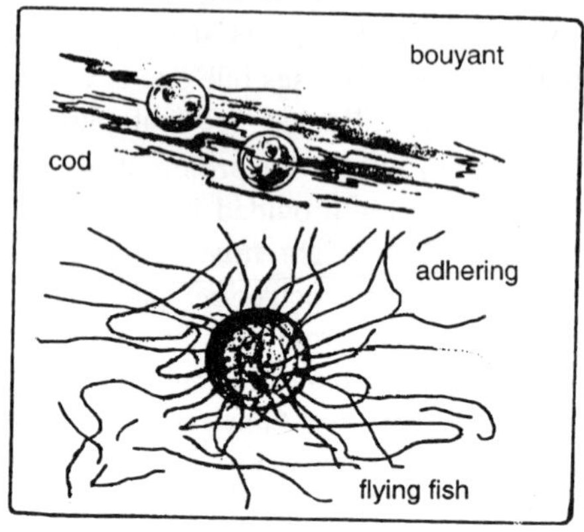

Laid in Masses

Placed in Another Animal

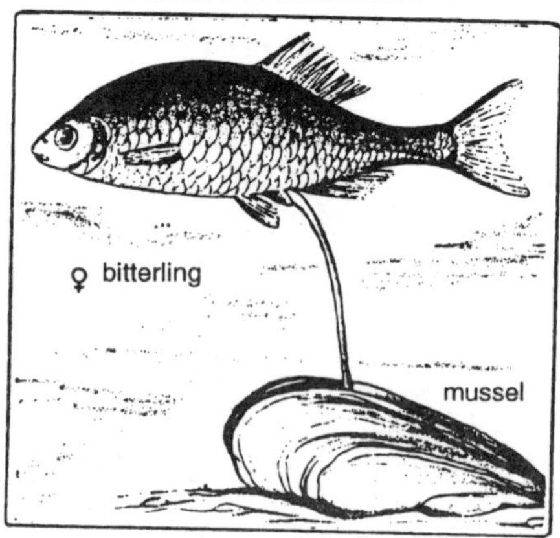

Figure 8.6 : Diversity in disposition of eggs and young in fishes.

nest-building fishes and is of great importance for successful reproduction. Once the site has been chosen, some fish will fight to the death to defend it. An outstanding example of this is the Siamese fighting fish, *Betta splendens,* well known to aquarium hobbyists as a species that defends its territory to the death. Many fishes, however, do not give special care to the eggs at all. In this instance, the eggs themselves are adapted to survival. They may float, either with or without a little filamentous appendage that will give them some attachment during development. Pelagic eggs which are transparent appear invisible to potential predators. The eggs may be in masses such as the hollow ropes of eggs of the perch *(Perca), so* that they are not scattered, do not fall into the mud, and have some protection by mass-size. The transparent egg balloon of the scorpionfish *(Scorpaena)* mentioned earlier, would be of some value in affording protection, as well as in providing for dispersal. The various attachment devices, as well as adhesiveness, also help the survival of eggs that are deserted by their parents.

Table 8.1 : Diversity in Nests and Spawning Places as Shown by Selected Examples from Fresh Waters of North America

I. Nest building species.

- A. Those exhibiting parental care.
  - 1. Nest a circular depression.
    - *a.* In bed of mud, silt, or sand, often in and among roots of aquatic flowering plants. Sunfishes (Centrarchidae) including the sunfishes proper *(Lepomis),* largemouth bass *(Micropterus salmoides),* crappies *(Pomoxis),* rock basses *(Ambloplites),* and warmouth *(Chaenobryttus);* bowfin *(Amia);* most bullheads *(Ictalurus).*
    - *b.* In gravel bottom. Smallmouth bass *(Micropterus dolomieui);* rarely, largemouth bass, rock basses.
  - 2. Nest excavated under a stone or other submerged object with the eggs attached to the underside of the roof of the nest. Johnny darter *(Etheostoma nigrum),* fantail darter *(Etheostoma flabellare),* stream-dwelling sculpins *(Cottus),* bluntnose and fathead minnows *(Pimephales).*
  - 3. Nest made of plant materials and spherical or mound shaped. Sticklebacks *(Culaea; Gasterosteus).*

4. Nest a tunnel. In bank, channel catfish *(Ictalurus punctatus);* in bottom, yellow bullhead *(Ictalurus natalis).*

B. Those deserting nests after spawning; nest a well-defined structure in gravelly bottom. Lampreys (Petromyzonidae); stream trouts including the brook trout *(Salvelinus fontinalis),* cutthroat trout *(Salmo clarki),* rainbow trout (S. *gairdneri),* and brown trout *(S. trutta);* salmons *(Oncorhynchus* and *Salmo salar);* creek chub *(Semotilus atromaculatus);* fallfish *(S. corporalis);* river chub *(Hybopsis micropogon);* hornyhead *(H. biguttata);* rainbow darter *(Etheostoma caeruleum).*

II. Species that do not build nests.

A. Scattering eggs, usually over aquatic plants, or their roots or remains. Northern pike *(Esox lucius);* carp *(Cyprinus carpio);* goldfish *(Carassius auratus);* golden shiner *(Notemigonus crysoleucas).*

B. Scattering eggs over shoals of sand, gravel, or boulders. Whitefishes and ciscoes *(Coregonus);* lake trout *(Salvelinus namaycush);* suckers *(Catostomus);* walleyes *(Stizostedion).*

C. Depositing eggs in single masses of definite form. A "rope" of eggs, yellow perch *(Perca flavescens).*

D. Spawning near surface without reference to vegetation or type of bottom. Brook silverside *(Labidesthes sicculus);* alewife *(Alosa pseudoharengus).*

Of those fishes that actually build nests, there are broadly two kinds: those that make a nest and then desert it and those that build one and stay with it to guard the site, eggs, and/or young. A few examples of fishes that build excavated nests then protectively fill them with stones interspersed with eggs are the creek chubs *(Semotilus),* the trouts *(Salmo),* and the salmons *(Oncorhynchus).* Some fishes that excavate a nest and then defend it against invaders are the bluntnose and fathead minnows *(Pimephales),* and the sun fishes and black basses (Centrarchidae). The latter family, of North American fresh waters, is characterised by this nest building and defensive habit in which the male is the aggressive partner-all except one genus, the Sacramento perch *(Archoplites),* which builds no nest. Bullheads *(Ictalurus)* dig a nest and stay on guard. The Siamese fighting fish *(Betta)* blows a surface bubble nest and defends it, as sticklebacks *(Culaea* and *Gasterosteus)* defend theirs made of twigs, leaves, and detritus on a substrate. The eggs of the bitterling *(Rhodeus)* are deposited in the mantle cavity of a freshwater mussel and those of the lumpsucker

(Careproctus), beneath the carapace of the Kamchatka crab. To carry egg protection to its highest degree, some fishes have evolved various types of internal incubation or gestation. Sometimes protection by the parent body is accorded the young by the male, as in the seahorses *(Hippocampus)* and pipefishes *(Syngnathus),* in which the eggs are placed by the female into the brood pouch of the male. In a Brazilian catfish *(Loricaria typus)* the male parent develops an enlarged lower lip to form a pouch in which labial incubation of the eggs takes place. The marine catfishes (Ariidae), including the gafftopsail catfish *(Bagre marinus)* and the sea catfish *(Galeichthys fells)* employ the mouth as an oral incubator. Each male parent carries from ten to thirty developing eggs in his mouth cavity where the hatchlings may still be found after they' have begun to feed independently. Oral incubation has also evolved in the cardinalfishes (Apogonidae), among others, that may thus carry a hundred or more eggs.

Truly internal incubation is that which occurs in livebearing fishes. Livebearing, with its many variations of embryonic and nutritive exchange, has evolved independently in several groups of fishes. Among the sharks, livebearing occurs in more than a dozen families. Most are nonplacental or ovoviviparous, but there are a few species that have well-defined nutritive and respiratory exchange structures among the genera *Triakis, Carcharhinus,* and *Mustelus* of the requiem sharks (Carcharhinidae), and *Sphyrna* of thc hammerhead sharks. Among the rays, skates, and relatives (Rajiformes), ovoviviparity is the rule, with the oviparous skates (Rajidae) being the exception. Interestingly, the oviparous sharks and relatives are bottom dwellers of relatively shoal waters and are never large, whereas the livebearing ones range from small to gigantic and are widely variable in their habits and distribution.

Livebearing among bony fishes is most highly developed in the topminnows (Cyprinodontiformes), especially in the families of the guppy *(Lebistes)* and mosquitofish *(Gambusia affinis),* Poeciliidae, Mexican livebearers (Goodeidae), the foureyefishes (Anablepidae), and jenynsiids (Jenynsiidae). It also appears in the halfbeaks (Hemiramphidae) and in several families of the perch-like fishes (Perciformes) : surfperches (Embiotocidae), clinids (Clinidae), eelpouts (Zoarcidae), brotulas (Brotulidae), and scorpionfishes and rockfishes (Scorpaenidae). In the latter it is universal among more than fifty species of *Sebastodes* of California coastal waters. The living coelacanth, *Latimeria, is* also ovoviviparous.

In the literature of ichthyology it has often been implied that the development of livebearing fishes takes place in a uterus. Some refinement of this generalisation is necessary. If it is allowed that a uterus is fundamentally an enlarged oviduct, the organ is present in livebearing sharks (Squaliformes). Among livebearing bony fishes (Osteichthyes), however, development takes place within the ovaries. In some, the development of the embryo is intrafollicular. Here the egg follicle is vascularised, dilated, and fluid-filled. The embryo, which is without an enveloping chorion, develops within the follicle of its original egg and ovulation and birth are simultaneous. This is the situation in the livebearers (Poeciliidae) and other families including the halfbeaks (Hemiramphidae) and the foureyefishes (Anablepidae). In some fishes, however, the egg may be fertilised while still in its follicle but completes development after leaving the ovigerous follicle within the cavity of the ovary; examples of this include the surfperches (Embiotocidae) and the families Jenynsiidae and Goodeidae. In some cyprinodontiforms, more than one brood may be present simultaneously in an ovary-a condition termed superfetation.

Adding to care of eggs, we should think for a moment of the care of the hatchlings which come from them. Hatchlings are actively defended by many fishes, for example, by members of the catfish family (Ictaluridae) and of the sunfish (Centrarchidae) and stickleback (Gasterosteidae) families. Actual herding of the young into places of shelter and away from enemies, even the kinds of antics that lead enemies away from the young, are sometimes practiced by fishes. The antics of the loon *(Gavia)* and other water birds, in distracting menaces to their young, are well known to naturalists. A largemouth bass *(Micropterus salmoides)* and a bowfin *(Amia) will* behave in much the same way, creating diversionary splashes and so forth in an opposite direction while the little school of black young moves away. Some African tilapias are called "mouthbrooders" because the young when they are hatched escape at time of danger into the oral cavity of either sex.

DEVELOPMENT

A hundred years ago, descriptive aspects of development were exciting the biological sciences. Now, although descriptions of the embryology and postnatal development have still not been made for most species of vertebratescertainly not for most fishes-excitement of discovery centers in learning about the mechanisms of the developmental processes, including the interaction of genetic and environmental forces in determining various characteristics of the individual.

In its broadest sense, development is an endless process, continuing from generation to generation. For the individual, the start is fertilisation of the egg; attainment of maturity and reproduction afford continuity; and the end of development is death. Spermatogenesis and oogenesis are thus nearterminal stages, preparatory for the next generation. For convenience, however, the successive periods of development may be regarded as: *(a)* early embryonic; (b) transitional embryonic or larval; and (c) postembryonic. Because of predation, limitations of food supply, and greater sensitivity of embryos and larvae to changes in chemical and physical properties of their environment, these stages of development sustain greater mortalities than later stages in many egg-laying fishes.

The early embryonic period starts when the egg is fertilised by a sperm and ends when the embryo has attained the generalised organ systems as they appear in common among all fish embryos.

The transitional phase of embryonic development involves the transformation of the generalised organ systems and body form of the early embryonic stage into those resembling the adult. Thus the definitive body form is attained at or near the end of this stage. During this period of development, two kinds of larvae, free-living and non-free-living, are represented among fishes. Free-living larvae are characterised by having their existence outside of protective embryonic structures. Non-free-living larvae complete their transition inside the egg membranes or within the body of either female or male parent. In species with indirect development, involving metamorphosis, the completion of metamorphosis ends this stage. In some fishes a late embryonic phase covers those developmental changes that occur after definitive body form has appeared. In live-born fishes, the period may be passed within the reproductive tract of the mother. In species with free-living larvae, adult shape is attained either soon after hatching or after a passage through several stages of development in which the young bear relatively little resemblance to the adult (e.g., convict surgeonfish, *Acanthurus triostegus)*.

Postembryonic development has juvenile, sexually mature adult, and senescent phases. In a juvenile the organ systems, particularly the reproductive system, complete their development. The body form is essentially that of an adult, although some differential growth of parts may go on. As an adult, the individual has the typical mature form of the species and is capable of reproduction. Adulthood passes into old age, senescence and degeneration. In old age, the reproductive potential declines and/or ends, other activities lessen, and various bodily parts slowly degenerate. The period terminates in death.

Embryonic Development

Impregnation

For present purposes we may think of embryonic development as beginning at the time of impregnation. In the process of impregnation, an egg is penetrated by a sperm. Activation of the egg results from this entry. Almost immediately there is a cortical reaction in the egg resulting in a block to the entry of additional spermatozoa. In some eggs the micropyle, which is the principal port of sperm entry through the vitelline membrane (often called the chorion) surrounding fish eggs, is sealed by material extruded from large vesicles, the cortical alveoli in the egg cortex. The entire vitelline membrane or chorion may also be made impervious to sperm entry. As a result, the process of polyspermy which is found in some other animals is not common in fishes. After completion of the cortical reaction, the vitelline membrane becomes the fertilisation membrane and the process of water hardening goes on in externally laid eggs. Independent of impregnation, the egg absorbs water and its chorion is at least temporarily adhesive because of the osmotic processes involved. When water uptake ends, the egg is turgid and firm to the touch.

At various periods of development fish eggs are especially sensitive to handling. They are quite often tender until the eyed stage is reached. Jolts, jars, and rapid changes of temperature are all particularly harmful to them during this early sensitive period. Once pigment has appeared in the iris of the eyes, the embryos are called eyed eggs. In this condition they are quite hardy and can even be transported for long distances if proper conditions, including those of moisture and temperature, are maintained. They remain in a fairly rugged condition until a few days before hatching. Then, perhaps because of softening of the eggshell in preparation for emefgence, the embryos again become tender.

Fertilisation

Actual fertilisation of an egg is not complete until the nucleus of the egg cell and that of the sperm have made union in the cytoplasm of the egg. Fusion of the male and female pronuclei from sperm and egg respectively completes the process of fertilisation. Now the chromosomes which carry hereditary factors, called genes, are

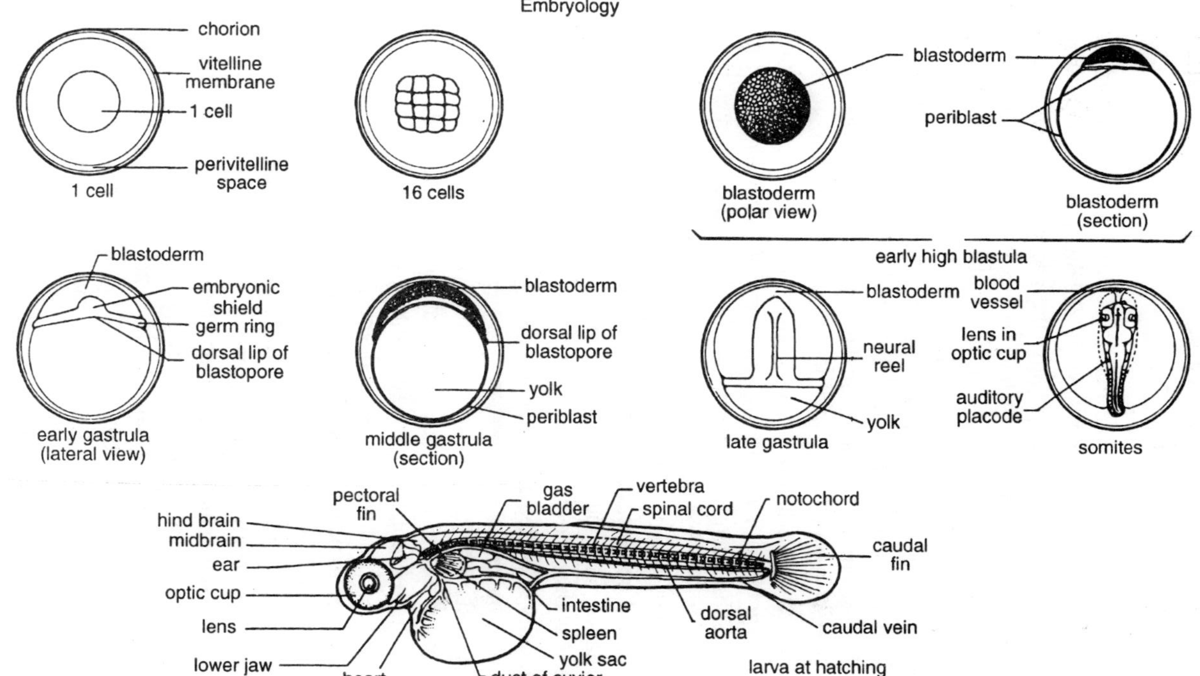

Figure 8.7 : Representative stages in the development of a bony fish, the mummiochog (Fundulus heteroclitus).

brought together in a process called amphimixis and the genes from both parents begin to exert their effects on the development of the embryo.

Cleavage

Both the processes and the results of embryonic development are similar among the major fish groups although detailed information about them is spotty. For all groups, the processes include cleavage, which is the division of the egg into successively smaller cells called blastomeres. Cleavage ultimately results in the formation of a transitory blastula stage, which consists of a single layer of cells called blastomeres arranged as the rim of a sphere containing a central cavity, the blastocoel, or spread out as a flat plate on the upper surface of a large yolk mass. In the blastula, some areas of cells are already destined to form certain organs. Gastrulation follows blastulation and converts the embryo to a two-layered stage consisting of an outer epiblast and an inner hypoblast. The antero-posterior axis becomes established at this time. Subsequently the gastrula elongates and the major organ-forming areas are extended. The foregoing processes establish the primitive generalised body form. Their rate is under genetic and environmental control, and details differ among major groups of fishes in accordance with the amount and distribution of yolk in relation to cytoplasm within the egg cell.

Typically the cleavage of chordate eggs is either holoblastic or meroblastic, but many intermediate types exist. In the holoblastic type, the entire cell divides into two, then four in meridional planes. The third cleavage plane is latitudinal, resulting in eight cells of nearly equal size. In meroblastic cleavage, only a small disc-like part of the egg divides to produce the blastoderm. The yolk material remains undivided, later to be enveloped by growing tissues. A classical example of holoblastic cleavage is seen in the cephalochordate amphioxus *(Branch iostoma)*. Here the cells divide nearly equally and result in a neat blastula which is a hollow ball of cells; the cavity within the blastula is the segmentation cavity or blastocoel. The blastula in turn gastrulates simply by having one of its sides invaginate to yield a gastrula resembling a rubber ball pushed in at one side. In the lampreys (Petromyzonidae) there is more yolk than in amphioxus. The eggs exhibit a polarity with more yolk at the lower or vegetal pole, an egg-cell condition termed telolecithal. As a result of yolk impedance, the cleavage is not as ideally holoblastic as in the amphioxus egg but is still classified as a holoblastic type. The yolk-free cells of the upper or animal pole (the micromeres) divide faster than those of the lower or vegetal pole of the egg (the macromeres), and overgrow the macromeres. Gastrulation is by convergence of cells toward the blastopore at the posterior end of the embryo, and subsequent turning-in of the micromeres at the border between them and the macromeres. Cleavage is somewhat holoblastic in the South American lungfish *(Lepidosiren),* the sturgeons *(Acipenser),* the gars *(Lepisosteus),* and the bowfin *(Amia),* but it exhibits various meroblastic features. The early development of these fishes is thus intermediate between holoblastic and meroblastic cleavage types.

Meroblastic cleavage is the prevalent form of egg division in fishes and other vertebrates. In the primitive hagfishes (Myxinidae) and in the sharks and relatives (Chondrichthyes), the egg is strongly telolecithal and cleavage is restricted to a small disc of cytoplasm at the animal pole. The same is true for the bony fishes (Osteichthyes) other than those described above. Cleavages result in the formation of two kinds of cells, blastoderm and periblast.

The blastoderm cells are distinct and produce the embryo. The periblast or trophoblast cells lie between the yolk and cells of the blastoderm and cover the entire yolk mass, having originated from the most marginal and outlying blastomeres. They become syncytial and are involved in mobilisation of yolk reserves. The blastulae of the hagfishes, of the sharks and their relatives, and of the teleosts (Actinopterygii) are disc-like and hence called blastodiscs. The blastocoele or segmentation cavity is between the outer blastoderm and the central periblast that covers the yolk beneath the blastoderm.

Through work in experimental embryology, it has become possible to identify, already in the blastula, regions destined to give rise to specific organs. The areas of presumptive gut-forming, notochordal, neural and epidermal cells have been thus established, as has a region of potential mesoderm. The term mesoderm refers to the middle germ layer of the embryo. It is formed during gastrulation when the hypoblast subdivides into mesoderm and endoderm. The outer germ layer of the early gastrula, the epiblast, is termed the ectoderm in the three-layered embryo.

Gastrulation

The, process of gastrulation is generally similar among most fishes with meroblastic cleavage. There are, however, some differences among bony fishes, for example, in the bowfin *(Amia),* as well as between bony fishes and the sharks and their relatives (Chondrichthyes).

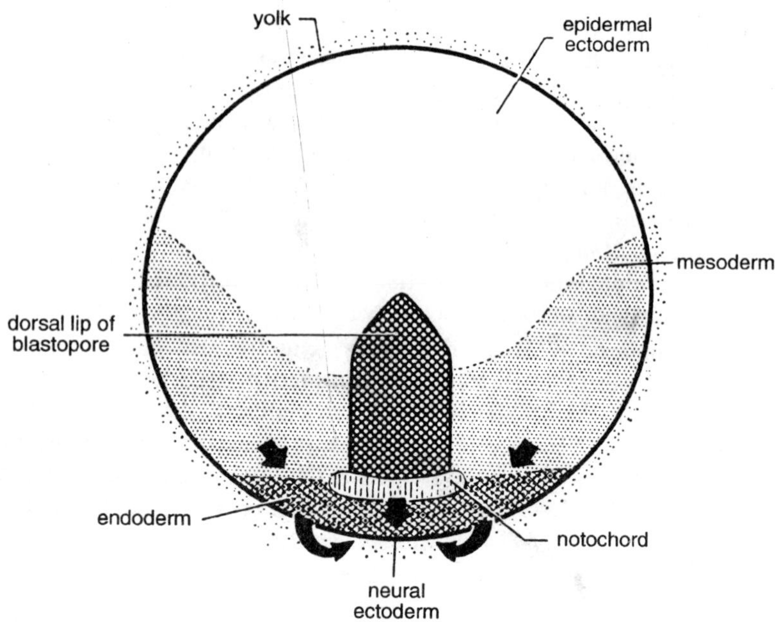

Figure 8.8 : Presumptive organ-forming areas of the blastoderm of the mummichog (Fundulus heteroclitus). Arrows indicate direction of cell movement.

A brief description of the basic process must suffice here. As the time for gastrulation comes on, the under-rim of the blastodisc thickens to form the "randwulst" or marginal ridge which comes to have an inner layer termed the germ ring. Caudally, the ring is thickest and is recognised as the embryonic shield. At the beginning of gastrulation, the presumptive endodermal cells at the caudal edge of the embryonic shield turn inward beneath the blastoderm and then stream forward beneath it to become the endodermal portion of the hypoblast. For this incursion of cells, an opening, the blastopore, appears at the tail end of the embryonic shield. Also moving inward over the dorsal lip of the blastopore are cells of the prechordal plate and notochord that establish the embryonic axis. Presumptive mesodermal cells also migrate inward over the lip of the blastopore and position themselves on both sides of the embryonic axis beneath the ectoderm. Initially the involuted cell mass is termed the hypoblast. During subsequent proliferation and delamination, the endodermal and mesodermal components of the hypoblast become organised to establish the primordia of the internal organ systems.

While the embolic processes of involution of cells go on, cellular overgrowth of the yolk (epiboly) also proceeds. In epiboly, randwulst and germ ring cells not involved in involution, accompanied by presumptive ectodermal cells, grow to cover the yolk mass outwardly as epiblast; the periblast provides through growth an inner covering of the yolk mass. The initial periblast cells and epiblast in covering the yolk form the yolk sac. As epiboly is proceeding, the presumptive neural plate, forerunner of the central nervous system, transforms into a ridge-like keel along the ectodermal middorsal line. This solid keel, in contact with notochordal cells beneath, gradually becomes overgrown by epidermis and eventually becomes tubulated to form the hollow neural tube.

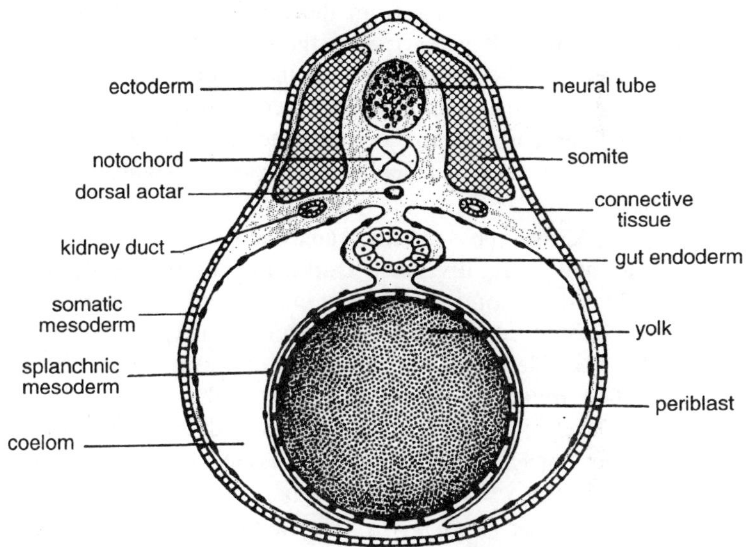

Figure 8.9 : Diagrammatic cross section of a fish embryo removed from egg shell during late embryonic stage to show how yolk sac is covered.

Gastrulation is complete in bony fishes when the yolk mass has been overgrown. During this process, some of the mesodermal tissue along both sides of the notochord has become organised into segments, the somites. Externally one can distinguish the boundary between embryo and yolk sac.

Organogenesis and Hatching

Subsequent to gastrulation in fishes, early embryonic development continues until a primitive vertebrate body form is established. The body at this time is more or less cylindrical and basically bilaterally symmetrical. Head and pharyngeal regions project from the yolk mass anteriorly; the trunk lies over the yolk; and the tail projects posteriorly. Five fundamental organ-forming tubes have appeared and have consistent relationships to the axial notochord.

Tubulation accounts for five tubes of tissue: epidermal (body-covering), neural, endodermal (primitive gut), and two mesodermal (somitic and coelomic cavities). These five tubes of tissue, plus the unspecialised mesenchymal cells of chiefly mesodermal origin, have key roles in organogenesis.

The ectodermal tube becomes the body covering or epidermis. Included in its derivatives among fishes are the outer layer of teeth, the olfactory epithelium and nerve, the lens of the eye, and the inner ear. The neural tube, arising from the sunken neural keel, is also ectodermal in origin. It gives rise to the brain and spinal cord and remaining parts of the eye, including the retina and optic nerve. Associated neural crest cells, among other things, provide assorted ganglia of the nervous system and contribute to the mesenchyme. The pigment cells known as melanophores appear to originate from the neural crest.

The mesodermal tubes of the trunk soon segregate into dorsal, intermediate, and lateral divisions. Dorsal mesoderm is early divided into the block-like somites already mentioned. Each somite is subdivided into three regions called sclerotome, myotome, and dermatome. The sclerotome contributes to the formation of the axial skeleton. The myotome develops into trunk musculature, the appendicular skeleton, and the appendages and their muscles. The dermatome develops into the connective tissue of the dermis of the skin, and its derivatives including scales.

Intermediate mesoderm gives rise to the kidneys, part of the gonads, and their ducts. Lateral mesoderm splits into inner and outer layers enclosing the coelomic cavity. Linings of the pericardial and peritoneal cavities come from lateral mesoderm, as do the heart, blood, trunk blood vessels, and all layers of the gut except its epithelial lining, which comes from endoderm. Mesenchyme in the head contributes to outer layers of the eye, the head skeleton, head musculature, and the dentinal layers of the teeth.

Endoderm supplies the primary sex cells (the germ tract) to the gonads in fishes and makes up the inner epithelium of the entire digestive tract and its associated glands. It also contributes to the development of several endocrine glands that are derived from the primitive gut, including the thyroid and the ultimobranchial bodies.

Hatching results from a softening of the chorion by enzymes or other chemical substances that are secreted from ectodermal glands usually on the anterior surface or from endodermal glands in the pharynx. The activity of the embryo assists in breaking through the chorion. Because the chemical composition of the chorion is not well understood, there is doubt about the mode of action of the enzymes (chorionases). A considerable part of the nutrient material in the chorion may be utilised by the embryo via the perivitelline fluid.

Recapitulation by Stages

When embryos and larvae of different species of either the same size and age are compared, most often the developmental stages are different. Comparisons are most useful when based on equal developmental stage. Even so, problems arise because of difficulties in defining these developmental stages. Nevertheless, in reviewing embryonic development, we may follow a sequence of stages of demonstrated utility in making comparisons among species, at least among oviparous ones, as derived for the mummichog *(Fundulus heteroclitus)*, a killifish (Cyprinodontidae). Approximate age in hours in the mummichog is cited for eggs held at 27°C (77°F):

Stage 1. The unfertilised egg.

Stage 2. One-celled embryo, the newly fertilised egg (1 hour).

Stage 3. Two-celled embryo (1½ hours).

Stage 4. Four-celled embryo (2 hours).

Stage 5. Eight-celled embryo (2½ hours).

Stage 6. Sixteen-celled embryo (3 hours).

Stage 7. Thirty-two-celled embryo (3½ hours).

Stage 8. Early high blastula. The blastoderm is elevated into a domed, cap-like structure and the periblast is being established (5½hours).

Stage 9. Late high blastula. Cells of blastoderm are smaller than in Stage 8.

Stage 10. Flat blastula. Blastoderm has become a flattened disc, rather than an elevated bulge as formerly (8 hours).

Stage 11. Expanding blastula. Blastoderm has begun to grow over yolk.

Stage 12. Early gastrula. Germ ring and embryonic shield formed and blastopore just opening (16 hours).

Stage 13. Middle gastrula. About half of yolk covered by blastoderm, neural keel just visible in midline of embryonic shield (19 hours).

Stage 14. Late gastrula. More than half of yolk covered by blastoderm, embryonic shield narrowing, neural keel more clearly visible (22 hours).

Stage 15. Closure of blastopore. Little embryonic differentiation except formation of rudiments of central nervous system and perhaps of optic vesicles and first somites (26 hours).

Stage 16. Expansion of forebrain begins (for formation of the optic vesicles). Three primary vesicles visible in brain.

Stage 17. Formation of cavity in optic vesicles, mesodermal segmentation provides one to four somites.

Stage 18. Formation of the auditory placode. Somites range from four to fourteen and extra-embryonic coelom appears as a cavity developed by yolk-sac epithelium (36 hours).

Stage 19. Cavity appears in neural cord behind brain. Ectoderm thickens to form lens of eye and olfactory pit. Somites range from fourteen to twenty (42 hours).

Stage 20. Expansion of midbrain to form optic lobes. Somites range from twenty to twenty-five. Melanophores appear about neural cord and are present over yolk. Pericardium established. Heart visible, tubular, and pulsing. Location of pectoral fin becoming visible by concentration of cells.

Stage 21. Motility. Muscular contractions come at about twenty-eight somites.

Stage 22. Circulation. Circulation starts with about thirty-five somites. Forebrain walls forming cerebral hemispheres.

Stage 23. Otoliths appear in ear. Melanophores appear in pericardium and blood flows in yolk vessels.

Stage 24. Pectoral-fin bud pointed.

Stage 25. Formation of urinary vesicle (as an outgrowth of hindgut).

Stage 26. Formation of liver and peritoneal cavity.

Stage 27. Pectoral fin becomes rounded.

Stage 28. Pigmentation of peritoneal walls.

Stage 29. Circulation established in pectoral fin. Fin becomes motile.

Stage 30. Rays appear in caudal fin. Lower jaw is formed.

Stage 31. Formation of gas bladder (diverticulum of gut). Eyes and jaws become motile.

Stage 32. Hatching. Some yolk is still present (11 days).

Stage 33. Pigmentation and growth of gas bladder.

Stage 34. Yolk absorption completed (12 days or more).

In subsequent larval development, skin and scales have yet to complete their differentiation, the axial skeleton must still be finished, the dorsal and anal fins and adult kidneys are yet to form, and definitive bodily proportions and pigmentation are still to be realised.

Nutritive and Respiratory Relationships

Embryos of the many different kinds of fishes have different nutritive and respiratory relationships depending on whether the young are born alive or hatch from externally laid eggs.

The embryos of egg layers essentially are totally yolk dependent for their food. Many species have one or more oil droplets in or about the yolk mass, which serve both as a potential source of energy and as a flotation or righting organ. Supplemental dissolved organic material and mineral salts may be absorbed directly from the water. Respiration and excretion of eggs laid in water is through the semipermeable egg membranes and shell.

For the embryos of all livebearers, there is nutritive and respiratory exchange *in utero* or *in ovario,* even when a placental arrangement is lacking, as in the ovoviviparous sharks or bony fishes. In these latter groups the uterine or ovarian juices make a physiological union between mother and embryo. The exchange is more intimate, however, among placental viviparous fishes. Histologically, the uterine lining varies from a mucus-secreting layer of cuboidal epithelial cells in species that depend on egg yolk for nourishment *(Acanthias vulgaris)* through forms with moderately folded serous-secreting linings *(Torpedo)* to those with uterine linings beset with villi (trophonemata) of varying lengths and complexities that secrete an abundance of fat *(Trygon violacea)*. Yolk is mainly digested within the intestine of embryos which depend on it for nourishment *(Squalus acanthias)*. True viviparity with yolk sac placenta is confined to certain species of two families of sharks (Carcharhinidae and Sphyrnidae). Embryos of *Sphyrna tiburo* develop elaborate circulatory networks in the walls of their yolk sacs, which become greatly folded. The uterine mucosa becomes folded, and the interdigitation of these two series of folds with a thinning of their epithelia bring maternal and fetal circulations into close proximity. The embryo is attached to the placenta by a modified yolk stalk (umbilical cord).

Among viviparous bony fishes, embryonic nutrition and respiration may successively involve several sources and organs, some of them quite special. In the Mexican livebearers (Goodeidae), the embryonal yolk sac, pericardial sac, and trophotaeniae variously take over the tasks. Trophotaeniae are finger-like extensions of rectal tissue. In the jenynsiid topminnows (Jenynsiidae), similar relationships exist between embryo and ovarian cavity and tissues, but trophotaeniae are replaced by trophonemas which are wormlike extensions of the ovarian wall that grow into the gill chambers of the embryo. In embryonic nutrition, the brotulas (Brotulidae) involve the yolk sac, body covering, gills, and finally the gut proper of the embryo through ingestion of the intraovarian fluid and by feeding on ovarian tissues, including egg cells. In the jenynsiids, cannibalism is practiced by some embryos on others within the same intraovarian cavity.

In the eggs of trouts and salmons (Salmoninae), the carbohydrate level increases gradually during development, although the glucose level falls temporarily at hatching. Glycogen is probably synthesised near hatching and is stored in the liver of the embryo near the end of the yolk-sac stage. Fat is used as a fuel, possibly 70 to 80 percent being consumed over the whole period of development. Fat contributes an estimated 68 calories per egg by the time the yolk is absorbed. Protein contributes some 45.9 calories to the process. Among various trour' ·,,.60 to 63 percent of the available protein in the yolk supplies energy and 16 to 40 percent of the fat. The energetic requirement of development is some 80 calories, and metabolic combustion uses 114 calories. The efficiency is high; about 40 percent of the energy originally incorporated in the egg is converted into growth of the embryo; the remainder, about 60 percent, goes for such activities as osmoregulation, secretion, circulation, and movements.

Transitional or Larval Development

Although in common usage a young fish is an embryo until birth or hatching, as indicated previously there are, according to species, varying amounts of subsequent development to be undergone in the earliest free-living stages. Generally, developmental stages prior to the adult stage but following hatching or birth are termed larval, and the(. during this period are called larvae. Larval fishes are also called fry. The period of larval life may range from a few moments to sever 1 years in length. The ammocete larvae of the sea lamprey (Petromyzon marinus), for example, require 5 or more years before metamorphosis.

In fishes, larval development is commonly, though not universally, divisible into prolarval and postlarval stages. Prolarvae are distinguished by the presence of the yolk sac and are commonly called sac fry by

Lamprey Metamorphosis

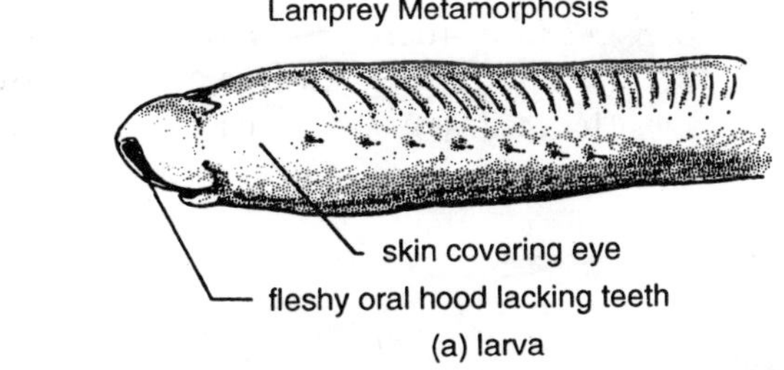

(a) larva

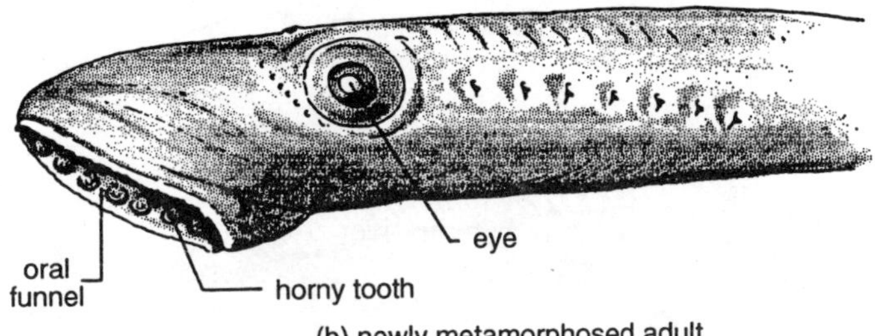

(b) newly metamorphosed adult

Figure 8.10 : Metamorphosis in the sea lamprey (Petromyzon marinus) : (a) nearly fullgrown larva; (b) recently metamorphosed adult. Both are drawn to the same scale.

fish culturists. In some species, when the yolk sac disappears the little fish is a diminutive adult, commonly called an alevin or an advanced fry by fish culturists. Such a direct development is chara eristic of the trouts and salmons (Salmoninae), the North American freshwater catfishes (Ictaluridae), the hagfishes (Myxinidae), and the sharks and their relatives (Chondrichthyes), as examples. In other fishes, a distinct postlarval stage follows the prolarval, sac-fry one. A postlarva must undergo a metamorphosis to lose larval structures and gain adult features. For many "species recog it oon of the postlarval stage is highly subjective. An indirect development has been recognised in many families including, among others, the South American lungfishes (Lepidosirenidae), lampreys (Petromyzonidae), bowfin (Amiidae), herrings (Clupeidae), suckers (Catostomidae), carps (Cyprinidae), sunfishes (Centrarchidae), goosefishes (Lophiidae), molas (Molidae), deepsea anglerfishes (Ceratiidae, rockfishes (Scorpaenidae), and eels (Anguillidae). Differences between postlarvae and adults in some fishes are so trenchant that they have resulted from time to time in taxonomic confusion. Thus larvae of the eel *(Anguilla)* were once described in the genus *Leptocephalus* and those of the lampreys (Petromyzonidae) in the separate genus *Ammocetes!*

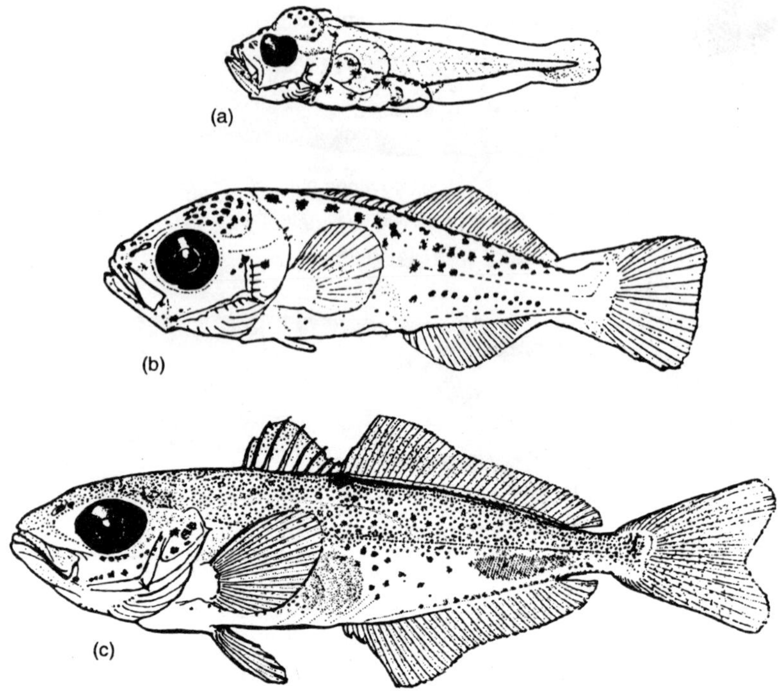

Figure 8.11 : Post-hatching development of a bony fish, the jack mackerel (Trachurus symmetricus). (a) Larva of 4.9 mm. lacking pelvic fin and development of rays in dorsal and anal fins. (b) Larva of 10 mm. with pelvic fin just appearing but with rays established in other fins; note changes in pigmentation over (a). (c) Juvenile of 28 mm. showing establishment of adult body form and further development of pigmentation over (a) and (b).

Features of postlarvae are many and diverse. They range from distinctive pigmentation as in many carps (Cyprinidae) and transparency as in the leptocephalus postlarvae of the eel *(Anguilla),* through profoundly anatomical features. Many marine fishes have a very sharp transition from a pelagic postlarva to a bottom-dwelling juvenile and adult. Syck c ange is nowhere better shown than in the shift from pelagic bilaterally symmetrical, upright young of the flatfishes (Pleuronectiformes) to bottom dwellers after metamorphosis, with both eyes on one side of the head, swimming with the eyeless side down. Other striking postlarvae have elaborate flotational devices. For example, the pelvic fins of larval goosefishes *(Lophius)* are enlarged, and the molas (Molidae) have bodily spines (as well as an evident caudal fin) which are lost at metamorphosis. The skin balloons outward

sometimes attainment of elevated relative robustness or fatness. Generally males undergo senescence and die at earlier ages than females. The period of old age also typically shows a slowing of activity with accompanying changes in feeding, distributional, and other habits. The period may be long in nature, but it may be prolonged in captivity. For example, the northern pike (*Esox lucius)* has been recorded as approaching the age of 25 years in nature, but, in captivity, it has lived 75 years. Senescence may set in when the individual is very small, as in the tiny Philippine goby *(Mistichthys luzonensis)* which seldom exceeds half an inch in length. It may also come when the individual is relatively young as it must do in another goby *(Latrunculus pellucidus),* the life-span of which is only a year.

In wild populations of fishes, death comes soonest to the fastest growing individuals.

Rates and Factors of Development

In practical fish culture the period of embryonic development, or the developmental period, extends from impregnation to the average hatching date of a batch of eggs or birth of a brood. This interval varies considerably according to species and even somewhat among the offspring of individual parents within species. Its duration is strongly influenced by environmental conditions. By definition, the developmental period ends when half of the eggs in a batch have hatched. The hatching (or birth) period is the interval of time over which a complement of eggs actually hatches (or young emerge from the female parent), from the first to the last.

Many forces affect the whole developmental process. They influence success and determine failure. They also affect rate and determine form and structure. Outstanding among environmental factors is temperature. For example, temperature affects rate of development or the fraction of the developmental process achieved per day. This rate has been expressed simply as the reciprocal of the developmental period in days. The larger the fraction, the faster the development. Thus a fish with a developmental period of *88* days has a rate of 1/88, whereas one developing in *9* days would have 1/9. To provide even more ready comparison among species, this rate may also be expressed as a decimal value (for example, 0.011 or 0.111 respectively), for the fractions just considered.

Temperature

Both the developmental period and the hatching period are generally shorter at higher temperatures than at lower ones. Interestingly, many species normally develop in nature under temperature conditions which are not optimal as determined by laboratory experiments. species differ in their temperature optima and tolerances during development; but for all kinds there are temperatures which are too low and too high for development to proceed. Extremes or sudden changes may be lethal. The range of temperatures over which normal development can proceed may be wide. The mummichog *(Fundulus heteroclitus),* for example, develops at temperatures from 12°C (53.6°F) to 27°C (80.6°F) with 2 percent or fewer abnormalities.

How developmental rate differs among species is shown in the following. At an average temperature of 43°F (6.1°C) during the developmental period, brown trout *(Salmo trutta)* require 88 days; brook trout *(Salvelinus fontinalis),* 80; sockeye salmon *(Oncorhynchus nerka), 75;* rainbow trout *(Salmo gairdneri),* 61; white perch *(Morone americana),* 20; cod *(Gadus morhua),* 14; and pollock *(Pollachius virens), 9.*

There has long been interest in the direct relationship of temperature to development of individuals within a species. Such a relationship was determined more than a hundred years ago for the brook trout *(Salvelinus fontinalis).* For this fish, the developmental period decreases with rise of average water temperature (in degrees Fahrenheit) as follows: 370, 165 days; 41°, 103 days; 48°, 56 days; 50°, 47 days; and 54°, 32 days. An early generalisation for eggs of the brook trout was that the developmental period was *50* days for an average water temperature of 50°F, and that, for each degree warmer or cooler, 5 days, respectively, less or more, would be required. For a long time, this scheme was the rule of thumb for making estimates at trout hatcheries.

Subsequently it has been realised that the time necessary to reach a definite stage of development multiplied by temperature is a constant *(K)* in the formula $yT = K$ as variously proposed in the past. When the developmental period is plotted as the abscissa and temperature as the ordinate, the resultant figure is a rectangular hyperbola (Figure *10.12).* In the foregoing temperature equation, y is time to reach a certain stage of development following fertilisation (for example, hatching), and T is the temperature at which the development is taking place. The corresponding intercept equation for rate of development is $v=Ko+KIT$ where v is the velocity or rate (that is, *1000* times the reciprocal of y) and K_0 and $K,$ are in order the constants for intercept (the biological zero) and slope. A plot of rate against temper-

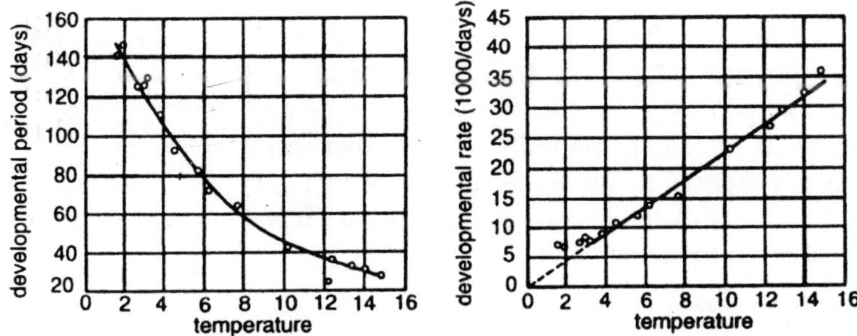

Figure 8.12 : (a) Effect of average water temperature on the developmental period (fertilization to mean date of hatching) in the brook trout (Salvelinus fontinalis).
(b) Rate of development (1000/developmental period in days) in relation to average temperature during developmental period.

ature gives a straight line useful in predicting deviations; however, this plot is curvilinear over wide ranges of temperature-simple linearity applies over a narrow range.

Temperature has an effect on the length of the hatching period, as well as on the total developmental interval. An example of the influence of temperature on the hatching period of different samples from a single lot of eggs of the rainbow trout *(Salmo gairdneri)* held at different temperatures follows.

Temperature		Hatching Period to
°F	°C	Nearest Day
42.3	5.7	14
46.8	8.2	8
55.2	12.9	6
63.7	17.6	3

The rate of hatching within the hatching period may be expected to follow a bell-shaped curve of distribution. It is at the peak of this curve under ordinary situations when 50 percent of the eggs will have hatched and the developmental period will have ended.

Temperature affects the efficiency with which food stored in the egg is converted into body weight of the embryo. In an experiment with hatchlings of the Atlantic salmon *(Salmo salar)*, gains in weight of the larvae were compared with losses in weight of the yolk sac in fish kept in a series of tanks differing in temperature from 0.2 to 16°C. Efficiency of yolk conversion into fish flesh was low and constant at about 42 percent in the coldest tanks. It started to increase at 5°C and reached a maximum efficiency of about 60 percent in the warmest tank, when calculated for a 10-day interval.

Light

Light can influence the developing fish embryo, but existing data are equivocal. Although some experiments with trout and salmon (Salmoninae) eggs have indicated that exposure to fluorescent light results in increased mortality, other experiments show no effect on embryos exposed to low levels of fluorescent lights (which are normally used in hatcheries). Juvenile cutthroat trout *(Salmo clarki)* even showed the fastest growth in light.

Dissolved Gases

Gases dissolved in the water are also factors in development, especially of the eggs of oviparous fishes. It may be expected that both optimum concentrations and intolerable extremes of dissolved oxygen exist and vary according to species. The generally accepted range for development lies between 4 and 12 parts per million (ppm) of dissolved oxygen. Coldwater, stream-spawning species have higher requirements than sluggish or stagnant, warmwater ones. Conditions and requirements of deepsea fishes are quite unknown, primarily because of difficulties in obtaining healthy fish from the deep ocean and holding and experimenting on them under high barometric pressures.

Oxygen pressure can influence the number of meristic elements. In the brown trout *(Salmo trutta)*, decreasing oxygen pressure during embryonic development produces an increase in the number of vertebrae.

At least two gases in water are toxic to fishes and their embryos-carbon dioxide and ammonia. Crude information suggests that under ordinary conditions of dissolved oxygen, concentrations of carbon dioxide up to 8 or 9 ppm have little effect on development. In concentrations from 10 to 30 ppm gradual impairment of the process may be expected, whereas concentrations greater than 30 ppm may arrest development and lead to death. Interestingly, raising carbon dioxide pressures during embryonic development decreases the number of vertebrae in the brown trout *(Salmo trutta)*. Very high levels, between 55 and 80 ppm, increase the mortality of posteyed embryos and also increase the number of deformed larvae. Ammonia can be toxic at low concentrations. Water quality specialists recognise 1.5 ppm as tolerable for aquatic life but express concern over greater concentrations. Water supersaturated with nitrogen such as occurs below some dams can result in gasbubble disease in fry.

Salinity

Salinity can de damaging to the eggs of freshwater fishes and, vice versa, fresh water can harm eggs of marine species. This is clearly shown in the barriers which either fresh or marine waters constitute for fishes primarily restricted to the other environment. If the salt content of water is intolerably high, the eggs of freshwater fishes immersed therein would lose water and die by shriveling. Similarity, if marine fish eggs are placed in fresh water they may imbibe water and burst. Salinity also has a selective role in development of some structures, as shown experimentally in the effect of its variation on the number of bony plates that develop on the sides of the threespine stickleback *(Gasterosteus aculeatus)*.

Endocrines

The importance of endocrine factors in development is well established. Of particular interest are the roles of the pituitary and thyroid in metamorphosis.

Amount of Yolk

The amount of yolk present in the egg has a relationship to the rate of development. Ordinarily in fishes it seems that the larger the amount of yolk, the slower the rate of development. Among animals generally, great parental dependence seems to be associated with relative slowness in development but it is not known if this applies to fishes.

FIXATION OF MERISTIC ELEMENTS

Rate of development also has something to do with the determination of meristic elements such as vertebrae, fin rays, gill rakers, and numbers of scale rows. For example, the golden shiner *(Notemigonus crysoleucas), a* North American minnow, in the North has more meristic elements than it does at the southern extent of its range near the Gulf of Mexico. Because less energy is expended in metabolism at the lower temperatures of the northern latitudes, more of the nutritive material of the egg may be available for synthesis of meristic elements. It is likely that environmental factors change the relation between growth and differentiation. If differentiation is late, more tissue is available to be differentiated, leading to a higher count. Analysed experimentally, the mechanism of such control is found to be more complicated than when judged from field studies. In the brown trout *(Salmo trutta)*, subsamples of fertilised eggs from one pair of parents (both with the same number of vertebrae) were treated with low, intermediate, and high constant temperatures during development. The outcome in numbers of vertebrae was lowest at the intermediate temperature and highest at temperatures above and below.

In the dorsal, anal, and pectoral fins the highest number of rays appeared at the intermediate temperature but was less at both the higher and lower temperatures. In the same group of experiments it was learned that the numbers of vertebrae and anal-fin elements are fixed in the embryonic stages before the eye is completed, thus well before hatching; the dorsal and pectoral rays are set later, as is the number of scales in the lateral line of the rainbow trout *(Salmo gairdneri)*.

BIOGENETIC THEORY

No discussion of the embryology of fishes would be complete without mention of the evidence that embryos afford in support of the biogenetic theory, which holds that ontogeny recapitulates phylogeny. Possible support for biogenesis is well illustrated by development of a heterocercal condition of the tail skeleton, among other features, in fishes. Heterocercal termination of vertebrae appears in embryos of many kinds of fishes, regardless of whether or not the definitive tail is heterocercal or some other such as the predominant homocercal type. This sequence in development has been held as evidence that the heterocercal condition is the more primitive on the ground that ontogeny or embryonic development is recapitulating phylogeny or evolution of the structure.

Index

F

G

H

I

T

U

V

X

Y

Z